OLIVIER HUMBLET

SIGNAL PROCESSING
FOR COMMUNICATIONS

About the cover photograph

Autumn leaves in the Gorges de l'Areuse, by Adrien Vetterli.

Besides being a beautiful picture, this photograph also illustrates a basic signal processing concept. The exposure time is on the order of a second, as can be seen from the fuzziness of the swirling leaves; in other words, the photograph is the average, over a one-second interval, of the light intensity and of the color at each point in the image. In more mathematical terms, the light on the camera's film is a three-dimensional process, with two spatial dimensions (the focal plane) and one time dimension. By taking a photograph we are sampling this process at a particular time, while at the same time integrating (i.e. lowpass filtering) the process over the exposure interval (which can range from a fraction of a second to several seconds).

COMMUNICATION AND INFORMATION SCIENCES

SIGNAL PROCESSING FOR COMMUNICATIONS

Paolo Prandoni and Martin Vetterli

EPFL Press
A Swiss academic publisher distributed by CRC Press

 CRC Press
Taylor & Francis Group

Taylor and Francis Group, LLC
6000 Broken Sound Parkway, NW, Suite 300,
Boca Raton, FL 33487

Distribution and Customer Service
orders@crcpress.com

www.crcpress.com

Library of Congress Cataloging-in-Publication Data
A catalog record for this book is available from the Library of Congress.

Cover photograph credit: *Autumn leaves in the Gorges de l'Areuse*, © Adrien Vetterli, all rights reserved.

This book is published under the editorial direction of Professor Serge Vaudenay (EPFL).

The authors and publisher express their thanks to the Ecole polytechnique fédérale de Lausanne (EPFL) for its generous support towards the publication of this book.

is an imprint owned by Presses polytechniques et universitaires romandes, a Swiss academic publishing company whose main purpose is to publish the teaching and research works of the Ecole polytechnique fédérale de Lausanne.

Presses polytechniques et universitaires romandes
EPFL – Centre Midi
Post office box 119
CH-1015 Lausanne, Switzerland
E-Mail: ppur@epfl.ch
Phone: 021 / 693 21 30
Fax: 021 / 693 40 27

www.epflpress.org

© 2008, First edition, EPFL Press
ISBN 978-2-940222-20-9 (EPFL Press)
ISBN 978-1-4200-7046-0 (CRC Press)

Printed in Italy

All right reserved (including those of translation into other languages). No part of this book may be reproduced in any form – by photoprint, microfilm, or any other means – nor transmitted or translated into a machine language without written permission from the publisher.

To wine, women and song.
Paolo Prandoni

To my children, Thomas and Noémie, who might one day learn from this book the magical inner-workings of their mp3 player, mobile phone and other objects of the digital age.
Martin Vetterli

Preface

The present text evolved from course notes developed over a period of a dozen years teaching undergraduates the basics of signal processing for communications. The students had mostly a background in electrical engineering, computer science or mathematics, and were typically in their third year of studies at Ecole Polytechnique Fédérale de Lausanne (EPFL), with an interest in communication systems. Thus, they had been exposed to signals and systems, linear algebra, elements of analysis (e.g. Fourier series) and some complex analysis, all of this being fairly standard in an undergraduate program in engineering sciences.

The notes having reached a certain maturity, including examples, solved problems and exercises, we decided to turn them into an easy-to-use text on signal processing, with a look at communications as an application. But rather than writing one more book on signal processing, of which many good ones already exist, we deployed the following variations, which we think will make the book appealing as an undergraduate text.

1. Less formal: Both authors came to signal processing by way of an interest in music and think that signal processing is fun, and should be taught to be fun! Thus, choosing between the intricacies of z-transform inversion through contour integration (how many of us have ever done this after having taken a class in signal processing?) or showing the Karplus-Strong algorithm for synthesizing guitar sounds (which also intuitively illustrates issues of stability along the way), you can guess where our choice fell.

    While mathematical rigor is not the emphasis, we made sure to be precise, and thus the text is not approximate in its use of mathematics. Remember, we think signal processing to be mathematics applied to a fun topic, and not mathematics for its own sake, nor a set of applications without foundations.

2. More conceptual: We could have said "more abstract", but this sounds scary (and may seem in contradiction with point 1 above, which of course it is not). Thus, the level of mathematical abstraction is probably higher than in several other texts on signal processing, but it allows to think at a higher conceptual level, and also to build foundations for more advanced topics. Therefore we introduce vector spaces, Hilbert spaces, signals as vectors, orthonormal bases, projection theorem, to name a few, which are powerful concepts not usually emphasized in standard texts. Because these are geometrical concepts, they foster understanding without making the text any more complex. Further, this constitutes the foundation of modern signal processing, techniques such as time-frequency analysis, filter banks and wavelets, which makes the present text an easy primer for more advanced signal processing books. Of course, we must admit, for the sake of full transparency, that we have been influenced by our research work, but again, this has been fun too!

3. More application driven: This is an engineering text, which should help the student solve real problems. Both authors are engineers by training and by trade, and while we love mathematics, we like to see their "operational value". That is, does the result make a difference in an engineering application?

   Certainly, the masterpiece in this regard is C. Shannon's 1948 foundational paper on "The Mathematical Theory of Communication". It completely revolutionized the way communication systems are designed and built, and, still today, we mostly live in its legacy. Not surprisingly, one of the key results of signal processing is the sampling theorem for bandlimited functions (often attributed to Shannon, since it appears in the above-mentioned paper), the theorem which single-handedly enabled the digital revolution. To a mathematician, this is a simple corollary to Fourier series, and he/she might suggest many other ways to represent such particular functions. However, the strength of the sampling theorem and its variations (e.g. oversampling or quantization) is that it is an operational theorem, robust, and applicable to actual signal acquisition and reconstruction problems.

   In order to showcase such powerful applications, the last chapter is entirely devoted to developing an end-to-end communication system, namely a modem for communicating digital information (or bits) over an analog channel. This real-world application (which is present in all modern communication devices, from mobile phones to ADSL boxes) nicely brings together many of the concepts and designs studied in the previous chapters.

Being less formal, more abstract and application-driven seems almost like moving simultaneously in several and possibly opposite directions, but we believe we came up with the right balancing act. Ultimately, of course, the readers and students are the judges!

A last and very important issue is the online access to the text and supplementary material. A full html version together with the unavoidable errata and other complementary material is available at www.sp4comm.org. A solution manual is available to teachers upon request.

As a closing word, we hope you will enjoy the text, and we welcome your feedback. Let signal processing begin, and be fun!

Martin Vetterli and Paolo Prandoni

Spring 2008, Paris and Grandvaux

Acknowledgements

The current book is the result of several iterations of a yearly signal processing undergraduate class and the authors would like to thank the students in Communication Systems at EPFL who survived the early versions of the manuscript and who greatly contributed with their feedback to improve and refine the text along the years. Invaluable help was also provided by the numerous teaching assistants who not only volunteered constructive criticism but came up with a lot of the exercices which appear at the end of each chapter (and their relative solutions). In no particular order: Andrea Ridolfi provided insightful mathematical remarks and also introduced us to the wonders of PsTricks while designing figures. Olivier Roy and Guillermo Barrenetxea have been indefatigable ambassadors between teaching and student bodies, helping shape exercices in a (hopefully) more user-friendly form. Ivana Jovanovic, Florence Bénézit and Patrick Vandewalle gave us a set of beautiful ideas and pointers thanks to their recitations on choice signal processing topics. Luciano Sbaiz always lent an indulgent ear and an insightful answer to all the doubts and worries which plague scientific writers. We would also like to express our personal gratitude to our families and friends for their patience and their constant support; unfortunately, to do so in a proper manner, we should resort to a lyricism which is sternly frowned upon in technical textbooks and therefore we must confine ourselves to a simple "thank you".

Contents

Preface vii

Chapter 1 What Is Digital Signal Processing? 1
    1.1 Some History and Philosophy 2
        1.1.1 Digital Signal Processing under the Pyramids 2
        1.1.2 The Hellenic Shift to Analog Processing 4
        1.1.3 "Gentlemen: *calculemus!*" 5
    1.2 Discrete Time 7
    1.3 Discrete Amplitude 10
    1.4 Communication Systems 12
    1.5 How to Read this Book 17
    Further Reading 18

Chapter 2 Discrete-Time Signals 19
    2.1 Basic Definitions 19
        2.1.1 The Discrete-Time Abstraction 21
        2.1.2 Basic Signals 23
        2.1.3 Digital Frequency 25
        2.1.4 Elementary Operators 26
        2.1.5 The Reproducing Formula 27
        2.1.6 Energy and Power 27
    2.2 Classes of Discrete-Time Signals 28
        2.2.1 Finite-Length Signals 29
        2.2.2 Infinite-Length Signals 30
    Examples 33
    Further Reading 36
    Exercises 36

Chapter 3 Signals and Hilbert Spaces 37
    3.1 Euclidean Geometry: a Review 38
    3.2 From Vector Spaces to Hilbert Spaces 41
        3.2.1 The Recipe for Hilbert Space 42
        3.2.2 Examples of Hilbert Spaces 45
        3.2.3 Inner Products and Distances 46

3.3 Subspaces, Bases, Projections.................................47
    3.3.1 Definitions...48
    3.3.2 Properties of Orthonormal Bases........................49
    3.3.3 Examples of Bases..51
3.4 Signal Spaces Revisited 53
    3.4.1 Finite-Length Signals....................................53
    3.4.2 Periodic Signals 53
    3.4.3 Infinite Sequences 54
Further Reading..55
Exercises 55

Chapter 4 Fourier Analysis   59

4.1 Preliminaries 60
    4.1.1 Complex Exponentials 61
    4.1.2 Complex Oscillations? Negative Frequencies? 61
4.2 The DFT (Discrete Fourier Transform) 63
    4.2.1 Matrix Form..64
    4.2.2 Explicit Form..64
    4.2.3 Physical Interpretation..................................67
4.3 The DFS (Discrete Fourier Series) 71
4.4 The DTFT (Discrete-Time Fourier Transform) 72
    4.4.1 The DTFT as the Limit of a DFS 75
    4.4.2 The DTFT as a Formal Change of Basis 77
4.5 Relationships between Transforms 81
4.6 Fourier Transform Properties 83
    4.6.1 DTFT Properties 83
    4.6.2 DFS Properties 85
    4.6.3 DFT Properties 86
4.7 Fourier Analysis in Practice 90
    4.7.1 Plotting Spectral Data 91
    4.7.2 Computing the Transform: the FFT 93
    4.7.3 Cosmetics: Zero-Padding 94
    4.7.4 Spectral Analysis 95
4.8 Time-Frequency Analysis 98
    4.8.1 The Spectrogram 98
    4.8.2 The Uncertainty Principle...............................100
4.9 Digital Frequency vs. Real Frequency 101
Examples..102
Further Reading...105
Exercises 106

Chapter 5 Discrete-Time Filters   109

5.1 Linear Time-Invariant Systems 109
5.2 Filtering in the Time Domain 111
    5.2.1 The Convolution Operator................................111
    5.2.2 Properties of the Impulse Response 113

	5.3 Filtering by Example – Time Domain	115
	5.3.1 FIR Filtering	115
	5.3.2 IIR Filtering	117
	5.4 Filtering in the Frequency Domain	121
	5.4.1 LTI "Eigenfunctions"	121
	5.4.2 The Convolution and Modulation Theorems	122
	5.4.3 Properties of the Frequency Response	123
	5.5 Filtering by Example – Frequency Domain	126
	5.6 Ideal Filters	129
	5.7 Realizable Filters	133
	5.7.1 Constant-Coefficient Difference Equations	134
	5.7.2 The Algorithmic Nature of CCDEs	135
	5.7.3 Filter Analysis and Design	136
	Examples	136
	Further Reading	143
	Exercises	143
Chapter 6	**The Z-Transform**	**147**
	6.1 Filter Analysis	148
	6.1.1 Solving CCDEs	148
	6.1.2 Causality	149
	6.1.3 Region of Convergence	150
	6.1.4 ROC and System Stability	152
	6.1.5 ROC of Rational Transfer Functions and Filter Stability	152
	6.2 The Pole-Zero Plot	152
	6.2.1 Pole-Zero Patterns	153
	6.2.2 Pole-Zero Cancellation	154
	6.2.3 Sketching the Transfer Function from the Pole-Zero Plot	155
	6.3 Filtering by Example – Z-Transform	156
	Examples	157
	Further Reading	159
	Exercises	159
Chapter 7	**Filter Design**	**165**
	7.1 Design Fundamentals	165
	7.1.1 FIR versus IIR	166
	7.1.2 Filter Specifications and Tradeoffs	168
	7.2 FIR Filter Design	171
	7.2.1 FIR Filter Design by Windowing	171
	7.2.2 Minimax FIR Filter Design	179
	7.3 IIR Filter Design	190
	7.3.1 All-Time Classics	191
	7.4 Filter Structures	195
	7.4.1 FIR Filter Structures	196
	7.4.2 IIR Filter Structures	197

		7.4.3 Some Remarks on Numerical Stability 200

        7.5 Filtering and Signal Classes 200
              7.5.1 Filtering of Finite-Length Signals 200
              7.5.2 Filtering of Periodic Sequences 201
        Examples 204
        Further Reading 208
        Exercises 208

Chapter 8 Stochastic Signal Processing   **217**
        8.1 Random Variables 217
        8.2 Random Vectors 219
        8.3 Random Processes 221
        8.4 Spectral Representation of Stationary Random Processes 223
              8.4.1 Power Spectral Density 224
              8.4.2 PSD of a Stationary Process 225
              8.4.3 White Noise 227
        8.5 Stochastic Signal Processing 227
        Examples 229
        Further Reading 232
        Exercises 233

Chapter 9 Interpolation and Sampling   **235**
        9.1 Preliminaries and Notation 236
        9.2 Continuous-Time Signals 237
        9.3 Bandlimited Signals 239
        9.4 Interpolation 240
              9.4.1 Local Interpolation 241
              9.4.2 Polynomial Interpolation 243
              9.4.3 Sinc Interpolation 245
        9.5 The Sampling Theorem 247
        9.6 Aliasing 250
              9.6.1 Non-Bandlimited Signals 250
              9.6.2 Aliasing: Intuition 251
              9.6.3 Aliasing: Proof 253
              9.6.4 Aliasing: Examples 255
        9.7 Discrete-Time Processing of Analog Signals 260
              9.7.1 A Digital Differentiator 260
              9.7.2 Fractional Delays 261
        Examples 262
        Appendix 266
        Further Reading 268
        Exercises 269

Chapter 10 A/D and D/A Conversions   **275**
        10.1 Quantization 275
              10.1.1 Uniform Scalar Quantization 278
              10.1.2 Advanced Quantizers 282

10.2 A/D Conversion 283
10.3 D/A Conversion 286
Examples 287
Further Reading 290
Exercises 290

Chapter 11 Multirate Signal Processing — 293
11.1 Downsampling 294
    11.1.1 Properties of the Downsampling Operator 294
    11.1.2 Frequency-Domain Representation 295
    11.1.3 Examples 297
    11.1.4 Downsampling and Filtering 302
11.2 Upsampling 304
    11.2.1 Upsampling and Interpolation 306
11.3 Rational Sampling Rate Changes 310
11.4 Oversampling 311
    11.4.1 Oversampled A/D Conversion 311
    11.4.2 Oversampled D/A Conversion 314
Examples 319
Further Reading 322
Exercises 322

Chapter 12 Design of a Digital Communication System — 327
12.1 The Communication Channel 328
    12.1.1 The AM Radio Channel 329
    12.1.2 The Telephone Channel 330
12.2 Modem Design: The Transmitter 331
    12.2.1 Digital Modulation and the Bandwidth Constraint 331
    12.2.2 Signaling Alphabets and the Power Constraint 339
12.3 Modem Design: the Receiver 347
    12.3.1 Hilbert Demodulation 348
    12.3.2 The Effects of the Channel 350
12.4 Adaptive Synchronization 353
    12.4.1 Carrier Recovery 353
    12.4.2 Timing Recovery 356
Further Reading 365
Exercises 365

Index — 367

Chapter 1

What Is Digital Signal Processing?

A *signal*, technically yet generally speaking, is a a formal description of a phenomenon evolving over time or space; by *signal processing* we denote any manual or "mechanical" operation which modifies, analyzes or otherwise manipulates the information contained in a signal. Consider the simple example of ambient temperature: once we have agreed upon a formal model for this physical variable – Celsius degrees, for instance – we can record the evolution of temperature over time in a variety of ways and the resulting data set represents a temperature "signal". Simple processing operations can then be carried out even just by hand: for example, we can plot the signal on graph paper as in Figure 1.1, or we can compute derived parameters such as the average temperature in a month.

Conceptually, it is important to note that signal processing operates on *an abstract representation* of a physical quantity and not on the quantity itself. At the same time, the *type* of abstract representation we choose for the physical phenomenon of interest determines the nature of a signal processing unit. A temperature regulation device, for instance, is not a signal processing system as a whole. The device does however contain a signal processing core in the feedback control unit which converts the instantaneous *measure* of the temperature into an ON/OFF trigger for the heating element. The physical nature of this unit depends on the temperature model: a simple design is that of a mechanical device based on the dilation of a metal sensor; more likely, the temperature signal is a voltage generated by a thermocouple and in this case the matched signal processing unit is an operational amplifier.

Finally, the adjective "digital" derives from *digitus*, the Latin word for finger: it concisely describes a world view where everything can be ultimately represented as an integer number. Counting, first on one's fingers and then

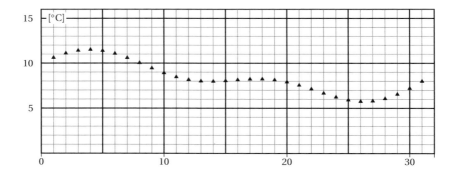

Figure 1.1 Temperature measurements over a month.

in one's head, is the earliest and most fundamental form of abstraction; as children we quickly learn that counting does indeed bring disparate objects (the proverbial "apples and oranges") into a common modeling paradigm, i.e. their cardinality. Digital signal processing is a flavor of signal processing in which everything *including time* is described in terms of integer numbers; in other words, the abstract representation of choice is a one-size-fit-all countability. Note that our earlier "thought experiment" about ambient temperature fits this paradigm very naturally: the measuring instants form a countable set (the days in a month) and so do the measures themselves (imagine a finite number of ticks on the thermometer's scale). In digital signal processing the underlying abstract representation is always the set of natural numbers regardless of the signal's origins; as a consequence, the physical nature of the processing device will also always remain the same, that is, a general digital (micro)processor. The extraordinary power and success of digital signal processing derives from the inherent universality of its associated "world view".

1.1 Some History and Philosophy

1.1.1 Digital Signal Processing under the Pyramids

Probably the earliest recorded example of digital signal processing dates back to the 25th century BC. At the time, Egypt was a powerful kingdom reaching over a thousand kilometers south of the Nile's delta. For all its latitude, the kingdom's populated area did not extend for more than a few kilometers on either side of the Nile; indeed, the only inhabitable areas in an otherwise desert expanse were the river banks, which were made fertile by the yearly flood of the river. After a flood, the banks would be left covered with a thin layer of nutrient-rich silt capable of supporting a full agricultural cycle. The floods of the Nile, however, were[1] a rather capricious meteorological phenomenon, with scant or absent floods resulting in little or no yield from the land. The pharaohs quickly understood that, in order to preserve stability, they would have to set up a grain buffer with which to compensate for the unreliability of the Nile's floods and prevent potential unrest in a famished population during "dry" years. As a consequence, studying and predicting the trend of the floods (and therefore the expected agricultural yield) was of paramount importance in order to determine the operating point of a very dynamic taxation and redistribution mechanism. The floods of the Nile were meticulously recorded by an array of measuring stations called "nilometers" and the resulting data set can indeed be considered a full-fledged digital signal defined on a time base of twelve months. The Palermo Stone, shown in the left panel of Figure 1.2, is a faithful record of the data in the form of a table listing the name of the current pharaoh alongside the yearly flood level; a more modern representation of an flood data set is shown on the left of the figure: bar the references to the pharaohs, the two representations are perfectly equivalent. The Nile's behavior is still an active area of hydrological research today and it would be surprising if the signal processing operated by the ancient Egyptians on their data had been of much help in anticipating for droughts. Yet, the Palermo Stone is arguably the first recorded digital signal which is still of relevance today.

 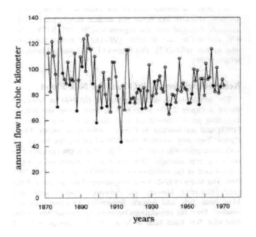

Figure 1.2 Representations of flood data for the river Nile: circa 2500 BC (left) and 2000 AD (right).

---

[1] The Nile stopped flooding Egypt in 1964, when the Aswan dam was completed.

1.1.2 The Hellenic Shift to Analog Processing

"Digital" representations of the world such as those depicted by the Palermo Stone are adequate for an environment in which quantitative problems are simple: counting cattle, counting bushels of wheat, counting days and so on. As soon as the interaction with the world becomes more complex, so necessarily do the models used to interpret the world itself. Geometry, for instance, is born of the necessity of measuring and subdividing land property. In the act of splitting a certain quantity into parts we can already see the initial difficulties with an integer-based world view;[2] yet, until the Hellenic period, western civilization considered natural numbers and their ratios all that was needed to describe nature in an operational fashion. In the 6th century BC, however, a devastated Pythagoras realized that the the side and the diagonal of a square are incommensurable, i.e. that $\sqrt{2}$ is *not* a simple fraction. The discovery of what we now call irrational numbers "sealed the deal" on an abstract model of the world that had already appeared in early geometric treatises and which today is called *the continuum*. Heavily steeped in its geometric roots (i.e. in the infinity of points in a segment), the continuum model postulates that time and space are an uninterrupted flow which can be divided arbitrarily many times into arbitrarily (and infinitely) small pieces. In signal processing parlance, this is known as the "analog" world model and, in this model, integer numbers are considered primitive entities, as rough and awkward as a set of sledgehammers in a watchmaker's shop.

In the continuum, the infinitely big and the infinitely small dance together in complex patterns which often defy our intuition and which required almost two thousand years to be properly mastered. This is of course not the place to delve deeper into this extremely fascinating epistemological domain; suffice it to say that the apparent incompatibility between the digital and the analog world views appeared right from the start (i.e. from the 5th century BC) in Zeno's works; we will appreciate later the immense import that this has on signal processing in the context of the sampling theorem.

Zeno's paradoxes are well known and they underscore this unbridgeable gap between our intuitive, integer-based grasp of the world and a model of

---

[2] The layman's aversion to "complicated" fractions is at the basis of many counting systems other than the decimal (which is just an accident tied to the number of human fingers). Base-12 for instance, which is still so persistent both in measuring units (hours in a day, inches in a foot) and in common language ("a dozen") originates from the simple fact that 12 happens to be divisible by 2, 3 and 4, which are the most common number of parts an item is usually split into. Other bases, such as base-60 and base-360, have emerged from a similar abundance of simple factors.

the world based on the continuum. Consider for instance the dichotomy paradox; Zeno states that if you try to move along a line from point A to point B you will never in fact be able to reach your destination. The reasoning goes as follows: in order to reach B, you will have to first go through point C, which is located mid-way between A and B; but, even before you reach C, you will have to reach D, which is the midpoint between A and C; and so on *ad infinitum*. Since there is an infinity of such intermediate points, Zeno argues, moving from A to B requires you to complete an infinite number of tasks, which is humanly impossible. Zeno of course was well aware of the empirical evidence to the contrary but he was brilliantly pointing out the extreme trickery of a model of the world which had not yet formally defined the concept of infinity. The complexity of the intellectual machinery needed to solidly counter Zeno's argument is such that even today the paradox is food for thought. A first-year calculus student may be tempted to offhandedly dismiss the problem by stating

$$\sum_{n=1}^{\infty} \frac{1}{2^n} = 1 \qquad (1.1)$$

but this is just a void formalism begging the initial question if the underlying notion of the continuum is not explicitly worked out.[3] In reality Zeno's paradoxes cannot be "solved", since they cease to be paradoxes once the continuum model is fully understood.

1.1.3 "Gentlemen: *calculemus!*"

The two competing models for the world, digital and analog, coexisted quite peacefully for quite a few centuries, one as the tool of the trade for farmers, merchants, bankers, the other as an intellectual pursuit for mathematicians and astronomers. Slowly but surely, however, the increasing complexity of an expanding world spurred the more practically-oriented minds to pursue science as a means to solve very tangible problems *besides* describing the motion of the planets. Calculus, brought to its full glory by Newton and Leibniz in the 17th century, proved to be an incredibly powerful tool when applied to eminently practical concerns such as ballistics, ship routing, mechanical design and so on; such was the faith in the power of the new science that Leibniz envisioned a not-too-distant future in which all human disputes, including problems of morals and politics, could be worked out with pen and paper: "gentlemen, *calculemus*". If only.

---

[3] An easy rebuttal of the bookish *reductio* above is asking to explain why $\sum 1/n$ diverges while $\sum 1/n^2 = \pi^2/6$ (Euler, 1740).

As Cauchy unsurpassably explained later, everything in calculus is a limit and therefore everything in calculus is a celebration of the power of the continuum. Still, in order to apply the calculus machinery to the real world, the real world has to be modeled as something calculus understands, namely a function of a real (i.e. *continuous*) variable. As mentioned before, there are vast domains of research well behaved enough to admit such an *analytical* representation; astronomy is the first one to come to mind, but so is ballistics, for instance. If we go back to our temperature measurement example, however, we run into the first difficulty of the analytical paradigm: we now need to model our measured temperature as a function of continuous time, which means that the value of the temperature should be available at *any* given instant and not just once per day. A "temperature function" as in Figure 1.3 is quite puzzling to define if all we have (and if all we *can* have, in fact) is just a set of empirical measurements reasonably spaced in time. Even in the rare cases in which an analytical model of the phenomenon is available, a second difficulty arises when the *practical* application of calculus involves the use of functions which are only available in tabulated form. The trigonometric and logarithmic tables are a typical example of how a continuous model needs to be made countable again in order to be put to real use. Algorithmic procedures such as series expansions and numerical integration methods are other ways to bring the analytic results within the realm of the practically computable. These parallel tracks of scientific development, the "Platonic" ideal of analytical results and the slide rule reality of practitioners, have coexisted for centuries and they have found their most durable mutual peace in digital signal processing, as will appear shortly.

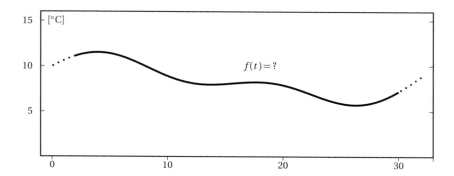

Figure 1.3 Temperature "function" in a continuous-time world model.

1.2 Discrete Time

One of the fundamental problems in signal processing is to obtain a permanent record of the signal itself. Think back of the ambient temperature example, or of the floods of the Nile: in both cases a description of the phenomenon was gathered by a naive *sampling* operation, i.e. by measuring the quantity of interest at regular intervals. This is a very intuitive process and it reflects the very natural act of "looking up the current value and writing it down". Manually this operation is clearly quite slow but it is conceivable to speed it up mechanically so as to obtain a much larger number of measurements per unit of time. Our measuring machine, however fast, still will never be able to take an *infinite* amount of samples in a finite time interval: we are back in the clutches of Zeno's paradoxes and one would be tempted to conclude that a true analytical representation of the signal is impossible to obtain.

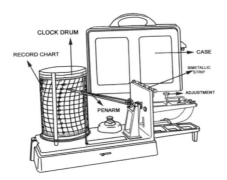

Figure 1.4 A thermograph.

At the same time, the history of applied science provides us with many examples of recording machines capable of providing an "analog" image of a physical phenomenon. Consider for instance a thermograph: this is a mechanical device in which temperature deflects an ink-tipped metal stylus in contact with a slowly rolling paper-covered cylinder. Thermographs like the one sketched in Figure 1.4 are still currently in use in some simple weather stations and they provide a chart in which a temperature function as in Figure 1.3 is duly plotted. Incidentally, the principle is the same in early sound recording devices: Edison's phonograph used the deflection of a steel pin connected to a membrane to impress a "continuous-time" sound wave as a groove on a wax cylinder. The problem with these analog recordings is that they are not abstract signals but *a conversion of a physical phenomenon into another physical phenomenon:* the temperature, for instance, is converted into the amount of ink on paper while the sound pressure wave is converted into the physical depth of the groove. The advent of electronics did not change the concept: an audio tape, for instance, is obtained by converting a pressure wave into an electrical current and then into a magnetic deflection. The fundamental consequence is that, for analog signals, a different signal processing system needs to be designed *explicitly* for each specific form of recording.

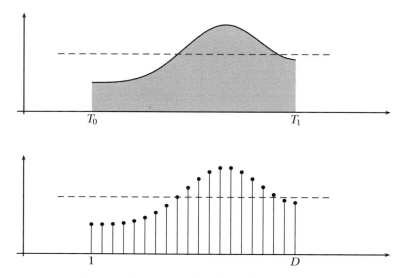

Figure 1.5 Analytical and empirical averages.

Consider for instance the problem of computing the average temperature over a certain time interval. Calculus provides us with the exact answer $\bar{C}$ *if* we know the elusive "temperature function" $f(t)$ over an interval $[T_0, T_1]$ (see Figure 1.5, top panel):

$$\bar{C} = \frac{1}{T_1 - T_0} \int_{T_0}^{T_1} f(t) \, dt \tag{1.2}$$

We can try to reproduce the integration with a "machine" adapted to the particular representation of temperature we have at hand: in the case of the thermograph, for instance, we can use a planimeter as in Figure 1.6, a manual device which computes the area of a drawn surface; in a more modern incarnation in which the temperature signal is given by a thermocouple, we can integrate the voltage with the RC network in Figure 1.7. In both cases, in spite of the simplicity of the problem, we can instantly see the practical complications and the degree of specialization needed to achieve something as simple as an average for an analog signal.

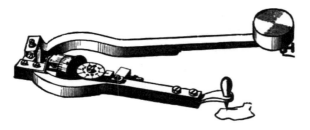

Figure 1.6 The planimeter: a mechanical integrator.

Now consider the case in which all we have is a set of daily measurements $c_1, c_2, \ldots, c_D$ as in Figure 1.1; the "average" temperature of our measurements over D days is simply:

$$\hat{C} = \frac{1}{D} \sum_{n=1}^{D} c_n \tag{1.3}$$

(as shown in the bottom panel of Figure 1.5) and this is an elementary sum of D terms which anyone can carry out by hand and which does not depend on how the measurements have been obtained: wickedly simple! So, obviously, the question is: "How different (if at all) is $\hat{C}$ from $\bar{C}$?" In order to find out we can remark that if we accept the existence of a temperature function $f(t)$ then the measured values c_n are *samples* of the function taken one day apart:

$$c_n = f(nT_s)$$

(where T_s is the duration of a day). In this light, the sum (1.3) is just the Riemann approximation to the integral in (1.2) and the question becomes an assessment on how good an approximation that is. Another way to look at the problem is to ask ourselves how much information we are discarding by only keeping samples of a continuous-time function.

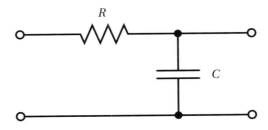

Figure 1.7 The RC network: an electrical integrator.

The answer, which we will study in detail in Chapter 9, is that in fact the continuous-time function and the set of samples *are perfectly equivalent representations* – provided that the underlying physical phenomenon "doesn't change too fast". Let us put the proviso aside for the time being and concentrate instead on the good news: first, the analog and the digital world can perfectly coexist; second, we actually possess a *constructive* way to move between worlds: the *sampling theorem*, discovered and rediscovered by many at the beginning of the 20th century[4], tells us that the continuous-time function can be obtained from the samples as

$$f(t) = \sum_{n=-\infty}^{\infty} c_n \frac{\sin(\pi(t - nT_s)/T_s)}{\pi(t - nT_s)/T_s} \qquad (1.4)$$

So, in theory, once we have a set of measured values, we can build the continuous-time representation and use the tools of calculus. In reality things are even simpler: if we plug (1.4) into our analytic formula for the average (1.2) we can show that the result is a simple sum like (1.3). So we don't need to explicitly go "through the looking glass" back to continuous-time: the tools of calculus have a discrete-time equivalent which we can use directly.

The equivalence between the discrete and continuous representations only holds for signals which are sufficiently "slow" with respect to how fast we sample them. This makes a lot of sense: we need to make sure that the signal does not do "crazy" things between successive samples; only if it is smooth and well behaved can we afford to have such sampling gaps. Quantitatively, the sampling theorem links the speed at which we need to repeatedly measure the signal to the maximum frequency contained in its spectrum. Spectra are calculated using the Fourier transform which, interestingly enough, was originally devised as a tool to break periodic functions into a *countable* set of building blocks. Everything comes together.

1.3 Discrete Amplitude

While it appears that the time continuum has been tamed by the sampling theorem, we are nevertheless left with another pesky problem: the precision of our measurements. If we model a phenomenon as an analytical function, not only is the argument (the time domain) a continuous variable but so is the function's value (the codomain); a practical measurement, however, will never achieve an infinite precision and we have another paradox

---

[4] Amongst the credited personnel: Nyquist, Whittaker, Kotel'nikov, Raabe, Shannon and Someya.

on our hands. Consider our temperature example once more: we can use a mercury thermometer and decide to write down just the number of degrees; maybe we can be more precise and note the half-degrees as well; with a magnifying glass we could try to record the tenths of a degree – but we would most likely have to stop there. With a more sophisticated thermocouple we could reach a precision of one hundredth of a degree and possibly more but, still, we would have to settle on a maximum number of decimal places. Now, if we know that our measures have a fixed number of digits, the set of all possible measures is actually countable and we have effectively mapped the codomain of our temperature function onto the set of integer numbers. This process is called *quantization* and it is the method, together with sampling, to obtain a fully digital signal.

In a way, quantization deals with the problem of the continuum in a much "rougher" way than in the case of time: we simply accept a loss of precision with respect to the ideal model. There is a very good reason for that and it goes under the name of *noise*. The mechanical recording devices we just saw, such as the thermograph or the phonograph, give the illusion of analytical precision but are in practice subject to severe mechanical limitations. Any analog recording device suffers from the same fate and even if electronic circuits can achieve an excellent performance, in the limit the unavoidable thermal agitation in the components constitutes a noise floor which limits the "equivalent number of digits". Noise is a fact of nature that cannot be eliminated, hence our acceptance of a finite (i.e. countable) precision.

Figure 1.8 Analog and digital computers.

Noise is not just a problem in measurement but also in processing. Figure 1.8 shows the two archetypal types of analog and digital computing devices; while technological progress may have significantly improved the speed of each, the underlying principles remain the same for both. An analog signal processing system, much like the slide rule, uses the displacement of physical quantities (gears or electric charge) to perform its task; each element in the system, however, acts as a source of noise so that complex or, more importantly, *cheap* designs introduce imprecisions in the final result (good slide rules used to be *very* expensive). On the other hand the abacus, working only with integer arithmetic, is a perfectly precise machine[5] even if it's made with rocks and sticks. Digital signal processing works with countable sequences of integers so that in a digital architecture no processing noise is introduced. A classic example is the problem of *reproducing* a signal. Before mp3 existed and file sharing became the bootlegging method of choice, people would "make tapes". When someone bought a vinyl record he would allow his friends to record it on a cassette; however, a "peer-to-peer" dissemination of illegally taped music never really took off because of the "second generation noise", i.e. because of the ever increasing hiss that would appear in a tape made from another tape. Basically only first generation copies of the purchased vinyl were acceptable quality on home equipment. With digital formats, on the other hand, duplication is really equivalent to copying down a (very long) list of integers and even very cheap equipment can do that without error.

Finally, a short remark on terminology. The amplitude accuracy of a set of samples is entirely dependent on the processing hardware; in current parlance this is indicated by the number of *bits per sample* of a given representation: compact disks, for instance, use 16 bits per sample while DVDs use 24. Because of its "contingent" nature, quantization is almost always ignored in the core theory of signal processing and all derivations are carried out as if the samples were real numbers; therefore, in order to be precise, we will almost always use the term *discrete-time* signal processing and leave the label "digital signal processing" (DSP) to the world of actual devices. Neglecting quantization will allow us to obtain very general results but care must be exercised: in the practice, actual implementations will have to deal with the effects of finite precision, sometimes with very disruptive consequences.

1.4 Communication Systems

Signals in digital form provide us with a very convenient abstract representation which is both simple and powerful; yet this does not shield us from the need to deal with an "outside" world which is probably best modeled by the analog paradigm. Consider a mundane act such as placing a call on a cell phone, as in Figure 1.9: humans are analog devices after all and they produce analog sound waves; the phone converts these into digital format,

---

[5] As long as we don't need to compute square roots; luckily, *linear* systems (which is what interests us) are made up only of sums and multiplications.

does some digital processing and then outputs an *analog* electromagnetic wave on its antenna. The radio wave travels to the base station in which it is demodulated, converted to digital format to recover the voice signal. The call, as a digital signal, continues through a switch and then is injected into an optical fiber *as an analog light wave*. The wave travels along the network and then the process is inverted until an analog sound wave is generated by the loudspeaker at the receiver's end.

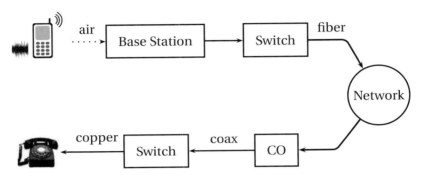

Figure 1.9 A prototypical telephone call and the associated transitions from the digital to the analog domain and back; processing in the blocks is done digitally while the links between blocks are over an analog medium.

Communication systems are in general a prime example of sophisticated interplay between the digital and the analog world: while all the processing is undoubtedly best done digitally, signal propagation in a medium (be it the the air, the electromagnetic spectrum or an optical fiber) is the domain of differential (rather than *difference*) equations. And yet, even when digital processing must necessarily hand over control to an analog interface, it does so in a way that leaves no doubt as to who's boss, so to speak: for, instead of transmitting an analog signal which is the reconstructed "real" function as per (1.4), we always transmit an analog signal which *encodes the digital representation of the data*. This concept is really at the heart of the "digital revolution" and, just like in the cassette tape example, it has to do with noise.

Imagine an analog voice signal $s(t)$ which is transmitted over a (long) telephone line; a simplified description of the received signal is

$$s_r(t) = \alpha s(t) + n(t)$$

where the parameter α, with $\alpha < 1$, is the *attenuation* that the signal incurs and where $n(t)$ is the noise introduced by the system. The noise function is of obviously unknown (it is often modeled as a Gaussian process, as we will see) and so, once it's added to the signal, it's impossible to eliminate it. Because of attenuation, the receiver will include an amplifier with gain G to restore the voice signal to its original level; with $G = 1/\alpha$ we will have

$$s_a(t) = G s_r(t) = s(t) + G n(t)$$

Unfortunately, as it appears, in order to regenerate the analog signal we also have amplified the noise G times; clearly, if G is large (i.e. if there is a lot of attenuation to compensate for) the voice signal end up buried in noise. The problem is exacerbated if many intermediate amplifiers have to be used in cascade, as is the case in long submarine cables.

Consider now a digital voice signal, that is, a discrete-time signal whose samples have been quantized over, say, 256 levels: each sample can therefore be represented by an 8-bit word and the whole speech signal can be represented as a very long sequence of binary digits. We now build an analog signal as a *two-level* signal which switches for a few instants between, say, -1 V and $+1$ V for every "0" and "1" bit in the sequence respectively. The received signal will still be

$$s_r(t) = \alpha s(t) + n(t)$$

but, to regenerate it, instead of linear amplification we can use nonlinear thresholding:

$$s_a(t) = \begin{cases} +1 & \text{if } s_r(t) \geq 0 \\ -1 & \text{if } s_r(t) < 0 \end{cases}$$

Figure 1.10 clearly shows that as long as the magnitude of the noise is less than α the two-level signal can be regenerated perfectly; furthermore, the regeneration process can be repeated as many times as necessary with no overall degradation.

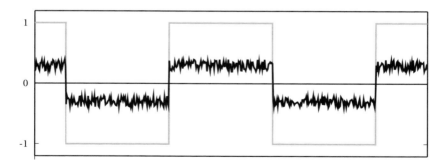

Figure 1.10 Two-level analog signal encoding a binary sequence: original signal $s(t)$ (light gray) and received signal $s_r(t)$ (black) in which both attenuation and noise are visible.

In reality of course things are a little more complicated and, because of the nature of noise, it is impossible to guarantee that some of the bits won't be corrupted. The answer is to use *error correcting codes* which, by introducing redundancy in the signal, make the sequence of ones and zeros robust to the presence of errors; a scratched CD can still play flawlessly because of the Reed-Solomon error correcting codes used for the data. Data coding is the core subject of Information Theory and it is behind the stellar performance of modern communication systems; interestingly enough, the most successful codes have emerged from group theory, a branch of mathematics dealing with the properties of closed sets of integer numbers. Integers again! Digital signal processing and information theory have been able to join forces so successfully because they share a common data model (the integer) and therefore they share the same architecture (the processor). Computer code written to implement a digital filter can dovetail seamlessly with code written to implement error correction; linear processing and nonlinear flow control coexist naturally.

A simple example of the power unleashed by digital signal processing is the performance of transatlantic cables. The first operational telegraph cable from Europe to North America was laid in 1858 (see Fig. 1.11); it worked for about a month before being irrecoverably damaged.[6] From then on, new materials and rapid progress in electrotechnics boosted the performance of each subsequent cable; the key events in the timeline of transatlantic communications are shown in Table 1.1. The first transatlantic *telephone* cable was laid in 1956 and more followed in the next two decades with increasing capacity due to multicore cables and better repeaters; the invention of the echo canceler further improved the number of voice channels for already deployed cables. In 1968 the first experiments in PCM digital telephony were successfully completed and the quantum leap was around the corner: by the end of the 70's cables were carrying over one order of magnitude more voice channels than in the 60's. Finally, the deployment of the first fiber optic cable in 1988 opened the door to staggering capacities (and enabled the dramatic growth of the Internet).

Finally, it's impossible not to mention the advent of *data compression* in this brief review of communication landmarks. Again, digital processing allows the coexistence of standard processing with sophisticated decision

---

[6] Ohm's law was published in 1861, so the first transatlantic cable was a little bit the proverbial cart before the horse. Indeed, the cable circuit formed an enormous RC equivalent circuit, i.e. a big lowpass filter, so that the sharp rising edges of the Morse symbols were completely smeared in time. The resulting intersymbol interference was so severe that it took hours to reliably send even a simple sentence. Not knowing how to deal with the problem, the operator tried to increase the signaling voltage ("crank up the volume") until, at 4000 V, the cable gave up.

Figure 1.11 Laying the first transatlantic cable.

Table 1.1 The main transatlantic cables from 1858 to our day.

Cable	Year	Type	Signaling	Capacity
	1858	Coax	telegraph	a few words per hour
	1866	Coax	telegraph	6-8 words per minute
	1928	Coax	telegraph	2500 characters per minute
TAT-1	1956	Coax	telephone	36 [48 by 1978] voice channels
TAT-3	1963	Coax	telephone	138 [276 by 1986] voice channels
TAT-5	1970	Coax	telephone	845 [2112 by 1993] voice channels
TAT-6	1976	Coax	telephone	4000 [10,000 by 1994] voice channels
TAT-8	1988	Fiber	data	280 Mbit/s ($\sim 40,000$ voice channels)
TAT-14	2000	Fiber	data	640 Gbit/s ($\sim 9,700,000$ voice channels)

logic; this enables the implementation of complex *data-dependent* compression techniques and the inclusion of psychoperceptual models in order to match the compression strategy to the characteristics of the human visual or auditory system. A music format such as mp3 is perhaps the first example to come to mind but, as shown in Table 1.2, all communication domains have been greatly enhanced by the gains in throughput enabled by data compression.

Table 1.2 Uncompressed and compressed data rates.

Signal	Uncompressed Rate	Common Rate
Music	4.32 Mbit/s (CD audio)	128 Kbit/s (MP3)
Voice	64 Kbit/s (AM radio)	4.8 Kbit/s (cellphone CELP)
Photos	14 MB (raw)	300 KB (JPEG)
Video	170 Mbit/s (PAL)	700 Kbit/s (DivX)

1.5 How to Read this Book

This book tries to build a largely self-contained development of digital signal processing theory *from within discrete time*, while the relationship to the analog model of the world is tackled only after all the fundamental "pieces of the puzzle" are already in place. Historically, modern discrete-time processing started to consolidate in the 50's when mainframe computers became powerful enough to handle the effective simulations of analog electronic networks. By the end of the 70's the discipline had by all standards reached maturity; so much so, in fact, that the major textbooks on the subject still in use today had basically already appeared by 1975. Because of its ancillary origin with respect to the problems of that day, however, discrete-time signal processing has long been presented as a tributary to much more established disciplines such as Signals and Systems. While historically justifiable, that approach is no longer tenable today for three fundamental reasons: first, the pervasiveness of digital storage for data (from CDs to DVDs to flash drives) implies that most devices today are designed for discrete-time signals to start with; second, the trend in signal processing devices is to move the analog-to-digital and digital-to-analog converters at the very beginning and the very end of the processing chain so that even "classically analog" operations such as modulation and demodulation are now done in discrete-time; third, the availability of numerical packages like Matlab provides a testbed for signal processing experiments (both academically and just for fun) which is far more accessible and widespread than an electronics lab (not to mention affordable).

The idea therefore is to introduce discrete-time signals as a self-standing entity (Chap. 2), much in the natural way of a temperature sequence or a series of flood measurements, and then to remark that the mathematical structures used to describe discrete-time signals are one and the same with the structures used to describe vector spaces (Chap. 3). Equipped with the geometrical intuition afforded to us by the concept of vector space, we can proceed to "dissect" discrete-time signals with the Fourier transform, which turns out to be just a change of basis (Chap. 4). The Fourier transform opens the passage between the time domain and the frequency domain and, thanks to this dual understanding, we are ready to tackle the concept of processing as performed by discrete-time linear systems, also known as filters (Chap. 5). Next comes the very practical task of designing a filter to order, with an eye to the subtleties involved in filter implementation; we will mostly consider FIR filters, which are unique to discrete time (Chaps 6 and 7). After a brief excursion in the realm of stochastic sequences (Chap. 8) we will finally build a bridge between our discrete-time world and the continuous-time models of physics and electronics with the concepts of sampling and interpolation (Chap. 9); and digital signals will be completely accounted for after a study of quantization (Chap. 10). We will finally go back to purely discrete time for the final topic, multirate signal processing (Chap. 11), before putting it all together in the final chapter: the analysis of a commercial voiceband modem (Chap. 12).

Further Reading

The Bible of digital signal processing was and remains *Discrete-Time Signal Processing*, by A. V. Oppenheim and R. W. Schafer (Prentice-Hall, last edition in 1999); exceedingly exhaustive, it is a must-have reference. For background in signals and systems, the eponymous *Signals and Systems*, by Oppenheim, Willsky and Nawab (Prentice Hall, 1997) is a good start.

Most of the historical references mentioned in this introduction can be integrated by simple web searches. Other comprehensive books on digital signal processing include S. K. Mitra's *Digital Signal Processing* (McGraw Hill, 2006) and *Digital Signal Processing*, by J. G. Proakis and D. K. Manolakis (Prentis Hall 2006). For a fascinating excursus on the origin of calculus, see D. Hairer and G. Wanner, *Analysis by its History* (Springer-Verlag, 1996). A more than compelling epistemological essay on the continuum is *Everything and More*, by David Foster Wallace (Norton, 2003), which manages to be both profound and hilarious in an unprecedented way.

Finally, the very prolific literature on current signal processing research is published mainly by the Institute of Electronics and Electrical Engineers (IEEE) in several of its transactions such as *IEEE Transactions on Signal Processing*, *IEEE Transactions on Image Processing* and *IEEE Transactions on Speech and Audio Processing*.

Chapter 2

Discrete-Time Signals

In this Chapter we define more formally the concept of the discrete-time signal and establish an associated basic taxonomy used in the remainder of the book. Historically, discrete-time signals have often been introduced as the discretized version of continuous-time signals, i.e. as the *sampled* values of analog quantities, such as the voltage at the output of an analog circuit; accordingly, many of the derivations proceeded within the framework of an underlying continuous-time reality. In truth, the discretization of analog signals is only part of the story, and a rather minor one nowadays. Digital signal processing, especially in the context of communication systems, is much more concerned with the *synthesis* of discrete-time signals rather than with sampling. That is why we choose to introduce discrete-time signals from an abstract and self-contained point of view.

2.1 Basic Definitions

A discrete-time signal is a complex-valued *sequence*. Remember that a sequence is defined as a complex-valued function of an integer index n, with $n \in \mathbb{Z}$; as such, it is a two-sided, infinite collection of values. A sequence can be defined analytically in closed form, as for example:

$$x[n] = (n \mod 11) - 5 \qquad (2.1)$$

shown as the "triangular" waveform plotted in Figure 2.1; or

$$x[n] = e^{j\frac{\pi}{20}n} \qquad (2.2)$$

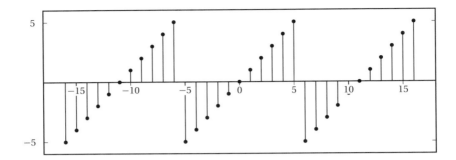

Figure 2.1 Triangular discrete-time wave.

which is a complex exponential of period 40 samples, plotted in Figure 2.2. An example of a sequence drawn from the real world is

$$x[n] = \text{The average Dow-Jones index in year } n \tag{2.3}$$

plotted in Figure 2.3 from year 1900 to 2002. Another example, this time of a random sequence, is

$$x[n] = \text{the } n\text{-th output of a random source } \mathcal{U}(-1,1) \tag{2.4}$$

a realization of which is plotted in Figure 2.4.

A few notes are in order:

- The dependency of the sequence's values on an integer-valued index n is made explicit by the use of *square brackets* for the functional argument. This is standard notation in the signal processing literature.

- The sequence index n is best thought of as a measure of *dimensionless time*; while it has no physical unit of measure, it imposes a chronological order on the values of the sequences.

- We consider complex-valued discrete-time signals; while physical signals can be expressed by real quantities, the generality offered by the complex domain is particularly useful in designing systems which *synthesize* signal, such as data communication systems.

- In graphical representations, when we need to emphasize the discrete-time nature of the signal, we resort to stem (or "lollipop") plots as in Figure 2.1. When the discrete-time domain is understood, we will often use a function-like representation as in Figure 2.3. In the latter case, each ordinate of the sequence is graphically connected to its

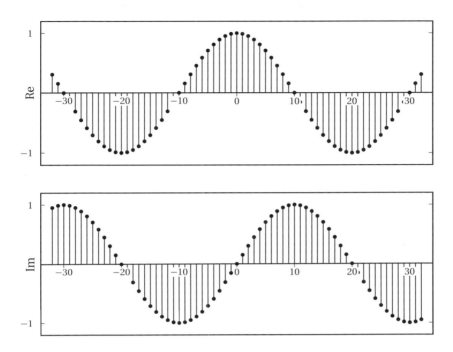

Figure 2.2 Discrete-time complex exponential $x[n] = e^{j\frac{\pi}{20}n}$ (real and imaginary parts).

neighbors, giving the illusion of a continuous-time function: while this makes the plot easier on the eye, it must be remembered that the signal is defined only over a *discrete* set.

2.1.1 The Discrete-Time Abstraction

While analytical forms of discrete-time signals such as the ones above are useful to illustrate the key points of signal processing and are absolutely necessary in the mathematical abstractions which follow, they are nonetheless just that, abstract examples. How does the notion of a discrete-time signal relate to the world around us? A discrete-time signal, in fact, captures our necessarily limited ability to take repeated accurate measurements of a physical quantity. We might be keeping track of the stock market index at the end of each day to draw a pencil and paper chart; or we might be measuring the voltage level at the output of a microphone 44,100 times per second (obviously not by hand!) to record some music via the computer's soundcard. In both cases we need "time to write down the value" and are therefore forced to neglect everything that happens between mea-

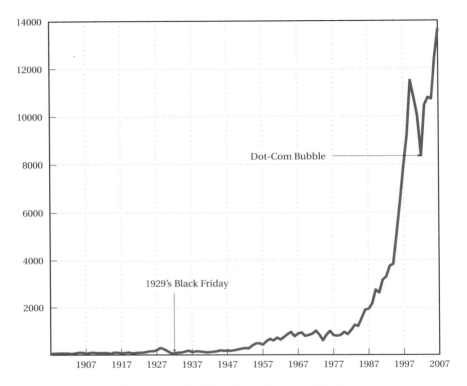

Figure 2.3 The Dow-Jones industrial index.

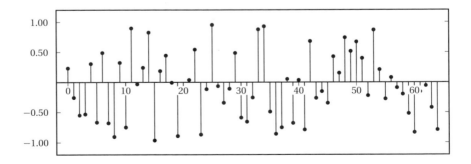

Figure 2.4 An example of random signal.

suring times. This "look and write down" operation is what is normally referred to as *sampling*. There are real-world phenomena which lend themselves very naturally and very intuitively to a discrete-time representation: the daily Dow-Jones index, for example, solar spots, yearly floods of the Nile, etc. There seems to be no irrecoverable loss in this neglect of intermediate values. But what about music, or radio waves? At this point it is not altogether clear from an intuitive point of view how a sampled measurement of these phenomena entail no loss of information. The mathematical proof of this will be shown in detail when we study the sampling theorem; for the time being let us say that "the proof of the cake is in the eating": just listen to your favorite CD!

The important point to make here is that, once a real-world signal is converted to a discrete-time representation, the underlying notion of "time between measurements" becomes completely abstract. All we are left with is a sequence of numbers, and all signal processing manipulations, with their intended results, are independent of the way the discrete-time signal is obtained. The power and the beauty of digital signal processing lies in part with its invariance with respect to the underlying physical reality. This is in stark contrast with the world of analog circuits and systems, which have to be realized in a version specific to the physical nature of the input signals.

2.1.2 Basic Signals

The following sequences are fundamental building blocks for the theory of signal processing.

Impulse. The discrete-time impulse (or discrete-time *delta function*) is potentially the simplest discrete-time signal; it is shown in Figure 2.5(a) and is defined as

$$\delta[n] = \begin{cases} 1 & n = 0 \\ 0 & n \neq 0 \end{cases} \quad (2.5)$$

Unit Step. The discrete-time unit step is shown in Figure 2.5(b) and is defined by the following expression:

$$u[n] = \begin{cases} 1 & n \geq 0 \\ 0 & n < 0 \end{cases} \quad (2.6)$$

The unit step can be obtained via a discrete-time integration of the impulse (see eq. (2.16)).

Exponential Decay. The discrete-time exponential decay is shown in Figure 2.5(c) and is defined as

$$x[n] = a^n u[n], \qquad a \in \mathbb{C}, |a| < 1 \tag{2.7}$$

The exponential decay is, as we will see, the free response of a discrete-time first order recursive filter. Exponential sequences are well-behaved only for values of a less than one in magnitude; sequences in which $|a| > 1$ are unbounded and represent an unstable behavior (their energy and power are both infinite).

Complex Exponential. The discrete-time complex exponential has already been shown in Figure 2.2 and is defined as

$$x[n] = e^{j(\omega_0 n + \phi)} \tag{2.8}$$

Special cases of the complex exponential are the real-valued discrete-time sinusoidal oscillations:

$$x[n] = \sin(\omega_0 n + \phi) \tag{2.9}$$
$$x[n] = \cos(\omega_0 n + \phi) \tag{2.10}$$

An example of (2.9) is shown in Figure 2.5(d).

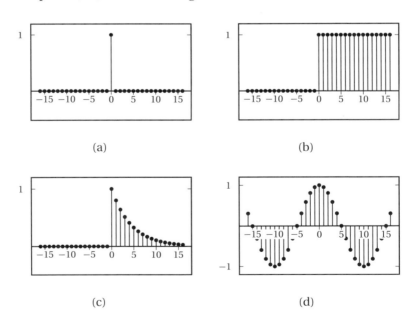

Figure 2.5 Basic signals. Impulse (a); unit step (b); decaying exponential (c); real-valued sinusoid (d).

2.1.3 Digital Frequency

With respect to the oscillatory behavior captured by the complex exponential, a note on the concept of "frequency" is in order. In the continuous-time world (the world of textbook physics, to be clear), where time is measured in seconds, the usual unit of measure for frequency is the Hertz which is equivalent to 1/second. In the discrete-time world, where the index n represents a dimensionless time, "digital" frequency is expressed in radians which is itself a dimensionless quantity.[1] The best way to appreciate this is to consider an algorithm to generate successive samples of a discrete-time sinusoid at a digital frequency ω_0:

$\omega \leftarrow 0;$	*initialization*
$\phi \leftarrow$ initial phase value;	
repeat	
$\quad x \leftarrow \sin(\omega + \phi);$	*compute next value*
$\quad \omega \leftarrow \omega + \omega_0;$	*update phase*
until done	

At each iteration,[2] the argument of the trigonometric function is incremented by ω_0 and a new output sample is produced. With this in mind, it is easy to see that *the highest frequency manageable by a discrete-time system is $\omega_{\max} = 2\pi$*; for any frequency larger than this, the inner 2π-periodicity of the trigonometric functions "maps back" the output values to a frequency between 0 and 2π. This can be expressed as an equation:

$$\sin(n(\omega + 2k\pi) + \phi) = \sin(n\omega + \phi) \tag{2.11}$$

for all values of $k \in \mathbb{Z}$. This 2π-equivalence of digital frequencies is a pervasive concept in digital signal processing and it has many important consequences which we will study in detail in the next Chapters.

---

[1] An angle measure in radians is dimensionless since it is defined in terms of the ratio of two lengths, the radius and the arc subtended by the measured angle on an arbitrary circle.

[2] Here is the algorithm written in C:
```
extern double omega0;
extern double phi;
static double omega = 0;
double GetNextValue()
{
    omega += omega0;
    return sin(omega + phi);
}
```

2.1.4 Elementary Operators

In this Section we present some elementary operations on sequences.

Shift. A sequence $x[n]$, shifted by an integer k is simply:

$$y[n] = x[n-k] \tag{2.12}$$

If k is positive, the signal is shifted "to the left", meaning that the signal has been *delayed*; if k is negative, the signal is shifted "to the right", meaning that the signal has been *advanced*. The delay operator can be indicated by the following notation:

$$\mathscr{D}_k\{x[n]\} = x[n-k]$$

Scaling. A sequence $x[n]$ scaled by a factor $\alpha \in \mathbb{C}$ is

$$y[n] = \alpha x[n] \tag{2.13}$$

If α is real, then the scaling represents a simple amplification or attenuation of the signal (when $\alpha > 1$ and $\alpha < 1$, respectively). If α is complex, amplification and attenuation are compounded with a phase shift.

Sum. The sum of two sequences $x[n]$ and $w[n]$ is their term-by-term sum:

$$y[n] = x[n] + w[n] \tag{2.14}$$

Please note that sum and scaling are linear operators. Informally, this means scaling and sum behave "intuitively":

$$\alpha(x[n] + w[n]) = \alpha x[n] + \alpha w[n]$$

or

$$\mathscr{D}_k\{x[n] + w[n]\} = x[n-k] + w[n-k]$$

Product. The product of two sequences $x[n]$ and $w[n]$ is their term-by-term product

$$y[n] = x[n]w[n] \tag{2.15}$$

Integration. The discrete-time equivalent of integration is expressed by the following running sum:

$$y[n] = \sum_{k=-\infty}^{n} x[k] \tag{2.16}$$

Intuitively, integration computes a non-normalized running average of the discrete-time signal.

Differentiation. A discrete-time approximation to differentiation is the first-order difference:[3]

$$y[n] = x[n] - x[n-1] \qquad (2.17)$$

With respect to Section 2.1.2, note how the unit step can be obtained by applying the integration operator to the discrete-time impulse; conversely, the impulse can be obtained by applying the differentiation operator to the unit step.

2.1.5 The Reproducing Formula

The signal reproducing formula is a simple application of the basic signal and signal properties that we have just seen and it states that

$$x[n] = \sum_{k=-\infty}^{\infty} x[k]\delta[n-k] \qquad (2.18)$$

Any signal can be expressed as a linear combination of suitably weighed and shifted impulses. In this case, the weights are nothing but the signal values themselves. While self-evident, this formula will reappear in a variety of fundamental derivations since it captures the "inner structure" of a discrete-time signal.

2.1.6 Energy and Power

We define the *energy* of a discrete-time signal as

$$E_x = \|x\|_2^2 = \sum_{n=-\infty}^{\infty} |x[n]|^2 \qquad (2.19)$$

(where the squared-norm notation will be clearer after the next Chapter). This definition is consistent with the idea that, if the values of the sequence represent a time-varying voltage, the above sum would express the total energy (in joules) dissipated over a 1Ω-resistor. Obviously, the energy is finite only if the above sum converges, i.e. if the sequence $x[n]$ is *square-summable*. A signal with this property is sometimes referred to as a *finite-energy signal*. For a simple example of the converse, note that a periodic signal which is not identically zero is *not* square-summable.

---

[3] We will see later that a more "correct" approximation to differentiation is given by a filter $H(e^{j\omega}) = j\omega$. For most applications, however, the first-order difference will suffice.

We define the *power* of a signal as the usual ratio of energy over time, taking the limit over the number of samples considered:

$$P_x = \lim_{N \to \infty} \frac{1}{2N} \sum_{-N}^{N-1} |x[n]|^2 \tag{2.20}$$

Clearly, signals whose energy is finite, have zero total power (i.e. their energy dilutes to zero over an infinite time duration). Exponential sequences which are not decaying (i.e. those for which $|a| > 1$ in (2.7)) possess infinite power (which is consistent with the fact that they describe an unstable behavior). Note, however, that many signals whose energy is infinite *do* have finite power and, in particular, periodic signals (such as sinusoids and combinations thereof). Due to their periodic nature, however, the above limit is undetermined; we therefore *define* their power to be simply the *average energy over a period*. Assuming that the period is N samples, we have

$$P_x = \frac{1}{N} \sum_{n=0}^{N-1} |x[n]|^2 \tag{2.21}$$

2.2 Classes of Discrete-Time Signals

The examples of discrete-time signals in (2.1) and (2.2) are two-sided, infinite sequences. Of course, in the practice of signal processing, it is impossible to deal with infinite quantities of data: for a processing algorithm to execute in a finite amount of time and to use a finite amount of storage, the input must be of finite length; even for algorithms that operate on the fly, i.e. algorithms that produce an output sample for each new input sample, an implicit limitation on the input data size is imposed by the necessarily limited life span of the processing device.[4] This limitation was all too apparent in our attempts to plot infinite sequences as shown in Figure 2.1 or 2.2: what the diagrams show, in fact, is just a meaningful and representative portion of the signals; as for the rest, the analytical description remains the only reference. When a discrete-time signal admits no closed-form representation, as is basically always the case with real-world signals, its finite time support arises naturally because of the finite time spent recording the signal: every piece of music has a beginning and an end, and so did every phone conversation. In the case of the sequence representing the Dow Jones index, for instance, we basically cheated since the index did not even exist for years before 1884, and its value tomorrow is certainly not known – so that

---

[4] Or, in the extreme limit, of the supervising engineer ...

the signal is not really a sequence, although it can be arbitrarily extended to one. More importantly (and more often), the finiteness of a discrete-time signal is explicitly imposed by design since we are interested in concentrating our processing efforts on a small portion of an otherwise longer signal; in a speech recognition system, for instance, the practice is to cut up a speech signal into small segments and try to identify the phonemes associated to each one of them.[5] A special case is that of periodic signals; even though these are bona-fide infinite sequences, it is clear that all information about them is contained in just one period. By describing one period (graphically or otherwise), we are, in fact, providing a full description of the sequence. The complete taxonomy of the discrete-time signals used in the book is the subject of the next Sections ans is summarized in Table 2.1.

2.2.1 Finite-Length Signals

As we just mentioned, a finite-length discrete-time signal of length N are just a collection of N complex values. To introduce a point that will reappear throughout the book, a finite-length signal of length N is entirely equivalent to a vector in $\mathbb{C}^N$. This equivalence is of immense import since all the tools of linear algebra become readily available for describing and manipulating finite-length signals. We can represent an N-point finite-length signal using the standard vector notation

$$\mathbf{x} = \begin{bmatrix} x_0 & x_1 & \ldots & x_{N-1} \end{bmatrix}^T$$

Note the transpose operator, which declares **x** as a *column* vector; this is the customary practice in the case of complex-valued vectors. Alternatively, we can (and often will) use a notation that mimics the one used for proper sequences:

$$x[n], \quad n = 0, \ldots, N-1$$

Here we *must* remember that, although we use the notation $x[n]$, $x[n]$ is *not defined* for values outside its support, i.e. for $n < 0$ or for $n \geq N$. Note that we can always obtain a finite-length signal from an infinite sequence by simply dropping the sequence values outside the indices of interest. Vector and sequence notations are equivalent and will be used interchangeably according to convenience; in general, the vector notation is useful when we want to stress the algorithmic or geometric nature of certain signal processing operations. The sequence notation is useful in stressing the algebraic structure of signal processing.

---

[5] Note that, in the end, phonemes are pasted together into words and words into sentences; therefore, for a complete speech recognition system, long-range dependencies become important again.

Finite-length signals are extremely convenient entities: their energy is always and, as a consequence, no stability issues arise in processing. From the computational point of view, they are not only a necessity but often the cornerstone of very efficient algorithmic design (as we will see, for instance, in the case of the FFT); one could say that all "practical" signal processing lives in $\mathbb{C}^N$. It would be extremely awkward, however, to develop the whole theory of signal processing only in terms of finite-length signals; the asymptotic behavior of algorithms and transformations for infinite sequences is also extremely valuable since a stability result proven for a general sequence will hold for all finite-length signals too. Furthermore, the notational flexibility which infinite sequences derive from their function-like definition is extremely practical from the point of view of the notation. We can immediately recognize and understand the expression $x[n-k]$ as a k-point shift of a sequence $x[n]$; but, in the case of finite-support signals, how are we to define such a shift? We would have to explicitly take into account the finiteness of the signal and the associated "border effects", i.e. the behavior of operations at the edges of the signal. For this reason, in most derivations which involve finite-length signal, these signals will be *embedded* into proper sequences, as we will see shortly.

2.2.2 Infinite-Length Signals

Aperiodic Signals. The most general type of discrete-time signal is represented by a generic infinite complex sequence. Although, as previously mentioned, they lie beyond our processing and storage capabilities, they are invaluably useful as a generalization in the limit. As such, they must be handled with some care when it comes to their properties. We will see shortly that two of the most important properties of infinite sequences con-

$$\tilde{x}[n] = \ldots, x_{N-2}, x_{N-1}, \overbrace{x_0, x_1, x_2, \ldots, x_{N-2}, x_{N-1}}^{\mathbf{x}}, x_0, x_1, \ldots$$

$$\updownarrow \rightarrow n = 0$$

$$\tilde{x}[n-1] = \ldots, x_{N-3}, x_{N-2}, \underbrace{x_{N-1}, x_0, x_1, x_2, \ldots, x_{N-2}}_{\mathbf{x}'}, x_{N-1}, x_0, \ldots$$

Figure 2.6 Equivalence between a right shift by one of a periodized signal and the circular shift of the original signal. $\mathbf{x}$ and $\mathbf{x}'$ are the length-N original signal and its right circular shift by one, respectively.

cern their summability: this can take the form of either *absolute summability* (stronger condition) or *square summability* (weaker condition, corresponding to finite energy).

Periodic Signals. A periodic sequence with period N is one for which

$$\tilde{x}[n] = \tilde{x}[n + kN], \qquad k \in \mathbb{Z} \tag{2.22}$$

The tilde notation $\tilde{x}[n]$ will be used whenever we need to explicitly stress a periodic behavior. Clearly an N-periodic sequence is completely defined by its N values over a period; that is, a periodic sequence "carries no more information" than a finite-length signal of length N.

Periodic Extensions. Periodic sequences are infinite in length, and yet their information is contained within a finite number of samples. In this sense, periodic sequences represent a first bridge between finite-length signals and infinite sequences. In order to "embed" a finite-length signal $x[n]$, $n = 0, \ldots, N-1$ into a sequence, we can take its periodized version:

$$\tilde{x}[n] = x[n \bmod N], \qquad n \in \mathbb{Z} \tag{2.23}$$

this is called the *periodic extension* of the finite length signal $x[n]$. This type of extension is the "natural" one in many contexts, for reasons which will be apparent later when we study the frequency-domain representation of discrete-time signals. Note that now an arbitrary shift of the periodic sequence corresponds to the periodization of a *circular shift* of the original finite-length signal. A circular shift by $k \in \mathbb{Z}$ is easily visualized by imagining a shift register; if we are shifting towards the right ($k > 0$), the values which pop out of the rightmost end of the shift register are pushed back in at the other end.[6] The relationship between the circular shift of a finite-length signal and the linear shift of its periodic extension is depicted in Figure 2.6. Finally, the energy of a periodic extension becomes infinite, while its power is simply the energy of the finite-length original signal scaled by $1/N$.

Finite-Support Signals. An infinite discrete-time sequence $\bar{x}[n]$ is said to have *finite support* if its values are zero for all indices outside of an interval; that is, there exist N and $M \in \mathbb{Z}$ such that

$$\bar{x}[n] = 0 \quad \text{for } n < M \text{ and } n > M + N - 1$$

Note that, although $\bar{x}[n]$ is an infinite sequence, the knowledge of M and of the N nonzero values of the sequence completely specifies the entire signal.

---

[6] For example, if $\mathbf{x} = [1\ 2\ 3\ 4\ 5]$, a right circular shift by 2 yields $\mathbf{x} = [4\ 5\ 1\ 2\ 3]$.

This suggests another approach to embedding a finite-length signal $x[n]$, $n = 0,\ldots,N-1$, into a sequence, i.e.

$$\bar{x}[n] = \begin{cases} x[n] & \text{if } 0 \le n < N-1 \\ 0 & \text{otherwise} \end{cases} \quad n \in \mathbb{Z} \quad (2.24)$$

where we have chosen $M = 0$ (but any other choice of M could be used). Note that, here, in contrast to the the periodic extension of $x[n]$, we are actually adding arbitrary information in the form of the zero values outside of the support interval. This is not without consequences, as we will see in the following Chapters. In general, we will use the bar notation $\bar{x}[n]$ for sequences defined as the finite support extension of a finite-length signal. Note that, now, the shift of the finite-support extension gives rise to a zero-padded shift of the signal locations between M and $M+N-1$; the dynamics of the shift are shown in Figure 2.7.

$$\bar{x}[n] = \ldots, 0, 0, \overbrace{x_0, x_1, x_2, \ldots, x_{N-2}, x_{N-1}}^{\mathbf{x}}, 0, 0, 0, 0, \ldots$$
$$\updownarrow \to n = 0$$
$$\bar{x}[n-1] = \ldots, 0, 0, \underbrace{0, x_0, x_1, x_2, \ldots, x_{N-3}, x_{N-2}}_{\mathbf{x}'}, x_{N-1}, 0, 0, \ldots$$

Figure 2.7 Relationship between the right shift by one of a finite-support extension and the zero padded shift of the original signal. $\mathbf{x}$ and $\mathbf{x}'$ are the length-N original signal and its zero-padded shift by one, respectively.

Table 2.1 Basic discrete-time signal types.

Signal Type	Notation	Energy	Power
Finite-Length	$x[n], \ n = 0, 1, \ldots, N-1$ $\mathbf{x}, \ \mathbf{x} \in \mathbb{C}^N$	$\sum_{n=0}^{N-1} \lvert x[n] \rvert^2$	undef.
Infinite-Length	$x[n], \ n \in \mathbb{Z}$	eq. (2.19)	eq. (2.20)
N-Periodic	$\tilde{x}[n], \ n \in \mathbb{Z}$ $\tilde{x}[n] = \tilde{x}[n+kN]$	∞	eq. (2.21)
Finite-Support	$\bar{x}[n], \ n \in \mathbb{Z}$ $\bar{x}[n] \ne 0$ for $M \le n \le M+N-1$	$\sum_{n=M}^{M+N-1} \lvert x[n] \rvert^2$	0

Examples

Example 2.1: Discrete-time in the Far West

The fact that the "fastest" digital frequency is 2π can be readily appreciated in old western movies. In classic scenarios there is always a sequence showing a stagecoach leaving town. We can see the spoked wagon wheels starting to turn forward faster and faster, then stop and then starting to turn backwards. In fact, each frame in the movie is a snapshot of a spinning disk with increasing angular velocity. The filming process therefore transforms the wheel's movement into a sequence of discrete-time positions depicting a circular motion with increasing frequency. When the speed of the wheel is such that the time between frames covers a full revolution, the wheel appears to be stationary: this corresponds to the fact that the maximum digital frequency $\omega = 2\pi$ is undistinguishable from the slowest frequency $\omega = 0$. As the speed of the real wheel increases further, the wheel on film starts to move backwards, which corresponds to a negative digital frequency. This is because a displacement of $2\pi + \alpha$ between successive frames is interpreted by the brain as a negative displacement of α: our intuition always privileges the most economical explanation of natural phenomena.

Example 2.2: Building periodic signals

Given a discrete-time signal $x[n]$ and an integer $N > 0$ we can always formally write

$$\tilde{y}[n] = \sum_{k=-\infty}^{\infty} x[n-kN]$$

The signal $\tilde{y}[n]$, if it exists, is an N-periodic sequence. The periodic signal $\tilde{y}[n]$ is "manufactured" by superimposing infinite copies of the original signal $x[n]$ spaced N samples apart. We can distinguish three cases:

(a) If $x[n]$ is finite-support and N is bigger than the size of the support, then the copies in the sum do not overlap; in the limit, if N is exactly equal to the size of the support then $\tilde{y}[n]$ corresponds to the periodic extension of $x[n]$ considered as a finite-length signal.

(b) If $x[n]$ is finite-support and N is smaller than the size of the support then the copies in the sum do overlap; for each n, the value of $\tilde{y}[n]$ is be the sum of at most a finite number of terms.

(c) If $x[n]$ has infinite support, then each value of $\tilde{y}[n]$ is be the sum of an infinite number of terms. Existence of $\tilde{y}[n]$ depends on the properties of $x[n]$.

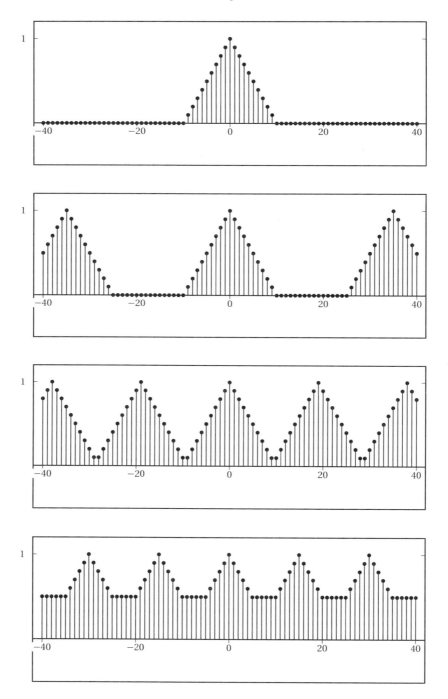

Figure 2.8 Periodization of a simple finite-support signal (support length $L = 19$); original signal (top panel) and periodized versions with $N = 35 > L$, $N = 19 = L$, $N = 15 < L$ respectively.

The first two cases are illustrated in Figure 2.8. In practice, the periodization of short sequences is an effective method to synthesize the sound of string instruments such as a guitar or a piano; used in conjunction with simple filters, the technique is known as the Karplus-Strong algorithm.

As an example of the last type, take for instance the signal $x[n] = \alpha^{-n} u[n]$. The periodization formula leads to

$$\tilde{y}[n] = \sum_{k=-\infty}^{\infty} \alpha^{-(n-kN)} u[n-kN] = \sum_{k=-\infty}^{\lfloor n/N \rfloor} \alpha^{-(n-kN)}$$

since $u[n-kN] = 0$ for $k \geq \lfloor n/N \rfloor$. Now write $n = mN + i$ with $m = \lfloor n/N \rfloor$ and $i = n \bmod N$. We have

$$\tilde{y}[n] = \sum_{k=-\infty}^{m} \alpha^{-(m-k)N-i} = \alpha^{-i} \sum_{h=0}^{\infty} \alpha^{-hN}$$

which exists and is finite for $|\alpha| > 1$; for these values of α we have

$$\tilde{y}[n] = \frac{\alpha^{-(n \bmod N)}}{1 - \alpha^{-N}} \tag{2.25}$$

which is indeed N-periodic. An example is shown in Figure 2.9.

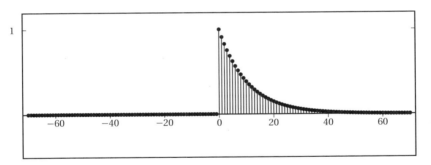

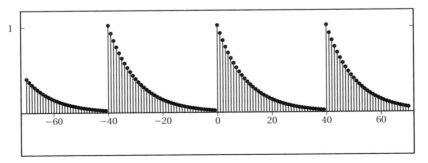

Figure 2.9 Periodization of $x[n] = 1.1^{-n} u[n]$ with $N = 40$; original signal (top panel) and periodized version (bottom panel).

Further Reading

For more discussion on discrete-time signals, see *Discrete-Time Signal Processing*, by A. V. Oppenheim and R. W. Schafer (Prentice-Hall, last edition in 1999), in particular Chapter 2.

Other books of interest include: B. Porat, A Course in *Digital Signal Processing* (Wiley, 1997) and R. L. Allen and D. W. Mills' *Signal Analysis* (IEEE Press, 2004).

Exercises

Exercise 2.1: Review of complex numbers.

(a) Let $s[n] := \dfrac{1}{2^n} + j\dfrac{1}{3^n}$. Compute $\sum_{n=1}^{\infty} s[n]$.

(b) Same question with $s[n] := \left(\dfrac{j}{3}\right)^n$.

(c) Characterize the set of complex numbers satisfying $z^* = z^{-1}$.

(d) Find 3 complex numbers $\{z_0, z_1, z_2\}$ which satisfy $z_i^3 = 1$, $i = 1, 2, 3$.

(e) What is the following infinite product $\prod_{n=1}^{\infty} e^{j\pi/2^n}$?

Exercise 2.2: Periodic signals. For each of the following discrete-time signals, state whether the signal is periodic and, if so, specify the period:

(a) $x[n] = e^{j\frac{n}{\pi}}$

(b) $x[n] = \cos(n)$

(c) $x[n] = \sqrt{\cos\left(\pi\dfrac{n}{7}\right)}$

(d) $x[n] = \sum_{k=-\infty}^{+\infty} y[n - 100k]$, with $y[n]$ absolutely summable.

Chapter 3

Signals and Hilbert Spaces

In the 17th century, algebra and geometry started to interact in a fruitful synergy which continues to the present day. Descartes's original idea of translating geometric constructs into algebraic form spurred a new line of attack in mathematics; soon, a series of astonishing results was produced for a number of problems which had long defied geometrical solutions (such as, famously, the trisection of the angle). It also spearheaded the notion of vector space, in which a geometrical point could be represented as an n-tuple of coordinates; this, in turn, readily evolved into the theory of linear algebra. Later, the concept proved useful in the opposite direction: many algebraic problems could benefit from our innate geometrical intuition once they were cast in vector form; from the easy three-dimensional visualization of concepts such as distance and orthogonality, more complex algebraic constructs could be brought within the realm of intuition. The final leap of imagination came with the realization that the concept of vector space could be applied to much more abstract entities such as infinite-dimensional objects and functions. In so doing, however, spatial intuition could be of limited help and for this reason, the notion of vector space had to be formalized in much more rigorous terms; we will see that the definition of Hilbert space is one such formalization.

Most of the signal processing theory which in this book can be usefully cast in terms of vector notation and the advantages of this approach are exactly what we have just delineated. Firstly of all, all the standard machinery of linear algebra becomes immediately available and applicable; this greatly simplifies the formalism used in the mathematical proofs which will follow and, at the same time, it fosters a good intuition with respect to the underlying principles which are being put in place. Furthermore, the vector notation creates a frame of thought which seamlessly links the more abstract results involving infinite sequences to the algorithmic reality involving finite-length signals. Finally, on the practical side, vector notation is the standard paradigm for numerical analysis packages such as Matlab; signal processing algorithms expressed in vector notation translate to working code with very little effort.

In the previous Chapter, we established the basic notation for the different classes of discrete-time signals which we will encounter time and again in the rest of the book and we hinted at the fact that a tight correspondence can be established between the concept of signal and that of vector space. In this Chapter, we pursue this link further, firstly by reviewing the familiar Euclidean space in finite dimensions and then by extending the concept of vector space to infinite-dimensional Hilbert spaces.

3.1 Euclidean Geometry: a Review

Euclidean geometry is a straightforward formalization of our spatial sensory experience; hence its cornerstone role in developing a basic intuition for vector spaces. Everybody is (or should be) familiar with Euclidean geometry and the natural "physical" spaces like $\mathbb{R}^2$ (the plane) and $\mathbb{R}^3$ (the three-dimensional space). The notion of *distance* is clear; *orthogonality* is intuitive and maps to the idea of a "right angle". Even a more abstract concept such as that of *basis* is quite easy to contemplate (the standard coordinate concepts of latitude, longitude and height, which correspond to the three orthogonal axes in $\mathbb{R}^3$). Unfortunately, immediate spatial intuition fails us for higher dimensions (i.e. for $\mathbb{R}^N$ with $N > 3$), yet the basic concepts introduced for $\mathbb{R}^3$ generalize easily to $\mathbb{R}^N$ so that it is easier to state such concepts for the higher-dimensional case and specialize them with examples for $N = 2$ or $N = 3$. These notions, ultimately, will be generalized even further to more abstract types of vector spaces. For the moment, let us review the properties of $\mathbb{R}^N$, the N-dimensional Euclidean space.

Vectors and Notation. A point in $\mathbb{R}^N$ is specified by an N-tuple of coordinates:[1]

$$\mathbf{x} = \begin{bmatrix} x_0 \\ x_1 \\ \vdots \\ x_{N-1} \end{bmatrix} = [x_0 \quad x_1 \quad \ldots \quad x_{N-1}]^T$$

---

[1] N-dimensional vectors are by default *column* vectors.

where $x_i \in \mathbb{R}$, $i = 0, 1, \ldots, N-1$. We call this set of coordinates a *vector* and the N-tuple will be denoted synthetically by the symbol $\mathbf{x}$; coordinates are usually expressed with respect to a "standard" orthonormal basis.[2] The vector $\mathbf{0} = [0 \ 0 \ \ldots \ 0]^T$, i.e. the null vector, is considered the origin of the coordinate system.

The generic n-th element in vector $\mathbf{x}$ is indicated by the subscript $\mathbf{x}_n$. In the following we will often consider a *set* of M arbitrarily chosen vectors in $\mathbb{R}^N$ and this set will be indicated by the notation $\{\mathbf{x}^{(k)}\}_{k=0\ldots M-1}$. Each vector in the set is indexed by the superscript $\cdot^{(k)}$. The n-th element of the k-th vector in the set is indicated by the notation $\mathbf{x}_n^{(k)}$

Inner Product. The inner product between two vectors $\mathbf{x}, \mathbf{y} \in \mathbb{R}^N$ is defined as

$$\langle \mathbf{x}, \mathbf{y} \rangle = \sum_{n=0}^{N-1} x_n y_n \tag{3.1}$$

We say that $\mathbf{x}$ and $\mathbf{y}$ are orthogonal, or $\mathbf{x} \perp \mathbf{y}$, when their inner product is zero:

$$\mathbf{x} \perp \mathbf{y} \iff \langle \mathbf{x}, \mathbf{y} \rangle = 0 \tag{3.2}$$

Norm. The norm of a vector is defined in terms of the inner product as

$$\|\mathbf{x}\|_2 = \sqrt{\sum_{n=0}^{N-1} x_n^2} = \langle \mathbf{x}, \mathbf{x} \rangle^{1/2} \tag{3.3}$$

It is easy to visualize geometrically that the norm of a vector corresponds to its length, i.e. to the distance between the origin and the point identified by the vector's coordinates. A remarkable property linking the inner product and the norm is the Cauchy-Schwarz inequality (the proof of which is nontrivial); given $\mathbf{x}, \mathbf{y} \in \mathbb{R}^N$ we can state that

$$|\langle \mathbf{x}, \mathbf{y} \rangle| \leq \|\mathbf{x}\|_2 \|\mathbf{y}\|_2$$

Distance. The concept of norm is used to introduce the notion of Euclidean *distance* between two vectors $\mathbf{x}$ and $\mathbf{y}$:

$$d(\mathbf{x}, \mathbf{y}) = \|\mathbf{x} - \mathbf{y}\|_2 = \sqrt{\sum_{n=0}^{N-1} (x_n - y_n)^2} \tag{3.4}$$

---

[2] The concept of basis will be defined more precisely later on; for the time being, consider a standard set of orthogonal axes.

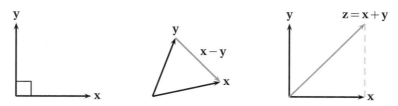

Figure 3.1 Elementary properties of vectors in $\mathbb{R}^2$: orthogonality of two vectors **x** and **y** (left); difference vector **x**−**y** (middle); sum of two orthogonal vectors **z** = **x**+**y**, also known as Pythagorean theorem (right).

From this, we can easily derive the Pythagorean theorem for N dimensions: if two vectors are orthogonal, $\mathbf{x} \perp \mathbf{y}$, and we consider the sum vector $\mathbf{z} = \mathbf{x} + \mathbf{y}$, we have

$$\|\mathbf{z}\|_2^2 = \|\mathbf{x}\|_2^2 + \|\mathbf{y}\|_2^2 \tag{3.5}$$

The above properties are graphically shown in Figure 3.1 for $\mathbb{R}^2$.

Bases. Consider a set of M arbitrarily chosen vectors in $\mathbb{R}^N$: $\{\mathbf{x}^{(k)}\}_{k=0...M-1}$. Given such a set, a key question is that of completeness: can *any* vector in $\mathbb{R}^N$ be written as a linear combination of vectors from the set? In other words, we ask ourselves whether, for any $\mathbf{z} \in \mathbb{R}^N$, we can find a set of M coefficients $\alpha_k \in \mathbb{R}$ such that $\mathbf{z}$ can be expressed as

$$\mathbf{z} = \sum_{k=0}^{M-1} \alpha_k \mathbf{x}^{(k)} \tag{3.6}$$

Clearly, M needs to be greater or equal to N, but what conditions does a set of vectors $\{\mathbf{x}^{(k)}\}_{k=0...M-1}$ need to satisfy so that (3.6) holds for any $\mathbf{z} \in \mathbb{R}^N$? There needs to be a set of M vectors that *span* $\mathbb{R}^N$, and it can be shown that this is equivalent to saying that the set must contain at least N *linearly independent* vectors. In turn, N vectors $\{\mathbf{y}^{(k)}\}_{k=0...N-1}$ are linearly independent if the equation

$$\sum_{k=0}^{N-1} \beta_k \mathbf{y}^{(k)} = 0 \tag{3.7}$$

is satisfied only when all the β_k's are zero. A set of N linearly independent vectors for $\mathbb{R}^N$ is called a *basis* and, amongst bases, the ones with mutually orthogonal vectors of norm equal to one are called *orthonormal bases*. For an orthonormal basis $\{\mathbf{y}^{(k)}\}$ we therefore have

$$\langle \mathbf{y}^{(k)}, \mathbf{y}^{(h)} \rangle = \begin{cases} 1 & \text{if } k = h \\ 0 & \text{otherwise} \end{cases} \tag{3.8}$$

Figure 3.2 reviews the above concepts in low dimensions.

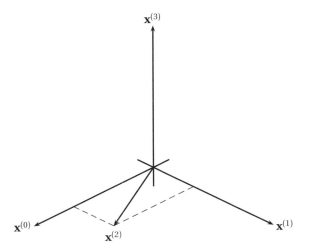

Figure 3.2 Linear independence and bases: $\mathbf{x}^{(0)}$, $\mathbf{x}^{(1)}$ and $\mathbf{x}^{(2)}$ are coplanar in $\mathbb{R}^3$ and, therefore, they do not form a basis; conversely, $\mathbf{x}^{(3)}$ and any two of $\{\mathbf{x}^{(0)}, \mathbf{x}^{(1)}, \mathbf{x}^{(2)}\}$ are linearly independent.

The standard orthonormal basis for $\mathbb{R}^N$ is the *canonical basis* $\{\boldsymbol{\delta}^{(k)}\}_{k=0...N-1}$ with

$$\boldsymbol{\delta}_n^{(k)} = \delta[n-k] = \begin{cases} 1 & \text{if } n = k \\ 0 & \text{otherwise} \end{cases}$$

The orthonormality of such a set is immediately apparent. Another important orthonormal basis for $\mathbb{R}^N$ is the normalized *Fourier basis* $\{\mathbf{w}^{(k)}\}_{k=0...N-1}$ for which

$$\mathbf{w}_n^{(k)} = \frac{1}{\sqrt{N}} e^{-j\frac{2\pi}{N}nk}$$

The orthonormality of the basis will be proved in the next Chapter.

3.2 From Vector Spaces to Hilbert Spaces

The purpose of the previous Section was to briefly review the elementary notions and spatial intuitions of Euclidean geometry. A thorough study of vectors in $\mathbb{R}^N$ and $\mathbb{C}^N$ is the subject of linear algebra; yet, the idea of vectors, orthogonality and bases is much more general, the basic ingredients being an inner product and the use of a square norm as in (3.3).

While the analogy between vectors in $\mathbb{C}^N$ and length-N signal is readily apparent, the question now hinges on how we are to proceed in order to generalize the above concepts to the class of infinite sequences. Intuitively, for instance, we can let N grow to infinity and obtain $\mathbb{C}^\infty$ as the Euclidean space for infinite sequences; in this case, however, much care must be exercised with expressions such as (3.1) and (3.3) which can diverge for sequences as simple as $x[n] = 1$ for all n. In fact, the proper generalization of $\mathbb{C}^N$ to an infinite number of dimensions is in the form of a particular vector space called *Hilbert space*; the structure of this kind of vector space imposes a set of constraints on its elements so that divergence problems, such as the one we just mentioned, no longer bother us. When we embed infinite sequences into a Hilbert space, these constraints translate to the condition that the corresponding signals have finite energy – which is a mild and reasonable requirement.

Finally, it is important to remember that the notion of Hilbert space is applicable to much more general vector spaces than $\mathbb{C}^N$; for instance, we can easily consider spaces of functions over an interval or over the real line. This generality is actually the cornerstone of a branch of mathematics called *functional analysis*. While we will not follow in great depth these kinds of generalizations, we will certainly point out a few of them along the way. The space of square integrable functions, for instance, will turn out to be a marvelous tool a few Chapters from now when, finally, the link between continuous—and discrete—time signals will be explored in detail.

3.2.1 The Recipe for Hilbert Space

A word of caution: we are now starting to operate in a world of complete abstraction. Here a vector is an entity *per se* and, while analogies and examples in terms of Euclidean geometry can be useful visually, they are, by no means, exhaustive. In other words: vectors are no longer just N-tuples of numbers; they can be anything. This said, a Hilbert space can be defined in incremental steps starting from a general notion of vector space and by supplementing this space with two additional features: the existence of an inner product and the property of completeness.

Vector Space. Consider a set of vectors V and a set of scalars S (which can be either $\mathbb{R}$ or $\mathbb{C}$ for our purposes). A vector space $H(V,S)$ is completely defined by the existence of a vector addition operation and a scalar multiplication operation which satisfy the following properties for any $\mathbf{x}, \mathbf{y}, \mathbf{z}, \in V$ and any $\alpha, \beta \in S$:

- Addition is commutative:

$$\mathbf{x}+\mathbf{y}=\mathbf{y}+\mathbf{x} \tag{3.9}$$

- Addition is associative:

$$(\mathbf{x}+\mathbf{y})+\mathbf{z}=\mathbf{x}+(\mathbf{y}+\mathbf{z}) \tag{3.10}$$

- Scalar multiplication is distributive:

$$\alpha(\mathbf{x}+\mathbf{y})=\alpha\mathbf{x}+\alpha\mathbf{y} \tag{3.11}$$

$$(\alpha+\beta)\mathbf{x}=\alpha\mathbf{x}+\beta\mathbf{x} \tag{3.12}$$

$$\alpha(\beta\mathbf{x})=(\alpha\beta)\mathbf{x} \tag{3.13}$$

- There exists a null vector $\mathbf{0}$ in V which is the additive identity so that $\forall \mathbf{x} \in V$:

$$\mathbf{x}+\mathbf{0}=\mathbf{0}+\mathbf{x}=\mathbf{x} \tag{3.14}$$

- $\forall \mathbf{x} \in V$ there exists in V an additive inverse $-\mathbf{x}$ such that

$$\mathbf{x}+(-\mathbf{x})=(-\mathbf{x})+\mathbf{x}=\mathbf{0} \tag{3.15}$$

- There exists an identity element "1" for *scalar* multiplication so that $\forall \mathbf{x} \in V$:

$$1\cdot\mathbf{x}=\mathbf{x}\cdot 1=\mathbf{x} \tag{3.16}$$

Inner Product Space. What we have so far is the simplest type of vector space; the next ingredient which we consider is the *inner product* which is essential to build a notion of *distance* between elements in a vector space. A vector space with an inner product is called an inner product space. An inner product for $H(V,S)$ is a function from $V \times V$ to S which satisfies the following properties for any $\mathbf{x}, \mathbf{y}, \mathbf{z}, \in V$:

- It is *distributive* with respect to vector addition:

$$\langle \mathbf{x}+\mathbf{y}, \mathbf{z} \rangle = \langle \mathbf{x}, \mathbf{z} \rangle + \langle \mathbf{y}, \mathbf{z} \rangle \tag{3.17}$$

- It possesses the *scaling property* with respect to scalar multiplication[3]:

$$\langle \mathbf{x}, \alpha \mathbf{y} \rangle = \alpha \langle \mathbf{x}, \mathbf{y} \rangle \tag{3.18}$$

$$\langle \alpha \mathbf{x}, \mathbf{y} \rangle = \alpha^* \langle \mathbf{x}, \mathbf{y} \rangle \tag{3.19}$$

- It is commutative within complex conjugation:

$$\langle \mathbf{x}, \mathbf{y} \rangle = \langle \mathbf{y}, \mathbf{x} \rangle^* \tag{3.20}$$

- The self-product is positive:

$$\langle \mathbf{x}, \mathbf{x} \rangle \geq 0 \tag{3.21}$$

$$\langle \mathbf{x}, \mathbf{x} \rangle = 0 \iff \mathbf{x} = \mathbf{0}. \tag{3.22}$$

From this definition of the inner product, a series of additional definitions and properties can be derived: first of all, orthogonality between two vectors is defined with respect to the inner product, and we say that the non-zero vectors $\mathbf{x}$ and $\mathbf{y}$ are orthogonal, or $\mathbf{x} \perp \mathbf{y}$, if and only if

$$\langle \mathbf{x}, \mathbf{y} \rangle = 0 \tag{3.23}$$

From the definition of an inner product, we can define the *norm* of a vector as (we will omit from now on the subscript 2 from the norme symbol):

$$\|\mathbf{x}\| = \langle \mathbf{x}, \mathbf{x} \rangle^{1/2} \tag{3.24}$$

In turn, the norm satisfies the *Cauchy-Schwartz inequality*:

$$\left| \langle \mathbf{x}, \mathbf{y} \rangle \right| \leq \|\mathbf{x}\| \cdot \|\mathbf{y}\| \tag{3.25}$$

with strict equality if and only if $\mathbf{x} = \alpha \mathbf{y}$. The norm also satisfies the *triangle inequality*:

$$\|\mathbf{x} + \mathbf{y}\| \leq \|\mathbf{x}\| + \|\mathbf{y}\| \tag{3.26}$$

with strict equality if and only if $\mathbf{x} = \alpha \mathbf{y}$ and $\alpha \in \mathbb{R}^+$. For orthogonal vectors, the triangle inequality becomes the famous Pythagorean theorem:

$$\|\mathbf{x} + \mathbf{y}\|^2 = \|\mathbf{x}\|^2 + \|\mathbf{y}\|^2 \qquad \text{for } \mathbf{x} \perp \mathbf{y} \tag{3.27}$$

Hilbert Space. A vector space $H(V, S)$ equipped with an inner product is called an inner product space. To obtain a Hilbert space, we need com-

---

[3] Note that in our notation, the left operand is conjugated.

pleteness. This is a slightly more technical notion, which essentially implies that convergent sequences of vectors in V have a limit that is also in V. To gain intuition, think of the set of rational numbers $\mathbb{Q}$ versus the set of real numbers $\mathbb{R}$. The set of rational numbers is incomplete, because there are convergent sequences in $\mathbb{Q}$ which converge to irrational numbers. The set of real numbers contains these irrational numbers, and is in that sense the completion of $\mathbb{Q}$. Completeness is usually hard to prove in the case of infinite-dimensional spaces; in the following it will be tacitly assumed and the interested reader can easily find the relevant proofs in advanced analysis textbooks. Finally, we will also only consider *separate* Hilbert spaces, which are the ones that admit orthonormal bases.

3.2.2 Examples of Hilbert Spaces

Finite Euclidean Spaces. The vector space $\mathbb{C}^N$, with the "natural" definition for the sum of two vectors $\mathbf{z} = \mathbf{x} + \mathbf{y}$ as

$$z_n = x_n + y_n \tag{3.28}$$

and the definition of the inner product as

$$\langle \mathbf{x}, \mathbf{y} \rangle = \sum_{n=0}^{N-1} x_n^* y_n \tag{3.29}$$

is a Hilbert space.

Polynomial Functions. An example of "functional" Hilbert space is the vector space $\mathbb{P}_N([0,1])$ of polynomial functions on the interval $[0,1]$ with maximum degree N. It is a good exercise to show that $\mathbb{P}_\infty([0,1])$ is not complete; consider for instance the sequence of polynomials

$$p_n(x) = \sum_{k=0}^{n} \frac{x^k}{k!}$$

This series converges as $p_n(x) \to e^x \notin \mathbb{P}_\infty([0,1])$.

Square Summable Functions. Another interesting example of functional Hilbert space is the space of *square integrable functions* over a finite interval. For instance, $L_2([-\pi, \pi])$ is the space of real or complex functions on the interval $[-\pi, \pi]$ which have a finite norm. The inner product over $L_2([-\pi, \pi])$ is defined as

$$\langle f, g \rangle = \int_{-\pi}^{\pi} f^*(t) g(t) \, dt \tag{3.30}$$

so that the norm of $f(t)$ is

$$\|f\| = \sqrt{\int_{-\pi}^{\pi} |f(t)|^2 \, dt} \qquad (3.31)$$

For $f(t)$ to belong to $L_2([-\pi,\pi])$ it must be $\|f\| < \infty$.

3.2.3 Inner Products and Distances

The inner product is a fundamental tool in a vector space since it allows us to introduce a notion of *distance* between vectors. The key intuition about this is a typical instance in which a geometric construct helps us to generalize a basic idea to much more abstract scenarios. Indeed, take the simple Euclidean space $\mathbb{R}^N$ and a given vector $\mathbf{x}$; for any vector $\mathbf{y} \in \mathbb{R}^N$ the inner product $\langle \mathbf{x}, \mathbf{y} \rangle$ is the measure of the *orthogonal projection* of $\mathbf{y}$ over $\mathbf{x}$. We know that the orthogonal projection defines the point on $\mathbf{x}$ which is closest to $\mathbf{y}$ and therefore this indicates how well we can approximate $\mathbf{y}$ by a simple scaling of $\mathbf{x}$. To illustrate this, it should be noted that

$$\langle \mathbf{x}, \mathbf{y} \rangle = \|\mathbf{x}\| \|\mathbf{y}\| \cos \theta$$

where θ is the angle between the two vectors (you can work out the expression in $\mathbb{R}^2$ to easily convince yourself of this; the result generalizes to any other dimension). Clearly, if the vectors are orthogonal, the cosine is zero and no approximation is possible. Since the inner product is dependent on the angular separation between the vectors, it represents a first rough measure of similarity between $\mathbf{x}$ and $\mathbf{y}$; in broad terms, it provides a measure of the difference in *shape* between vectors.

In the context of signal processing, this is particularly relevant since most of the times, we are interested in the difference in shape" between signals. As we have said before, discrete-time signals *are* vectors; the computation of their inner product will assume different names according to the processing context in which we find ourselves: it will be called *filtering*, when we are trying to approximate or modify a signal or it will be called *correlation* when we are trying to detect one particular signal amongst many. Yet, in all cases, it will still be an inner product, i.e. a *qualitative* measure of similarity between vectors. In particular, the concept of orthogonality between signals implies that the signals are perfectly distinguishable or, in other words, that their shape is completely different.

The need for a *quantitative* measure of similarity in some applications calls for the introduction of the Euclidean distance, which is derived from the inner product as

$$d(\mathbf{x},\mathbf{y}) = \langle \mathbf{x}-\mathbf{y}, \mathbf{x}-\mathbf{y} \rangle^{1/2} = \|\mathbf{x}-\mathbf{y}\| \qquad (3.32)$$

In particular, for $\mathbb{C}^N$ the Euclidean distance is defined by the expression:

$$d(\mathbf{x},\mathbf{y}) = \sqrt{\sum_{n=0}^{N-1} |x_n - y_n|^2} \tag{3.33}$$

whereas for $L_2([-\pi,\pi])$ we have

$$d(\mathbf{x},\mathbf{y}) = \sqrt{\int_{-\pi}^{\pi} |x(t) - y(t)|^2 \, dt} \tag{3.34}$$

In the practice of signal processing, the Euclidean distance is referred to as the *root mean square error*;[4] this is a global, quantitative goodness-of-fit measure when trying to approximate signal **y** with **x**.

Incidentally, there are other types of distance measures which do not rely on a notion of inner product; for example in $\mathbb{C}^N$ we could define

$$d(\mathbf{x},\mathbf{y}) = \max_{0 \leq n < N} |\mathbf{x}_n - \mathbf{y}_n| \tag{3.35}$$

This distance is based on the supremum norm and is usually indicated by $\|\mathbf{x}-\mathbf{y}\|_\infty$; however, it can be shown that there is no inner product from which this norm can be derived and therefore no Hilbert space can be constructed where $\|\cdot\|_\infty$ is the natural norm. Nonetheless, this norm will reappear later, in the context of optimal filter design.

3.3 Subspaces, Bases, Projections

Now that we have defined the properties of Hilbert space, it is only natural to start looking at the consequent inner *structure* of such a space. The best way to do so is by introducing the concept of *basis*. You can think of a basis as the "skeleton" of a vector space, i.e. a structure which holds everything together; yet, this skeleton is flexible and we can twist it, stretch it and rotate it in order to highlight some particular structure of the space and facilitate access to particular information that we may be seeking. All this is accomplished by a linear transformation called a *change of basis*; to anticipate the topic of the next Chapter, we will see shortly that the Fourier transform is an instance of basis change.

Sometimes, we are interested in exploring in more detail a specific subset of a given vector space; this is accomplished via the concept of *subspace*. A subspace is, as the name implies, a restricted region of the global space,

---

[4] Almost always, the square distance is considered instead; its name is then the *mean square error*, or MSE.

with the additional property that it is *closed* under the usual vector operations. This implies that, once in a subspace, we can operate freely without ever leaving its confines; just like a full-fledged space, a subspace has its own skeleton (i.e. the basis) and, again, we can exploit the properties of this basis to highlight the features that interest us.

3.3.1 Definitions

Assume $H(V,S)$ is a Hilbert space, with V a vector space and S a set of scalars (i.e. $\mathbb{C}$).

Subspace. A subspace of V is defined as a subset $P \subseteq V$ that satisfies the following properties:

- Closure under addition, i.e.

$$\forall \mathbf{x}, \mathbf{y} \in P \Rightarrow \mathbf{x} + \mathbf{y} \in P \tag{3.36}$$

- Closure under scalar multiplication, i.e.

$$\forall \mathbf{x} \in P, \forall \alpha \in S \Rightarrow \alpha \mathbf{x} \in P \tag{3.37}$$

Clearly, V is a subspace of itself.

Span. Given an arbitrary set of M vectors $W = \{\mathbf{x}^{(m)}\}_{m=0,1,\ldots,M-1}$, the *span* of these vectors is defined as

$$\text{span}(W) = \left\{ \sum_{m=0}^{M-1} \alpha_m \mathbf{x}^{(m)} \right\}, \qquad \alpha_m \in S \tag{3.38}$$

i.e. the span of W is the set of all possible linear combinations of the vectors in W. The set of vectors W is called *linearly independent* if the following holds:

$$\sum_{m=0}^{M-1} \alpha_m \mathbf{x}^{(m)} = 0 \iff \alpha_m = 0 \qquad \text{for } m = 0, 1, \ldots, M-1 \tag{3.39}$$

Basis. A set of K vectors $W = \{\mathbf{x}^{(k)}\}_{k=0,1,\ldots,K-1}$ from a subspace P is a *basis* for that subspace if

- The set W is linearly independent.

- Its span covers P, i.e. $\text{span}(W) = P$.

Signals and Hilbert Spaces

The last statement affirms that any $\mathbf{y} \in P$ can be written as a linear combination of $\{\mathbf{x}^{(k)}\}_{k=0,1,\dots,K-1}$ or that, for all $\mathbf{y} \in P$, there exist K coefficients α_k such that

$$\mathbf{y} = \sum_{k=0}^{K-1} \alpha_k \mathbf{x}^{(k)} \qquad (3.40)$$

which is equivalently expressed by saying that the set W is *complete* in P.

Orthogonal/Orthonormal Basis. An orthonormal basis for a subspace P is a set of K basis vectors $W = \{\mathbf{x}^{(k)}\}_{k=0,1,\dots,K-1}$ for which

$$\langle \mathbf{x}^{(i)}, \mathbf{x}^{(j)} \rangle = \delta[i-j] \qquad 0 \le i,j < K \qquad (3.41)$$

which means orthogonality across vectors and unit norm. Sometimes, the set of vectors can be orthogonal but not normal (i.e. the norm of the vectors is not unitary). This is not a problem provided that we remember to include the appropriate normalization factors in the analysis and/or synthesis formulas. Alternatively, an lineary idependent set of vectors can be orthonormalized via the Gramm-Schmidt procedure, which can be found in any linear algebra textbook.

Among all bases, *orthonormal bases* are the most "beautiful" in a way because of their structure and their properties. One of the most important properties for finite-dimensional spaces is the following:

- A set of N orthogonal vectors in an N-dimensional subspace is a basis for the subspace.

In other words, in finite dimensions, once we find a full set of orthogonal vectors, we are sure that the set spans the space.

3.3.2 Properties of Orthonormal Bases

Let $W = \{\mathbf{x}^{(k)}\}_{k=0,1,\dots,K-1}$ be an orthonormal basis for a (sub)space P. Then the following properties (all of which are easily verified) hold:

Analysis Formula. The coefficients in the linear combination (3.40) are obtained simply as

$$\alpha_k = \langle \mathbf{x}^{(k)}, \mathbf{y} \rangle \qquad (3.42)$$

The coefficients $\{\alpha_k\}$ are called the *Fourier coefficients*[5] of the orthonormal expansion of $\mathbf{y}$ with respect to the basis W and (3.42) is called the Fourier *analysis formula*; conversely, Equation (3.40) is called the *synthesis formula*.

---

[5] Fourier coefficients often refer to the particular case of Fourier series. However, the term generally refers to coefficients in any orthonormal basis.

Parseval's Identity For an orthonormal basis, there is a norm conservation property given by *Parseval's identity*:

$$\|\mathbf{y}\|^2 = \sum_{k=0}^{K-1} |\langle \mathbf{x}^{(k)}, \mathbf{y} \rangle|^2 \qquad (3.43)$$

For physical quantities, the norm is dimensionally equivalent to a measure of energy; accordingly, Parseval's identity is also known as the *energy conservation formula*.

Bessel's Inequality. The generalization of Parseval's equality is *Bessel's inequality*. In our subspace P, consider a set of L orthonormal vectors $G \subset P$ (a set which is not necessarily a basis since it may be $L < K$), with $G = \{\mathbf{g}^{(l)}\}_{l=1,2,\ldots L-1}$; then the norm of any vector $\mathbf{y} \in P$ is lower bounded as:

$$\|\mathbf{y}\|^2 \geq \sum_{l=0}^{L-1} |\langle \mathbf{g}^{(l)}, \mathbf{y} \rangle|^2 \qquad (3.44)$$

and the lower bound is reached for all $\mathbf{y}$ if and only if the system G is complete, that is, if it is an orthonormal basis for P.

Best Approximations. Assume P is a subspace of V; if we try to approximate a vector $\mathbf{y} \in V$ by a linear combination of basis vectors from the subspace P, then we are led to the concepts of least squares approximations and orthogonal projections. First of all, we define the *best* linear approximation $\hat{\mathbf{y}} \in P$ of a general vector $\mathbf{y} \in V$ to be the approximation which minimizes the norm $\|\mathbf{y} - \hat{\mathbf{y}}\|$. Such approximation is easily obtained by projecting $\mathbf{y}$ onto an orthonormal basis for P, as shown in Figure 3.3. With W as our usual orthonormal basis for P, the projection is given by

$$\hat{\mathbf{y}} = \sum_{k=0}^{K-1} \langle \mathbf{x}^{(k)}, \mathbf{y} \rangle \mathbf{x}^{(k)} \qquad (3.45)$$

Define the approximation error as the vector $\mathbf{d} = \mathbf{y} - \hat{\mathbf{y}}$; it can be easily shown that:

- The error is orthogonal to the approximation, i.e. $\mathbf{d} \perp \hat{\mathbf{y}}$.

- The approximation minimizes the error square norm, i.e.

$$\arg\min_{\hat{\mathbf{y}} \in P} \|\mathbf{y} - \hat{\mathbf{y}}\|_2 = \mathbf{d} \qquad (3.46)$$

This approximation with an orthonormal basis has a key property: it can be *successively refined*. Assume you have the approximation with the first m terms of the orthonormal basis:

$$\hat{\mathbf{y}}_m = \sum_{k=0}^{m-1} \langle \mathbf{x}^{(k)}, \mathbf{y} \rangle \mathbf{x}^{(k)} \qquad (3.47)$$

and now you want to compute the $(m+1)$-term approximation. This is simply given by

$$\hat{\mathbf{y}}_{m+1} = \hat{\mathbf{y}}_m + \langle \mathbf{x}^{(m)}, \mathbf{y} \rangle \mathbf{x}^{(m)} \qquad (3.48)$$

While this seems obvious, it is actually a small miracle, since it does not hold for more general, non-orthonormal bases.

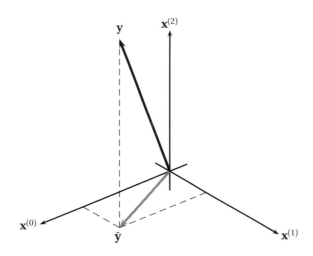

Figure 3.3 Orthogonal projection of the vector $\mathbf{y}$ onto the subspace W spanned by $\{\mathbf{x}^{(0)}, \mathbf{x}^{(1)}\}$, leading to the approximation $\hat{\mathbf{y}}$. This approximation minimizes the square norm $\|\mathbf{y} - \hat{\mathbf{y}}\|_2$ among all approximations belonging to W.

3.3.3 Examples of Bases

Considering the examples of 3.2.2, we have the following:

Finite Euclidean Spaces. For the simplest case of Hilbert spaces, namely $\mathbb{C}^N$, orthonormal bases are also the most intuitive since they contain exactly N mutually orthogonal vectors of unit norm. The classical example is the canonical basis $\{\boldsymbol{\delta}^{(k)}\}_{k=0...N-1}$ with

$$\delta_n^{(k)} = \delta[n-k] \qquad (3.49)$$

but we will soon study more interesting bases such as the Fourier basis $\{\mathbf{w}^{(k)}\}$, for which

$$\mathbf{w}_n^{(k)} = e^{j\frac{2\pi}{N}nk}$$

In $\mathbb{C}^N$, the analysis and synthesis formulas (3.42) and (3.40) take a particularly neat form. For any set $\{\mathbf{x}^{(k)}\}$ of N orthonormal vectors one can indeed arrange the conjugates of the basis vectors[6] as the successive rows of an $N \times N$ square matrix $\mathbf{M}$ so that each matrix element is the conjugate of the n-th element of the m-th basis vector:

$$\mathbf{M}_{mn} = \left(\mathbf{x}_n^{(m)}\right)^*$$

$\mathbf{M}$ is called a *change of basis* matrix. Given a vector $\mathbf{y}$, the set of expansion coefficient $\{\alpha_k\}_{k=0...N-1}$ can now be written *itself* as a vector[7] $\boldsymbol{\alpha} \in \mathbb{C}^N$. Therefore, we can rewrite the analysis formula (3.42) in matrix-vector form and we have

$$\boldsymbol{\alpha} = \mathbf{M}\mathbf{y} \tag{3.50}$$

The reconstruction formula (3.40) for $\mathbf{y}$ from the expansion coefficients, becomes, in turn,

$$\mathbf{y} = \mathbf{M}^H \boldsymbol{\alpha} \tag{3.51}$$

where the superscript denotes the Hermitian transpose (transposition and conjugation of the matrix). The previous equation shows that $\mathbf{y}$ is a linear combination of the columns of $\mathbf{M}^H$, which, in turn, are of course the vectors $\{\mathbf{x}^{(k)}\}$. The orthogonality relation (3.49) takes the following forms:

$$\mathbf{M}^H \mathbf{M} = \mathbf{I} \tag{3.52}$$
$$\mathbf{M}\mathbf{M}^H = \mathbf{I} \tag{3.53}$$

since left inverse equals right inverse for square matrices; this implies that $\mathbf{M}$ has orthonormal rows as well as orthonormal columns.

Polynomial Functions. A basis for $\mathbb{P}_N([0,1])$ is $\{x^k\}_{0 \leq k < N}$. This basis, however, is not an orthonormal basis. It can be transformed to an orthonormal basis by a standard Gramm-Schmidt procedure; the basis vector thus obtained are called *Legendre polynomials*.

---

[6] Other definitions may build $\mathbf{M}$ by stacking the *non*-conjugated basis vectors instead; the procedure is however entirely equivalent. Here we choose this definition in order to be consistent with the usual derivation of the Discrete Fourier Transform, which we will see in the next Chapter.

[7] This isomorphism is rather special and at the foundation of Linear Algebra. If the original vector space V is not $\mathbb{C}^N$, the analysis formula will always provide us with a vector of complex values, but this vector will *not* be in V.

Square Summable Functions. An orthonormal basis set for $L_2([-\pi,\pi])$ is the set $\{(1/\sqrt{2\pi})e^{jnt}\}_{n\in\mathbb{Z}}$. This is actually the classic Fourier basis for functions on an interval. Please note that, here, as opposed to the previous examples, the number of basis vectors is actually infinite. The orthogonality of these basis vectors is easily verifiable; their completeness, however, is rather hard to prove and this, unfortunately, is very much the rule for all non-trivial, infinite-dimensional basis sets.

3.4 Signal Spaces Revisited

We are now in a position to formalize our intuitions so far, with respect to the equivalence between discrete-time signals and vector spaces, with a particularization for the three main classes of signals that we have introduced in the previous Chapter. Note that in the following, we will liberally interchange the notations **x** and $x[n]$ to denote a sequence as a vector embedded in its appropriate Hilbert space.

3.4.1 Finite-Length Signals

The correspondence between the class of finite-length, length-N signals and $\mathbb{C}^N$ should be so immediate at this point that it does not need further comment. As a reminder, the canonical basis is the canonical basis for $\mathbb{C}^N$. The k-th canonical basis vector is often expressed in signal form as

$$\delta[n-k] \qquad n=0,\ldots,N-1, \quad k=0,\ldots,N-1$$

3.4.2 Periodic Signals

As we have seen, N-periodic signals are equivalent to length-N signals. The space of N-periodic sequences is therefore isomorphic to $\mathbb{C}^N$. In particular, the sum of two sequences considered as vectors is the standard pointwise sum for the elements:

$$z[n] = x[n] + y[n] \qquad n \in \mathbb{Z} \qquad (3.54)$$

while, for the inner product, we extend the summation over a period only:

$$\langle x[n], y[n]\rangle = \sum_{n=0}^{N-1} x^*[n]y[n] \qquad (3.55)$$

The canonical basis for the space of N-periodic sequences is the canonical basis for $\mathbb{C}^N$, because of the isomorphism; in general, any basis for $\mathbb{C}^N$ is also a basis for the space of N-periodic sequences. Sometimes, however, we will also consider an explicitly *periodized* version of the basis. For the canonical basis, in particular, the periodized basis is composed of N vectors of infinite-length $\{\tilde{\boldsymbol{\delta}}^{(k)}\}_{k=0...N-1}$ with

$$\tilde{\boldsymbol{\delta}}^{(k)} = \sum_{i=-\infty}^{\infty} \delta[n-k-iN]$$

Such a sequence is called a *pulse train*. Note that here we are abandoning mathematical rigor, since the norm of each of these basis vectors is infinite; yet the *pulse train*, if handled with care, can be a useful tool in formal derivations.

3.4.3 Infinite Sequences

In the case of infinite sequences, whose "natural" Euclidean space would appear to be $\mathbb{C}^\infty$, the situation is rather delicate. While the sum of two sequences can be defined in the usual way, by extending the sum for $\mathbb{C}^N$ to $\mathbb{C}^\infty$, care must be taken when evaluating the inner product. We have already pointed out that the formula:

$$\langle x[n], y[n] \rangle = \sum_{n=-\infty}^{\infty} x^*[n]y[n] \qquad (3.56)$$

can diverge even for simple constant sequences such as $x[n] = y[n] = 1$. A way out of this impasse is to restrict ourselves to $\ell_2(\mathbb{Z})$, the space of *square summable sequences*, for which

$$\|x\|^2 = \sum_{n \in \mathbb{Z}} |x[n]|^2 < \infty \qquad (3.57)$$

This is the space of choice for all the theoretical derivations involving infinite sequences. Note that these sequences are often called "of finite energy", since the square norm corresponds to the definition of energy as given in (2.19).

The canonical basis for $\ell_2(\mathbb{Z})$ is simply the set $\{\boldsymbol{\delta}^{(k)}\}_{k \in \mathbb{Z}}$; in signal form:

$$\boldsymbol{\delta}^{(k)} = \delta[n-k], \qquad n, k \in \mathbb{Z} \qquad (3.58)$$

This is an infinite set, and actually an infinite set of linearly independent vectors, since

$$\delta[n-k] = \sum_{l \in \mathbb{Z}/k} \alpha_l \delta[n-l] \qquad (3.59)$$

has no solution for any k. Note that, for an arbitrary signal $x[n]$ the analysis formula gives

$$\alpha_k = \langle \boldsymbol{\delta}^{(k)}, \mathbf{x} \rangle = \langle \delta[n-k], x[n] \rangle = x[k]$$

so that the reconstruction formula becomes

$$x[n] = \sum_{k=-\infty}^{\infty} \alpha_k \boldsymbol{\delta}^{(k)} = \sum_{k=-\infty}^{\infty} x[k]\delta[n-k]$$

which is the reproducing formula (2.18). The Fourier basis for $\ell_2(\mathbb{Z})$ will be introduced and discussed at length in the next Chapter.

As a last remark, note that the space of *all* finite-support signals, which is clearly a subset of $\ell_2(\mathbb{Z})$, does *not* form a Hilbert space. Clearly, the space is closed under addition and scalar multiplication, and the canonical inner product is well behaved since all sequences have only a finite number of nonzero values. However, the space is not complete; to clarify this, consider the following family of signals:

$$y_k[n] = \begin{cases} 1/n & |n| < k \\ 0 & \text{otherwise} \end{cases}$$

For k growing to infinity, the sequence of signals converges as $y_k[n] \to y[n] = 1/n$ for all n; while $y[n]$ is indeed in $\ell_2(\mathbb{Z})$, since

$$\sum_{n=0}^{\infty} \frac{1}{n^2} = \frac{\pi^2}{6}$$

$y[n]$ is clearly *not* a finite-support signal.

Further Reading

A comprehensive review of linear algebra, which contains all the concepts of Hilbert spaces but in finite dimensions, is the classic by G. Strang, *Linear Algebra and Its Applications* (Brooks Cole, 2005). For an introduction to Hilbert spaces, there are many mathematics books; we suggest N. Young, *An Introduction to Hilbert Space* (Cambridge University Press, 1988). As an alternative, a more intuitive and engineering-motivated approach is in the classic book by D. G. Luenberger, *Optimization by Vector Space Methods* (Wiley, 1969).

Exercises

Exercise 3.1: Elementary operators. An *operator* $\mathscr{H}$ is a transformation of a given signal and is indicated by the notation

$$y[n] = \mathscr{H}\{x[n]\}$$

For instance, the delay operator $\mathscr{D}$ is indicated as

$$\mathscr{D}\{x[n]\} = x[n-1]$$

and the one-step difference operator is indicated as

$$\mathscr{V}\{x[n]\} = x[n] - \mathscr{D}\{x[n]\} = x[n] - x[n-1] \tag{3.60}$$

A *linear* operator is one for which the following holds:

$$\begin{cases} \mathscr{H}\{ax[n]\} = a\mathscr{H}\{x[n]\} \\ \mathscr{H}\{x[n] + y[n]\} = \mathscr{H}\{x[n]\} + \mathscr{H}\{y[n]\} \end{cases}$$

(a) Show that the delay operator $\mathscr{D}$ is linear.

(b) Show that the differentiation operator $\mathscr{V}$ is linear.

(c) Show that the squaring operator $\mathscr{S}\{x[n]\} = x^2[n]$ is *not* linear.

In $\mathbb{C}^N$, any linear operator on a vector $\mathbf{x}$ can be expressed as a matrix-vector multiplication for a suitable $N \times N$ matrix $\mathbf{A}$. In $\mathbb{C}^N$, we define the delay operator as the left *circular* shift of a vector:

$$\mathscr{D}\{\mathbf{x}\} = [x_{N-1} \quad x_0 \quad x_1 \quad \ldots \quad x_{N-2}]^T$$

Assume $N = 4$ for convenience; it is easy to see that

$$\mathscr{D}\{\mathbf{x}\} = \begin{bmatrix} 0 & 0 & 0 & 1 \\ 1 & 0 & 0 & 0 \\ 0 & 1 & 0 & 0 \\ 0 & 0 & 1 & 0 \end{bmatrix} \mathbf{x} = \mathbf{D}\mathbf{x}$$

so that $\mathbf{D}$ is the matrix associated to the delay operator.

(d) Using the same definition for the one-step difference operator as in (3.60), write out the associated matrix for the operator in $\mathbb{C}^4$.

(e) Consider the following matrix:

$$\mathbf{A} = \begin{bmatrix} 1 & 0 & 0 & 0 \\ 1 & 1 & 0 & 0 \\ 1 & 1 & 1 & 0 \\ 1 & 1 & 1 & 1 \end{bmatrix}$$

Which operator do you think it is associated to? What does the operator do?

Exercise 3.2: Bases. Let $\{\mathbf{x}^{(k)}\}_{k=0,\ldots,N-1}$ be a basis for a subspace S. Prove that any vector $\mathbf{z} \in S$ is *uniquely* represented in this basis.
Hint: prove by contradiction.

Exercise 3.3: Vector spaces and signals.

(a) Show that the set of all ordered n-tuples $[a_1, a_2, \ldots, a_n]$ with the natural definition for the sum:

$$[a_1, a_2, \ldots, a_n] + [b_1, b_2, \ldots, b_n] = [a_1 + b_1, a_2 + b_2, \ldots, a_n + b_n]$$

and the multiplication by a scalar:

$$\alpha[a_1, a_2, \ldots, a_n] = [\alpha a_1, \alpha a_2, \ldots, \alpha a_n]$$

form a vector space. Give its dimension and find a basis.

(b) Show that the set of signals of the form $y(x) = a\cos(x) + b\sin(x)$ (for arbitrary a, b), with the usual addition and multiplication by a scalar, form a vector space. Give its dimension and find a basis.

(c) Are the four diagonals of a cube orthogonal?

(d) Express the discrete-time impulse $\delta[n]$ in terms of the discrete-time unit step $u[n]$ and conversely.

(e) Show that any function $f(t)$ can be written as the sum of an odd and an even function, i.e. $f(t) = f_o(t) + f_e(t)$ where $f_o(-t) = -f_o(t)$ and $f_e(-t) = f_e(t)$.

Exercise 3.4: The Haar basis. Consider the following change of basis matrix in $\mathbb{C}^8$, with respect to the standard orthonormal basis:

$$\mathbf{H} = \begin{bmatrix} 1 & -1 & 0 & 0 & 0 & 0 & 0 & 0 \\ 0 & 0 & 1 & -1 & 0 & 0 & 0 & 0 \\ 0 & 0 & 0 & 0 & 1 & -1 & 0 & 0 \\ 0 & 0 & 0 & 0 & 0 & 0 & 1 & -1 \\ 1 & 1 & -1 & -1 & 0 & 0 & 0 & 0 \\ 0 & 0 & 0 & 0 & 1 & 1 & -1 & -1 \\ 1 & 1 & 1 & 1 & -1 & -1 & -1 & -1 \\ 1 & 1 & 1 & 1 & 1 & 1 & 1 & 1 \end{bmatrix}$$

Note the pattern in the first four rows, in the next two, and in the last two.

(a) What is an easy way to prove that the rows in **H** do indeed form a basis? (*Hint: it is sufficient to show that they are linearly independent, i.e. that the matrix has full rank...*)

The basis described by **H** is called the *Haar basis* and it is one of the most celebrated cornerstones of a branch of signal processing called wavelet analysis. To get a feeling for its properties, consider the following set of numerical experiments (you can use Matlab or any other numerical package):

(c) Verify that $\mathbf{HH}^H$ is a diagonal matrix, which means that the vectors are orthogonal.

(d) Consider a constant signal $\{\mathbf{x} = [1 \quad 1 \quad \ldots \quad 1]$ and compute its coefficients in the Haar basis.

(e) Consider an alternating signal $\{\mathbf{x} = [+1 \quad -1 \quad +1 \quad \ldots \quad +1 \quad -1]$ and compute its coefficients in the Haar basis.

Chapter 4

Fourier Analysis

Fourier theory has a long history, from J. Fourier's early work on the transmission of heat to recent results on non-harmonic Fourier series. Fourier theory is a branch of harmonic analysis, and in that sense, a topic in pure and applied mathematics. At the same time, because of its usefulness in practical applications, Fourier analysis is a key tool in several engineering branches, and in signal processing in particular.

Why is Fourier analysis so important? To understand this, let us take a short philosophical detour. Interesting signals are time-varying quantities: you can imagine, for instance, the voltage level at the output of a microphone or the measured level of the tide at a particular location; in all cases, the variation of a signal, over time, implies that a transfer of energy is happening somewhere, and ultimately this is what we want to study. Now, a time-varying value which only *increases* over time is not only a physical impossibility but a recipe for disaster for whatever system is supposed to deal with it; fuses will blow, wires will melt and so on. Oscillations, on the other hand, are nature's and man's way of keeping things in motion without trespassing all physical bounds; from Maxwell's wave equation to the mechanics of the vocal cords, from the motion of an engine to the ebb and flow of the tide, oscillatory behavior is the recurring theme. Sinusoidal oscillations are the purest form of such a constrained motion and, in a nutshell, Fourier's immense contribution was to show that (at least mathematically) one could express any given phenomenon as the combined output of a number of sinusoidal "generators".

Sinusoids have another remarkable property which justifies their ubiquitous presence. Indeed, *any linear time-invariant transformation of a sinusoid is a sinusoid at the same frequency*: we express this by saying that sinusoidal oscillations are eigenfunctions of linear time-invariant systems. This is a formidable tool for the analysis and design of signal processing structures, as we will see in much detail in the context of discrete-time filters.

The purpose of the present Chapter is to introduce and analyze some key results on Fourier series and Fourier transforms in the context of discrete-time signal processing. It appears that, as we mentioned in the previous Chapter, the Fourier transform of a signal is a change of basis in an appropriate Hilbert space. While this notion constitutes an extremely useful unifying framework, we also point out the peculiarities of its specialization within the different classes of signal. In particular, for finite-length signals we highlight the eminently algebraic nature of the transform, which leads to efficient computational procedures; for infinite sequences, we will analyze some of its interesting mathematical subtleties.

4.1 Preliminaries

The Fourier transform of a signal is an alternative representation of the data in the signal. While a signal lives in the *time domain*,[1] its Fourier representation lives in the *frequency domain*. We can move back and forth at will from one domain to the other using the direct and inverse Fourier operators, since these operators are invertible.

In this Chapter we study three types of Fourier transform which apply to the three main classes of signals that we have seen so far:

- the Discrete Fourier Transform (DFT), which maps length-N signals into a set of N discrete frequency components;

- the Discrete Fourier Series (DFS), which maps N-periodic sequences into a set of N discrete frequency components;

- the Discrete-Time Fourier Transform (DTFT), which maps infinite sequences into the space of 2π-periodic function of a real-valued argument.

The frequency representation of a signal (given by a set of coefficients in the case of the DFT and DFS and by a frequency distribution in the case of the DTFT) is called the *spectrum*.

---

[1] *Discrete*-time, of course.

4.1.1 Complex Exponentials

The basic ingredient of all the Fourier transforms which follow, is the discrete-time complex exponential; this is a sequence of the form:

$$x[n] = A\,e^{j(\omega n + \phi)} = A\left[\cos(\omega n + \phi) + j\sin(\omega n + \phi)\right]$$

A complex exponential represents an oscillatory behavior; $A \in \mathbb{R}$ is the amplitude of the oscillation, ω is its frequency and ϕ is its initial phase. Note that, actually, a discrete-time complex exponential sequence is not always a periodic sequence; it is periodic only if $\omega = 2\pi(M/N)$ for some value of $M, N \in \mathbb{Z}$. The power of a complex exponential is equal to the average energy over a period, i.e. $|A|^2$, irrespective of frequency.

4.1.2 Complex Oscillations? Negative Frequencies?

In the introduction, we hinted at the fact that Fourier analysis allows us to decompose a physical phenomenon into oscillatory components. However, it may seem odd, that we have chosen to use complex oscillation for the analysis of real-world signals. It may seem even odder that these oscillations can have a negative frequency and that, as we will soon see in the context of the DTFT, the spectrum extends over to the negative axis.

The starting point in answering these legitimate questions is to recall that the use of complex exponentials is essentially a matter of convenience. One could develop a complete theory of frequency analysis for real signals using only the basic trigonometric functions. You may actually have noticed this if you are familiar with the Fourier Series of a real function; yet the notational overhead is undoubtedly heavy since it involves two separate sets of coefficients for the sine and cosine basis functions, plus a distinct term for the zero-order coefficient. The use of complex exponentials elegantly unifies these separate series into a single complex-valued sequence. Yet, one may ask again, what does it mean for the spectrum of a musical sound to be complex? Simply put, the complex nature of the spectrum is a compact way of representing two concurrent pieces of information which uniquely define each spectral component: its *frequency* and its *phase*. These two values form a two-element vector in $\mathbb{R}^2$ but, since $\mathbb{R}^2$ is isomorphic to $\mathbb{C}$, we use complex numbers for their mathematical convenience.

With respect negative frequencies, one must first of all consider, yet again, a basic complex exponential sequence such as $x[n] = e^{j\omega n}$. We can visualize its evolution over discrete-time as a series of points on the unit circle in the complex plane. At each step, the angle increases by ω, defining a counterclockwise circular motion. It is easy to see that a complex exponential sequence of frequency $-\omega$ is just the same series of points with the difference that the points move *clockwise* instead; this is illustrated in detail in Figure 4.1. If we decompose a *real* signal into complex exponentials, we will show that, for any given frequency value, the phases of the positive and negative components are always opposite in sign; as the two oscillations move in opposite directions along the unit circle, their complex part will always cancel out exactly, thus returning a purely real signal.[2]

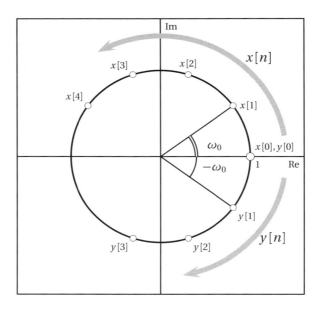

Figure 4.1 Complex exponentials as a series of points on the unit circle; $x[n] = e^{j\omega_0 n}$ and $y[n] = e^{-j\omega_0 n}$ for $\omega_0 = \pi/5$.

The final step in developing a comfortable feeling for complex oscillations comes from the realization that, in the *synthesis* of discrete-time signals (and especially in the case of communication systems) it is actually more convenient to work with complex-valued signals, themselves. Although the transmitted signal of a device like an ADSL box is a real signal, the internal representation of the underlying sequences is complex; therefore complex oscillations become a necessity.

---

[2] To anticipate a question which may appear later, the fact that modulation "makes negative frequencies appear in the positive spectrum" is really a consequence of the following very mundane formula:

$$\cos\alpha\cos\beta = \frac{1}{2}[\cos(\alpha+\beta)+\cos(\alpha-\beta)]$$

4.2 The DFT (Discrete Fourier Transform)

We now develop a Fourier representation for finite-length signals; to do so, we need to find a set of oscillatory signals of length N which contain a whole number of periods over their support. We start by considering a family of finite-length sinusoidal signals (indexed by an integer k) of the form

$$w_k[n] = e^{j\omega_k n}, \qquad n = 0, \ldots, N-1 \qquad (4.1)$$

where all the ω_k's are distinct frequencies which fulfill our requirements. To determine these frequency values, note that, in order for $w_k[n]$ to contain a whole number of periods over N samples, it must conform to

$$w_k[N] = w_k[0] = 1$$

which translates to

$$\left(e^{j\omega_k}\right)^N = 1$$

The above equation has N distinct solutions which are the N roots of unity $e^{j2\pi m/N}$, $m = 0, \ldots, N-1$; if we define the complex number

$$W_N = e^{-j\frac{2\pi}{N}}$$

then the family of N signals in (4.1) can be written as

$$w_k[n] = W_N^{-nk}, \qquad n = 0, \ldots, N-1 \qquad (4.2)$$

for each value of $k = 0, \ldots, N-1$. We can represent these N signals as a set of vectors $\{\mathbf{w}^{(k)}\}_{k=0,\ldots,N-1}$ in $\mathbb{C}^N$ with

$$\mathbf{w}^{(k)} = \begin{bmatrix} 1 & W_N^{-k} & W_N^{-2k} & \cdots & W_N^{-(N-1)k} \end{bmatrix}^T \qquad (4.3)$$

The real and imaginary parts of $\mathbf{w}^{(k)}$ for $N = 32$ and for some values of k are plotted in Figures 4.2 to 4.5.

We can verify that $\{\mathbf{w}^{(k)}\}_{k=0,\ldots,N-1}$ is a set of N orthogonal vectors and therefore a basis for $\mathbb{C}^N$; indeed we have (noting that $(W_N^{-k})^* = W_N^k$):

$$\langle \mathbf{w}^{(m)}, \mathbf{w}^{(n)} \rangle = \sum_{i=0}^{N-1} W_N^{(m-n)i} = \begin{cases} N & \text{for } m = n \\ \dfrac{1 - W_N^{(m-n)N}}{1 - W_N^{(m-n)}} = 0 & \text{for } m \neq n \end{cases} \qquad (4.4)$$

since $W_N^{iN} = 1$ for all $i \in \mathbb{Z}$. In more compact notation we can therefore state that

$$\langle \mathbf{w}^{(m)}, \mathbf{w}^{(n)} \rangle = N\,\delta[n-m] \qquad (4.5)$$

The vectors $\{\mathbf{w}^{(k)}\}_{k=0,\ldots,N-1}$ are the Fourier basis for $\mathbb{C}^N$, and therefore for the space of length-N signals. It is immediately evident that this basis is not orthonormal, since $\left\|\mathbf{w}^{(k)}\right\|^2 = N$, but that it could be made orthonormal simply by scaling the basis vectors by $(1/\sqrt{N})$. In signal processing practice, however, it is customary to keep the normalization factor explicit in the change of basis formulas; this is mostly due to computational reasons, as we will see later, but, for the sake of consistency with the mainstream literature, we will also follow this convention.

4.2.1 Matrix Form

The Discrete Fourier Transform (DFT) analysis and synthesis formulas can now be easily expressed in the familiar matrix notation as in Section 3.3.3: define an $N \times N$ square matrix $\mathbf{W}$ by stacking the conjugate of the basis vectors, i.e. $\mathbf{W}_{nk} = e^{-j(2\pi/N)nk} = W_N^{nk}$; from this we can state, for all vectors $\mathbf{x} \in \mathbb{C}^N$:

$$\mathbf{X} = \mathbf{W}\mathbf{x} \tag{4.6}$$

$$\mathbf{x} = \frac{1}{N}\mathbf{W}^H \mathbf{X} \tag{4.7}$$

(note the normalization factor in the reconstruction formula). Here, $\mathbf{X}$ is the set of Fourier coefficients in vector form, whose physical interpretation we will explore shortly. Note that the DFT preserves the energy of the finite-length signal: indeed Parseval's relation (3.43) becomes

$$\|\mathbf{x}\|_2 = \frac{1}{\sqrt{N}} \|\mathbf{X}\|_2 \tag{4.8}$$

(once again, note the explicit normalization factor).

4.2.2 Explicit Form

It is very common in the literature to explicitly write out the inner products in (4.6) and (4.7); this is both for historical reasons and to underscore the highly structured form of this transformation which, as we will see, is the basis for very efficient computational procedures. In detail, we have the analysis formula

$$X[k] = \sum_{n=0}^{N-1} x[n] W_N^{nk}, \qquad k = 0, \ldots, N-1 \tag{4.9}$$

and the dual synthesis formula

$$x[n] = \frac{1}{N} \sum_{k=0}^{N-1} X[k] W_N^{-nk}, \qquad n = 0, \ldots, N-1 \tag{4.10}$$

where we have used the standard convention of "lumping" the normalizing factor $(1/N)$ entirely within in the reconstruction sum (4.10).

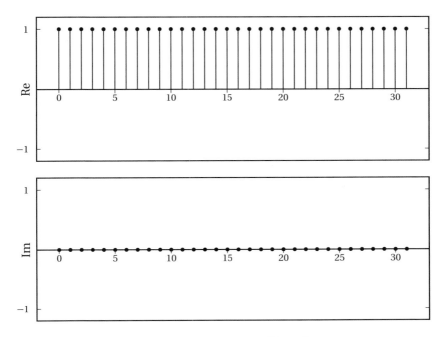

Figure 4.2 Basis vector $\mathbf{w}^{(0)} \in \mathbb{C}^{32}$.

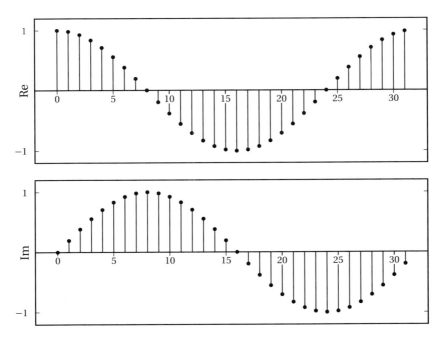

Figure 4.3 Basis vector $\mathbf{w}^{(1)} \in \mathbb{C}^{32}$.

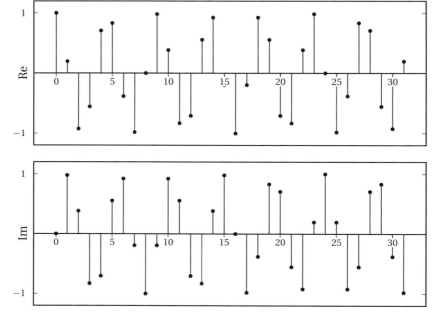

Figure 4.4 Basis vector $\mathbf{w}^{(7)} \in \mathbb{C}^{32}$.

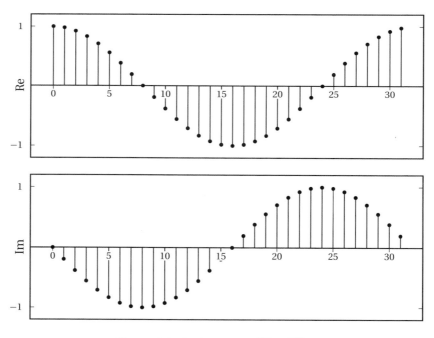

Figure 4.5 Basis vector $\mathbf{w}^{(31)} \in \mathbb{C}^{32}$.

4.2.3 Physical Interpretation

To return to the physical interpretation of the DFT, what we have just obtained is the decomposition of a finite-length signal into a set of N sinusoidal components; the magnitude and initial phase of each oscillator are given by the coefficients $X[k]$ in (4.9) (or, equivalently, by the vector elements $\mathbf{X}_k$ in (4.6)[3]). To stress the point again:

- take an array of N complex sinusoidal generators;
- set the frequency of the k-th generator to $(2\pi/N)k$;
- set the amplitude of the k-th generator to $|X[k]|$, i.e. to the magnitude of the k-th DFT coefficient;
- set the phase of the k-th generator to $\sphericalangle X[k]$, i.e. to the phase of the k-th DFT coefficient;
- start the generators at the same time and sum their outputs.

The first N output values of this "machine" are exactly $x[n]$.

If we look at this from the opposite end, each $X[k]$ shows "how much" oscillatory behavior at frequency $2\pi/k$, is contained in the signal; this is consistent with the fact that an inner product is a measure of similarity. The coefficients $X[k]$ are referred to as the *spectrum* of the signal. The square magnitude $|X[k]|^2$ is a measure (up to a scale factor N) of the signal's energy at the frequency $(2\pi/N)k$; the coefficients $X[k]$, therefore, show exactly how the global energy of the original signal is distributed in the frequency domain while Parseval's equality (4.8) guarantees that the result is consistent. The phase of each Fourier coefficient, indicated by $\sphericalangle X[k]$, specifies the initial phase of each oscillator for the reconstruction formula, i.e. the *relative alignment* of each complex exponential at the onset of the signal. While this does not influence the energy distribution in frequency, it does have a significant effect on the shape of the signal in the discrete-time domain as we will shortly see in more detail.

Some examples for signals in $\mathbb{C}^{64}$ are plotted in Figures 4.6–4.9. Figure 4.6 shows one of the simplest cases: indeed, the signal $x[n]$ is a sinusoid whose frequency coincides with that of one of the basis vectors (precisely, to that of $\mathbf{w}^{(4)}$) and, as a consequence of the orthogonality of the basis, only $X[4]$ and $X[60]$ are nonzero (this can be easily verified by decomposing the sinusoid as the sum of two appropriate basis functions). Figure 4.7 shows the same signal, but this time with a phase offset. The magnitude DFT

---

[3] From now on, the perfect equivalence between the notations $y[n]$ and $\mathbf{y}_n$ while dealing with a length-N signal will be taken for granted.

does not change, but the phase offset appears in the phase of the transform. Figure 4.8 shows the transform of a sinusoid whose frequency does *not* coincide with any of the basis frequencies. As a consequence, *all* of the basis

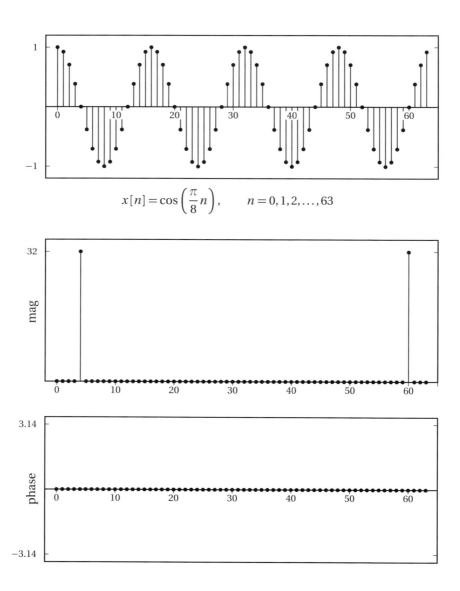

$$x[n] = \cos\left(\frac{\pi}{8}n\right), \quad n = 0, 1, 2, \ldots, 63$$

Figure 4.6 Signal and DFT (example).

vectors are needed to reconstruct the signal. Clearly, the magnitude is larger for frequencies closer to hot of the original signal's ($6\pi/64$ and $7\pi/64$ in this case); yet, to reconstruct $x[n]$ *exactly*, we need the contribution of the entire

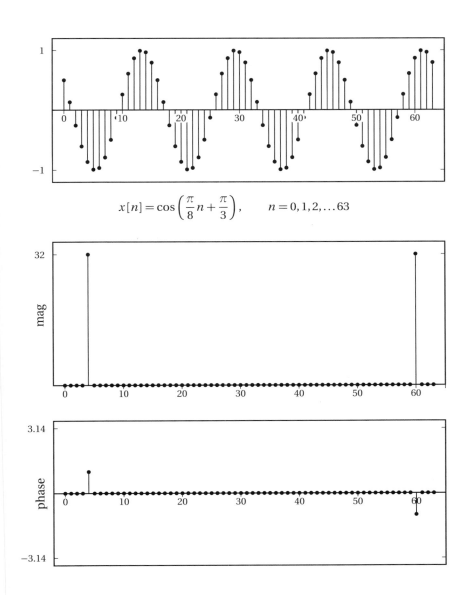

Figure 4.7 Signal and DFT (example cont.d).

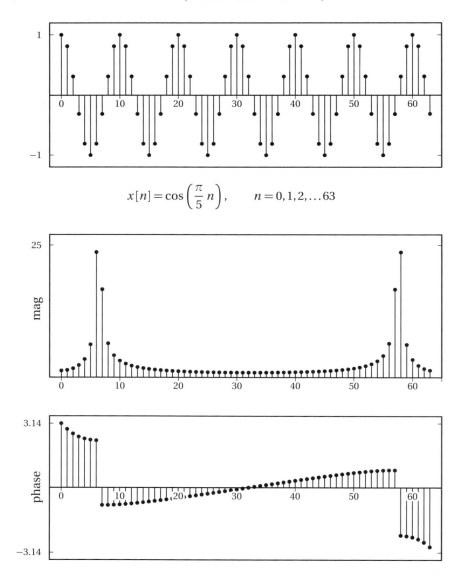

Figure 4.8 Signal and DFT (example cont.d).

basis. Finally, Figure 4.9 shows the DFT of a step signal. It can be shown (with a few trigonometric manipulations) that the DFT of a step signal is

$$X[k] = \frac{\sin\big((\pi/N)Mk\big)}{\sin\big((\pi/N)k\big)} e^{-j\frac{\pi}{N}(M-1)k}$$

where N is the length of the signal and M is the length of the step (in Figure 4.9 $N = 64$ and $M = 5$, for instance.)

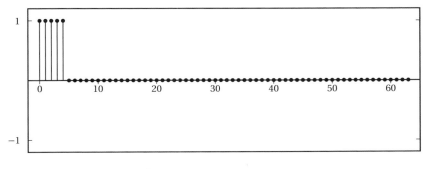

$$x[n] = \sum_{k=0}^{4} \delta[n-k], \qquad n = 0, 1, 2, \ldots, 63$$

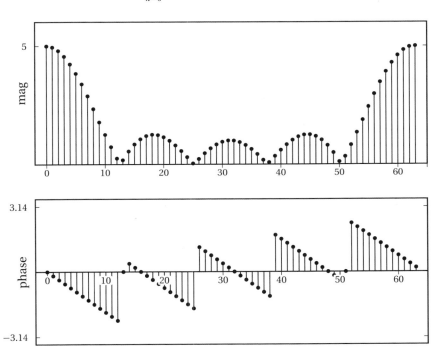

Figure 4.9 Signal and DFT (example cont.d).

4.3 The DFS (Discrete Fourier Series)

Consider the reconstruction formula in (4.10); what happens if we let the index n roam outside of the $[0, N-1]$ interval? Since $W_N^{(n+iN)k} = W_N^{nk}$ for all $i \in \mathbb{Z}$, we note that $x[n+iN] = x[n]$. In other words, the reconstruction formula in (4.10) implicitly defines a *periodic sequence* of period N. This is the reason why, earlier, we stated that periodic sequences are the natural way to embed a finite-length signal into a sequence: their Fourier representation is formally identical. This is not surprising since a) we have already established a correspondence between $\mathbb{C}^N$ and $\tilde{\mathbb{C}}^N$ and b) we are actually expressing a length-N sequence as a combination of N-periodic basis signals.

The Fourier representation of periodic sequences is called the Discrete Fourier Series (DFS), and its explicit analysis and synthesis formulas are the exact equivalent of (4.9) and (4.10), modified only with respect to the range of the indices. We have already seen that in (4.10), the reconstruction formula, n is now in $\mathbb{Z}$. Symmetrically, due to the N-periodicity of W_N^k, we can let the index k in (4.9) assume any value in $\mathbb{Z}$ too; this way, the DFS coefficients become an N-periodic sequence themselves and the DFS becomes a change of basis in $\tilde{\mathbb{C}}^N$ using the definition of inner product given in Section (3.4.2) and the formal periodic basis for $\tilde{\mathbb{C}}^N$:

$$\tilde{X}[k] = \sum_{n=0}^{N-1} \tilde{x}[n] W_N^{nk}, \qquad k \in \mathbb{Z} \qquad (4.11)$$

$$\tilde{x}[n] = \frac{1}{N} \sum_{k=0}^{N-1} \tilde{X}[k] W_N^{-nk}, \qquad n \in \mathbb{Z} \qquad (4.12)$$

4.4 The DTFT (Discrete-Time Fourier Transform)

We now consider a Fourier representation for infinite non-periodic sequences. From a purely mathematical point of view, the Discrete-Time Fourier Transform of a sequence $x[n]$ is defined as

$$X(e^{j\omega}) = \sum_{n=-\infty}^{\infty} x[n] e^{-j\omega n} \qquad (4.13)$$

The DTFT is therefore a complex-valued function of the *real* argument ω, and, as can be easily verified, it is periodic in ω with period 2π. The somewhat odd notation $X(e^{j\omega})$ is quite standard in the signal processing literature and offers several advantages:

- it stresses the basic periodic nature of the transform since, obviously, $e^{j(\omega+2\pi)} = e^{j\omega}$;

- regardless of context, it immediately identifies a function as the Fourier transform of a discrete-time sequence: for exemple $U(e^{j\lambda})$ is just as readily recognizable;

- it provides a convenient notational framework which unifies the Fourier transform and the z-transform (which we will see shortly).

The DTFT, when it exists, can be inverted via the integral

$$x[n] = \frac{1}{2\pi} \int_{-\pi}^{\pi} X(e^{j\omega}) e^{j\omega n} d\omega \qquad (4.14)$$

as can be easily verified by substituting (4.13) into 4.14) and using

$$\int_{-\pi}^{\pi} e^{-j\omega(n-k)} = 2\pi \delta[n-k]$$

In fact, due to the 2π-periodicity of the DTFT, the integral in (4.14) can be computed over *any* 2π-wide interval on the real line (i.e. between 0 and 2π, for instance). The relation between a sequence $x[n]$ and its DTFT $X(e^{j\omega})$ will be indicated in the general case by

$$x[n] \overset{\text{DTFT}}{\longleftrightarrow} X(e^{j\omega})$$

While the DFT and DFS were signal transformations which involved only a finite number of quantities, both the infinite summation and the real-valued argument, appearing in the DTFT, can create an uneasiness which overshadows the conceptual similarities between the transforms. In the following, we start by defining the mathematical properties of the DTFT and we try to build an intuitive feeling for this Fourier representation, both with respect to its physical interpretation and to its conformity to the "change of basis" framework, that we used for the DFT and DFS.

Mathematically, the DTFT is a transform operator which maps discrete-time sequences onto the space of 2π-periodic functions. Clearly, for the DTFT to exist, the sum in (4.13) must converge, i.e. the limit for $M \to \infty$ of the partial sum

$$X_M(e^{j\omega}) = \sum_{n=-M}^{M} x[n] e^{-j\omega n} \qquad (4.15)$$

must exist and be finite. Convergence of the partial sum in (4.15) is very easy to prove for *absolutely summable* sequences, that is for sequences satisfying

$$\lim_{M \to \infty} \sum_{n=-M}^{M} |x[n]| < \infty \qquad (4.16)$$

since, according to the triangle inequality,

$$|X_M(e^{j\omega})| \leq \sum_{n=-M}^{M} |x[n] e^{-j\omega n}| = \sum_{n=-M}^{M} |x[n]| \qquad (4.17)$$

For this class of sequences it can also be proved that the convergence of $X_M(e^{j\omega})$ to $X(e^{j\omega})$ is uniform and that $X(e^{j\omega})$ is continuous. While absolute summability is a sufficient condition, it can be shown that the sum in (4.15) is convergent also for all *square-summable* sequences, i.e. for sequences whose energy is finite; this is very important to us with respect to the discussion in Section 3.4.3 where we defined the Hilbert space $\ell_2(\mathbb{Z})$. In the case of square summability only, however, the convergence of (4.15) is no longer uniform but takes place only in the mean-square sense, i.e.

$$\lim_{M \to \infty} \int_{-\pi}^{\pi} \left| X_M(e^{j\omega}) - X(e^{j\omega}) \right|^2 d\omega = 0 \qquad (4.18)$$

Convergence in the mean square sense implies that, while the total energy of the error signal becomes zero, the pointwise values of the partial sum may never approach the values of the limit. One manifestation of this odd behavior is called the *Gibbs phenomenon*, which has important consequences in our approach to filter design, as we will see later. Furthermore, in the case of square-summable sequences, $X(e^{j\omega})$ is no longer guaranteed to be continuous.

As an example, consider the sequence:

$$x[n] = \begin{cases} 1 & \text{for } -N \leq n \leq N \\ 0 & \text{otherwise} \end{cases} \qquad (4.19)$$

Its DTFT can be computed as the sum[4]

$$\begin{aligned}
X(e^{j\omega}) &= \sum_{n=-N}^{N} e^{-j\omega n} \\
&= \sum_{n=1}^{N} e^{j\omega n} + \sum_{n=0}^{N} e^{-j\omega n} \\
&= \frac{1 - e^{-j\omega(N+1)}}{1 - e^{-j\omega}} + \frac{1 - e^{j\omega(N+1)}}{1 - e^{j\omega}} - 1 \\
&= e^{j\omega/2} \frac{1 - e^{-j\omega(N+1)}}{e^{j\omega/2} - e^{-j\omega/2}} + e^{-j\omega/2} \frac{1 - e^{j\omega(N+1)}}{e^{-j\omega/2} - e^{j\omega/2}} - 1 \\
&= \frac{e^{j\omega(N+\frac{1}{2})} - e^{-j\omega(N+\frac{1}{2})}}{e^{j\omega/2} - e^{-j\omega/2}} \\
&= \frac{\sin\left(\omega \left(N + \frac{1}{2}\right)\right)}{\sin(\omega/2)}
\end{aligned}$$

---

[4] Remember that $\sum_{n=0}^{N} x^n = \dfrac{1 - x^{N+1}}{1 - x}$.

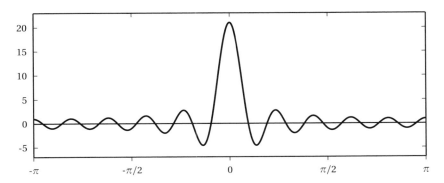

Figure 4.10 The DTFT of the signal in (4.19).

The DTFT of this particular signal turns out to be real (we will see later that this is a consequence of the signal's symmetry) and it is plotted in Figure 4.10. When, as is very often the case, the DTFT is complex-valued, the usual way to represent it graphically takes the magnitude and the phase separately into account. The DTFT is always a 2π-periodic function and the standard convention is to plot the interval from $-\pi$ to π. Larger intervals can be considered if the periodicity needs to be made explicit; Figure 4.11, for instance, shows five full periods of the same function.

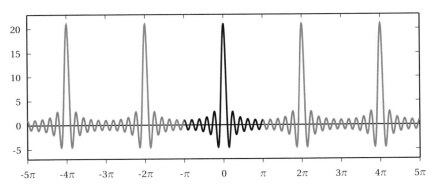

Figure 4.11 The DTFT of the signal in (4.19), with explicit periodicity.

4.4.1 The DTFT as the Limit of a DFS

A way to gain some intuition about the structure of the DTFT formulas is to consider the DFS of periodic sequences with larger and larger periods. Intuitively, as we look at the structure of the Fourier basis for the DFS, we can see that the number of basis vectors in (4.9) grows with the length N of the period and, consequently, the frequencies of the underlying complex exponentials become "denser" between 0 and 2π. We want to show that, in the limit, we end up with the reconstruction formula of the DTFT.

To do so, let us restrict ourselves to the domain of absolute summable sequences; for these sequences, we know that the sum in (4.13) exists. Now, given an absolutely summable sequence $x[n]$, we can always build an N-periodic sequence $\tilde{x}[n]$ as

$$\tilde{x}[n] = \sum_{i=-\infty}^{\infty} x[n+iN] \qquad (4.20)$$

for any value of N (see Example 2.2); this is guaranteed by the fact that the above sum converges for all $n \in \mathbb{Z}$ (because of the absolute summability of $x[n]$) so that all values of $\tilde{x}[n]$ are finite. Clearly, there is overlap between successive copies of $x[n]$; the intuition, however, is the following: since in the end we will consider very large values for N and since $x[n]$, because of absolute summability, decays rather fast with n, the resulting overlap of "tails" will be negligible. This can be expressed as

$$\lim_{N \to \infty} \tilde{x}[n] = x[n]$$

Now consider the DFS of $\tilde{x}[n]$:

$$\tilde{X}[k] = \sum_{n=0}^{N-1} \tilde{x}[n] e^{-j\frac{2\pi}{N}nk} = \sum_{i=-\infty}^{\infty} \left(\sum_{n=0}^{N-1} x[n+iN] e^{-j\frac{2\pi}{N}(n+iN)k} \right) \qquad (4.21)$$

where in the last term we have used (4.20), interchanged the order of the summation and exploited the fact that $e^{-j(2\pi/N)(n+iN)k} = e^{-j(2\pi/N)nk}$. We can see that, for every value of i in the outer sum, the argument of the inner sum varies between iN and $iN + N - 1$, i.e. non-overlapping intervals, so that the double summation can be simplified as

$$\tilde{X}[k] = \sum_{m=-\infty}^{\infty} x[m] e^{-j\frac{2\pi}{N}mk} \qquad (4.22)$$

and therefore

$$\tilde{X}[k] = X(e^{j\omega})\big|_{\omega=\frac{2\pi}{N}k} \qquad (4.23)$$

This already gives us a noteworthy piece of intuition: the DFS coefficients for the periodized signal are a discrete set of values of its DTFT (here considered solely as a formal operator) computed at multiples of $2\pi/N$. As N grows, the spacing between these frequency intervals narrows more and more so that, in the limit, the DFS converges to the DTFT.

To check that this assertion is consistent, we can now write the DFS reconstruction formula using the DFS values given to us by inserting (4.23) in (4.10):

$$\tilde{x}[n] = \frac{1}{N} \sum_{k=0}^{N-1} X(e^{j\frac{2\pi}{N}k}) e^{j\frac{2\pi}{N}nk} \tag{4.24}$$

By defining $\Delta = (2\pi/N)$, we can rewrite the above expression as

$$\tilde{x}[n] = \frac{1}{2\pi} \sum_{k=0}^{N-1} X(e^{j(k\Delta)}) e^{j(k\Delta)n} \Delta \tag{4.25}$$

and the summation is easily recognized as the Riemann sum with step Δ approximating the integral of $f(\omega) = X(e^{j\omega})e^{j\omega n}$ between 0 and 2π. As N goes to infinity (and therefore $\tilde{x}[n] \to x[n]$), we can therefore write

$$\tilde{x}[n] \to \frac{1}{2\pi} \int_0^{2\pi} X(e^{j\omega}) e^{j\omega n} \, d\omega \tag{4.26}$$

which is indeed the DTFT reconstruction formula (4.14).[5]

4.4.2 The DTFT as a Formal Change of Basis

We now show that, if we are willing to sacrifice mathematical rigor, the DTFT can be cast in the same conceptual framework we used for the DFT and DFS, namely as a basis change in a vector space. The following formulas are to be taken as nothing more than a set of purely symbolic derivations, since the mathematical hypotheses under which the results are well defined are far from obvious and are completely hidden by the formalism. It is only fair to say, however, that the following expressions represent a very handy and intuitive toolbox to grasp the essence of the duality between the discrete-time and the frequency domains and that they can be put to use very effectively to derive quick results when manipulating sequences.

One way of interpreting Equation (4.13) is to see that, for any given value ω_0, the corresponding value of the DTFT is the inner product in $\ell_2(\mathbb{Z})$ of the sequence $x[n]$ with the sequence $e^{j\omega_0 n}$; formally, at least, we are still performing a projection in a vector space akin to $\mathbb{C}^\infty$:

$$X(e^{j\omega}) = \langle e^{j\omega n}, x[n] \rangle$$

Here, however, the set of "basis vectors" $\{e^{j\omega n}\}_{\omega \in \mathbb{R}}$ is indexed by the real variable ω and is therefore uncountable. This uncountability is mirrored in

---

[5] Clearly (4.26) is equivalent to (4.14) in spite of the different integration limits since all the quantities under the integral sign are 2π-periodic and we are integrating over a period.

the inversion formula (4.14), in which the usual summation is replaced by an integral; in fact, the DTFT operator maps $\ell_2(\mathbb{Z})$ onto $L_2([-\pi,\pi])$ which is a space of 2π-periodic, square integrable functions. This interpretation preserves the physical meaning given to the inner products in (4.13) as a way to measure the frequency content of the signal at a given frequency; in this case the number of oscillators is infinite and their frequency separation becomes infinitesimally small.

To complete the picture of the DTFT as a change of basis, we want to show that, at least formally, the set $\{e^{j\omega n}\}_{\omega\in\mathbb{R}}$ constitutes an orthogonal "basis" for $\ell_2(\mathbb{Z})$.[6] In order to do so, we need to introduce a quirky mathematical entity called the Dirac delta functional; this is defined in an implicit way by the following formula

$$\int_{-\infty}^{\infty} \delta(t-\tau)f(t)\,dt = f(\tau) \tag{4.27}$$

where $f(t)$ is an arbitrary integrable function on the real line; in particular

$$\int_{-\infty}^{\infty} \delta(t)f(t)\,dt = f(0) \tag{4.28}$$

While no ordinary function satisfies the above equation, $\delta(t)$ can be interpreted as shorthand for a limiting operation. Consider, for instance, the family of parametric functions[7]

$$r_k(t) = k\,\text{rect}(kt) \tag{4.29}$$

which are plotted in Figure 4.12. For any continuous function $f(t)$ we can write

$$\int_{-\infty}^{\infty} r_k(t)f(t)\,dt = k\int_{-1/2k}^{1/2k} f(t)\,dt = f(\gamma)\big|_{\gamma\in[-1/2k,1/2k]} \tag{4.30}$$

where we have used the Mean Value theorem. Now, as k goes to infinity, the integral converges to $f(0)$; hence we can say that the limit of the series of functions $r_k(t)$ converges then to the Dirac delta. As already stated, the delta cannot be considered as a proper function, so the expression $\delta(t)$ outside of an integral sign has no mathematical meaning; it is customary however to associate an "idea" of function to the delta and we can think of it as being

---

[6] You can see here already why this line of thought is shaky unsafe: indeed, $e^{j\omega n} \notin \ell_2(\mathbb{Z})$!

[7] The rect function is discussed more exhaustively in Section 5.6 its definition is

$$\text{rect}(x) = \begin{cases} 1 & \text{for } |x| \leq 1/2 \\ 0 & \text{otherwise} \end{cases}$$

undefined for $t \neq 0$ and to have a value of ∞ at $t = 0$. This interpretation, together with (4.27), defines the so-called *sifting property* of the Dirac delta; this property allows us to write (outside of the integral sign):

$$\delta(t-\tau)f(t) = \delta(t-\tau)f(\tau) \tag{4.31}$$

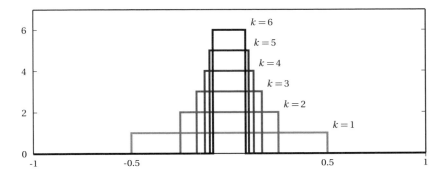

Figure 4.12 The Dirac delta as the limit of a family of rectangular functions.

The physical interpretation of the Dirac delta is related to quantities expressed as continuous *distributions* for which the most familiar example is probably that of a probability distribution (pdf). These functions represent a value which makes physical sense only over an interval of nonzero measure; the punctual value of a distribution is only an abstraction. The Dirac delta is the operator that extracts this punctual value from a distribution, in a sense capturing the essence of considering smaller and smaller observation intervals.

To see how the Dirac delta applies to our basis expansion, note that equation (4.27) is formally identical to an inner product over the space of functions on the real line; by using the definition of such an inner product we can therefore write

$$f(t) = \int_{-\infty}^{\infty} \langle \delta(s-\tau), f(s) \rangle \, \delta(t-\tau) \, d\tau \tag{4.32}$$

which is, in turn, formally identical to the reconstruction formula of Section 3.4.3. In reality, we are interested in the space of 2π-periodic functions, since that is where DTFTs live; this is easily accomplished by building a 2π-periodic version of the delta as

$$\tilde{\delta}(\omega) = 2\pi \sum_{k=-\infty}^{\infty} \delta(\omega - 2\pi k) \tag{4.33}$$

where the leading 2π factor is for later convenience. The resulting object is called a *pulse train*, similarly to what we built for the case of periodic sequences in $\tilde{\mathbb{C}}^N$. Using the pulse train and given any 2π-periodic function $f(\omega)$, the reconstruction formula (4.32) becomes

$$f(\omega) = \frac{1}{2\pi} \int_{\sigma}^{\sigma+2\pi} \langle \tilde{\delta}(\theta - \phi), f(\theta) \rangle \, \tilde{\delta}(\omega - \phi) \, d\phi \qquad (4.34)$$

for any $\sigma \in \mathbb{R}$.

Now that we have the delta notation in place, we are ready to start. First of all, we show the formal orthogonality of the basis functions $\{e^{j\omega n}\}_{\omega \in \mathbb{R}}$. We can write

$$\frac{1}{2\pi} \int_{-\pi}^{\pi} \tilde{\delta}(\omega - \omega_0) e^{j\omega n} d\omega = e^{j\omega_0 n} \qquad (4.35)$$

The left-hand side of this equation has the exact form of the DTFT reconstruction formula (4.14); hence we have found the fundamental relationship

$$e^{j\omega_0 n} \xleftrightarrow{\text{DTFT}} \tilde{\delta}(\omega - \omega_0) \qquad (4.36)$$

Now, the DTFT of a complex exponential $e^{j\sigma n}$ is, in our change of basis interpretation, simply the inner product $\langle e^{j\omega n}, e^{j\sigma n} \rangle$; because of (4.36) we can therefore express this as

$$\langle e^{j\omega n}, e^{j\sigma n} \rangle = \tilde{\delta}(\omega - \sigma) \qquad (4.37)$$

which is formally equivalent to the orthogonality relation in (4.5).

We now recall for the last time that the delta notation subsumes a limiting operation: the DTFT pair (4.36) should be interpreted as shorthand for the limit of the partial sums

$$s_k(\omega) = \sum_{n=-k}^{k} e^{-j\omega n}$$

(where we have chosen $\omega_0 = 0$ for the sake of example). Figure 4.13 plots $|s_k(\omega)|$ for increasing values of k (we show only the $[-\pi, \pi]$ interval, although of course the functions are 2π-periodic). The family of functions $s_k(\omega)$ is exactly equivalent to the family of functions $r_k(t)$ we saw in (4.29); they too become increasingly narrow while keeping a constant area (which turns out to be 2π). That is why we can simply state that $s_k(\omega) \to \tilde{\delta}(\omega)$.

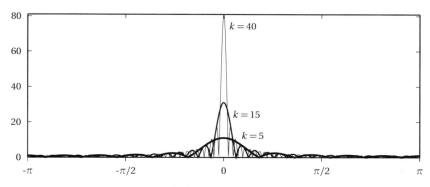

Figure 4.13 The sum $\left|\sum_{n=-k}^{k} e^{-j\omega n}\right|$ for different values of k.

From (4.36) we can easily obtain other interesting results: by setting $\omega_0 = 0$ and by exploiting the linearity of the DTFT operator, we can derive the DTFT of a constant sequence:

$$\alpha \xrightarrow{\text{DTFT}} \alpha \tilde{\delta}(\omega) \qquad (4.38)$$

or, using Euler's formulas, the DTFTs of sinusoidal functions:

$$\cos(\omega_0 n + \phi) \xrightarrow{\text{DTFT}} \frac{1}{2}\left[e^{j\phi}\tilde{\delta}(\omega - \omega_0) + e^{-j\phi}\tilde{\delta}(\omega + \omega_0)\right] \qquad (4.39)$$

$$\sin(\omega_0 n + \phi) \xrightarrow{\text{DTFT}} \frac{-j}{2}\left[e^{j\phi}\tilde{\delta}(\omega - \omega_0) - e^{-j\phi}\tilde{\delta}(\omega + \omega_0)\right] \qquad (4.40)$$

As we can see from the above examples, we are defining the DTFT for sequences which are not even square-summable; again, these transforms are purely a notational formalism used to capture a behavior, in the limit, as we showed before.

4.5 Relationships between Transforms

We can now show that, thanks to the delta formalism, the DTFT is the most general type of Fourier transform for discrete-time signals. Consider a length-N signal $x[n]$ and its N DFT coefficients $X[k]$; consider also the sequences obtained from $x[n]$ either by periodization or by building a finite-support sequence. The computation of the DTFTs of these sequences highlights the relationships linking the three types of discrete-time transforms that we have seen so far.

Periodic Sequences. Given a length-N signal $x[n]$, $n = 0, \ldots, N-1$, consider the associated N-periodic sequence $\tilde{x}[n] = x[n \bmod N]$ and its N DFS coefficients $X[k]$. If we try to write the analysis DTFT formula for $\tilde{x}[n]$ we have

$$\tilde{X}(e^{j\omega}) = \sum_{n=-\infty}^{\infty} \tilde{x}[n] e^{-j\omega n}$$

$$= \sum_{n=-\infty}^{\infty} \left(\frac{1}{N} \sum_{k=0}^{N-1} X[k] e^{j\frac{2\pi}{N}nk} \right) e^{-j\omega n} \quad (4.41)$$

$$= \frac{1}{N} \sum_{k=0}^{N-1} X[k] \left(\sum_{n=-\infty}^{\infty} e^{j\frac{2\pi}{N}nk} e^{-j\omega n} \right) \quad (4.42)$$

where in (4.41) we have used the DFS reconstruction formula. Now we recognize in the last term important to recognize the last terms of (4.42) as the DTFT of a complex exponential of frequency $(2\pi/N)k$; we can therefore write

$$\tilde{X}(e^{j\omega}) = \frac{1}{N} \sum_{k=0}^{N-1} X[k] \tilde{\delta}\left(\omega - \frac{2\pi}{N}k\right) \quad (4.43)$$

which is the relationship between the DTFT and the DFS. If we restrict ourselves to the $[-\pi, \pi]$ interval, we can see that the DTFT of a periodic sequence is a series of regularly spaced deltas placed at the N roots of unity and whose amplitude is proportional to the DFS coefficients of the sequence. In other words, *the DTFT is uniquely determined by the DFS and vice versa.*

Finite-Support Sequences. Given a length-N signal $x[n]$, $n = 0, \ldots, N-1$ and its N DFT coefficients $X[k]$, consider the associated finite-support sequence

$$\bar{x}[n] = \begin{cases} x[n] & 0 \leq n < N \\ 0 & \text{otherwise} \end{cases}$$

from which we can easily derive the DTFT of $\bar{x}$ as

$$\bar{X}(e^{j\omega}) = \sum_{k=0}^{N-1} X[k] \Lambda\left(\omega - \frac{2\pi}{N}k\right) \quad (4.44)$$

with

$$\Lambda(\omega) = \frac{1}{N} \sum_{m=0}^{N-1} e^{-j\omega m}$$

What the above expression means, is that the DTFT of the finite support sequence $\bar{x}[n]$ is again *uniquely defined by the N DFT coefficients of the* *finite-length signal* $x[n]$ and it can be obtained by a type of Lagrangian interpolation. As in the previous case, the values of DTFT at the roots of unity are equal to the DFT coefficients; note, however, that the transform of a finite support sequence is very different from the DTFT of a periodized sequence. The latter, in accordance with the definition of the Dirac delta, is defined only in the limit and for a finite set of frequencies; the former is just a (smooth) interpolation of the DFT.

4.6 Fourier Transform Properties

4.6.1 DTFT Properties

The DTFT possesses the following properties.

Symmetries and Structure. The DTFT of a time-reversed sequence is

$$x[-n] \stackrel{\text{DTFT}}{\longleftrightarrow} X(e^{-j\omega}) \qquad (4.45)$$

while, for the complex conjugate of a sequence we have

$$x^*[n] \stackrel{\text{DTFT}}{\longleftrightarrow} X^*(e^{-j\omega}) \qquad (4.46)$$

For the very important case of a *real* sequence $x[n] \in \mathbb{R}$, property 4.46 implies that the DTFT is conjugate-symmetric:

$$X(e^{j\omega}) = X^*(e^{-j\omega}) \qquad (4.47)$$

which leads to the following special symmetries for real signals:

- The magnitude of the DTFT is symmetric:

$$|X(e^{j\omega})| = |X(e^{-j\omega})| \qquad (4.48)$$

- The phase of the DTFT is antisymmetric:

$$\sphericalangle X(e^{j\omega}) = -\sphericalangle X(e^{-j\omega}) \qquad (4.49)$$

- The real part of the DTFT is symmetric:

$$\text{Re}\{X(e^{j\omega})\} = \text{Re}\{X(e^{-j\omega})\} \qquad (4.50)$$

- The imaginary part of the DTFT is antisymmetric:

$$\text{Im}\{X(e^{j\omega})\} = -\text{Im}\{X(e^{-j\omega})\} \qquad (4.51)$$

Finally, if $x[n]$ is real and symmetric, then the DTFT is real:

$$x[n]\in\mathbb{R},\ x[-n]=x[n] \iff X(e^{j\omega})\in\mathbb{R} \tag{4.52}$$

while, for real antisymmetric signals we have that the DTFT is purely imaginary:

$$x[n]\in\mathbb{R},\ x[-n]=-x[n] \iff \operatorname{Re}\{X(e^{j\omega})\}=0 \tag{4.53}$$

Linearity and Shifts. The DTFT is a linear operator:

$$\alpha x[n]+\beta y[n] \stackrel{\text{DTFT}}{\longleftrightarrow} \alpha X(e^{j\omega})+\beta Y(e^{j\omega}) \tag{4.54}$$

A shift in the discrete-time domain leads to multiplication by a phase term in the frequency domain:

$$x[n-n_0] \stackrel{\text{DTFT}}{\longleftrightarrow} e^{-j\omega n_0} X(e^{j\omega}) \tag{4.55}$$

while multiplication of the signal by a complex exponential (i.e. signal *modulation* by a complex "carrier" at frequency ω_0) leads to

$$e^{j\omega_0 n} x[n] \stackrel{\text{DTFT}}{\longleftrightarrow} X(e^{j(\omega-\omega_0)}) \tag{4.56}$$

which means that the spectrum is shifted by ω_0. This last result is known as the *modulation theorem*.

Energy Conservation. The DTFT satisfies the *Plancherel-Parseval* equality:

$$\langle x[n], y[n] \rangle = \frac{1}{2\pi} \langle X(e^{j\omega}), Y(e^{j\omega}) \rangle \tag{4.57}$$

or, using the respective definitions of inner product for $\ell_2(\mathbb{Z})$ and $L_2([-\pi,\pi])$:

$$\sum_{n=-\infty}^{\infty} x^*[n] y[n] = \frac{1}{2\pi} \int_{-\pi}^{\pi} X^*(e^{j\omega}) Y(e^{j\omega})\, d\omega \tag{4.58}$$

(note the explicit normalization factor $1/2\pi$). The above equality specializes into *Parseval's theorem* as

$$\sum_{n=-\infty}^{\infty} |x[n]|^2 = \frac{1}{2\pi} \int_{-\pi}^{\pi} |X(e^{j\omega})|^2\, d\omega \tag{4.59}$$

which establishes the conservation of energy property between the time and the frequency domains.

4.6.2 DFS Properties

The DTFT properties we have just seen extend easily to the Fourier Transform of periodic signals. The easiest way to obtain the particularizations which follow is to apply relationship (4.43) to the results of the previous Section.

Symmetries and Structure. The DFS of a time-reversed sequence is

$$\tilde{x}[-n] \overset{\text{DFS}}{\longleftrightarrow} \tilde{X}[-k] \qquad (4.60)$$

while, for the complex conjugate of a sequence we have

$$\tilde{x}^*[n] \overset{\text{DFS}}{\longleftrightarrow} \tilde{X}^*[-k] \qquad (4.61)$$

For *real* periodic sequences, the following special symmetries hold (see (4.47)–(4.53)):

$$\tilde{X}[k] = \tilde{X}^*[-k] \qquad (4.62)$$
$$|\tilde{X}[k]| = |\tilde{X}[-k]| \qquad (4.63)$$
$$\measuredangle \tilde{X}[k] = -\measuredangle \tilde{X}[-k] \qquad (4.64)$$
$$\text{Re}\{\tilde{X}[k]\} = \text{Re}\{\tilde{X}[-k]\} \qquad (4.65)$$
$$\text{Im}\{\tilde{X}[k]\} = -\text{Im}\{\tilde{X}[-k]\} \qquad (4.66)$$

Finally, if $\tilde{x}[n]$ is real and symmetric, then the DFS is real:

$$\tilde{x}[n] = \tilde{x}[-n] \iff \tilde{X}[k] \in \mathbb{R} \qquad (4.67)$$

while, for real antisymmetric signals, we can state that the DFS is purely imaginary.

Linearity and Shifts. The DFS is a linear operator, since it can be expressed as a matrix-vector product. A shift in the discrete-time domain leads to multiplication by a phase term in the frequency domain:

$$\tilde{x}[n-n_0] \overset{\text{DFS}}{\longleftrightarrow} W_N^{kn_0} \tilde{X}[k] \qquad (4.68)$$

while multiplication of the signal by a complex exponential of frequency multiple of $2\pi/N$ leads to of a shift in frequency:

$$W_N^{-nk_0} \tilde{x}[n] \overset{\text{DFS}}{\longleftrightarrow} \tilde{X}[k-k_0] \qquad (4.69)$$

Energy Conservation. We have already seen the energy conservation property in the context of basis expansion. Here, we simply recall Parseval's theorem, which states

$$\sum_{n=0}^{N-1} |\tilde{x}[n]|^2 = \frac{1}{N} \sum_{k=0}^{N-1} |\tilde{X}[k]|^2 \qquad (4.70)$$

4.6.3 DFT Properties

The properties of the DFT are obviously the same as those for the DFS, given the formal equivalence of the transforms. The only detail is how to interpret shifts, index reversal and symmetries for finite, length-N vectors; this is easily solved by considering the fact that the equivalence DFT-DFS translates in the time domain to a homomorphism between a length-N signal and its associated N-periodic extension to an infinite sequence. A shift, for instance, can be applied to the periodized version of the signal and the resulting shifted length N signal is given by the values of the shifted sequence from 0 to $N-1$, as previously explained in Section 2.2.2.

Mathematically, this means that all shifts and time reversals of a length-N signal are operated *modulo N*. Consider a length-N signal:

$$\begin{bmatrix} x[0] & x[1] & \ldots & x[N-1] \end{bmatrix}^T = x[n] \tag{4.71}$$

Its time-reversed version is

$$\begin{bmatrix} x[0] & x[N-1] & x[N-2] & \ldots & x[2] & x[1] \end{bmatrix}^T = x[-n \bmod N] \tag{4.72}$$

as we can easily see by considering the underlying periodic extension (note that $x[0]$ remains in place!) A shift by k corresponds to a circular shift:

$$\begin{bmatrix} x[k] & x[k-1] & \ldots & x[0] & x[N-1] & \ldots & x[k-1] \end{bmatrix}^T$$
$$= x[(n-k) \bmod N] \tag{4.73}$$

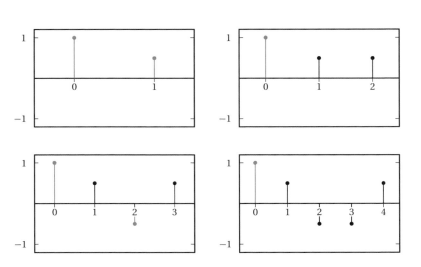

Figure 4.14 Examples of finite-length symmetric signals for $N = 2, 3, 4, 5$. Unconstrained values are drawn in gray.

The concept of symmetry can be reinterpreted as follows: a symmetric signal is equal to its time reversed version; therefore, for a length-N signal to be symmetric, the first member of (4.71) must equal the first member of (4.72), that is

$$x[k] = x[N-k], \qquad k = 1, 2, \ldots, \lfloor (N-1)/2 \rfloor \qquad (4.74)$$

Note that, in the above definition, the index k runs from 1 of $\lfloor (N-1)/2 \rfloor$; this means that symmetry does not place any constraint on the value of $x[0]$ and, similarly, the value of $x[N/2]$ is also unconstrained for even-length signals. Figure 4.14 shows some examples of symmetric length-N signals for different values of N. Of course the same definition can be used for anti-symmetric signals with just a change of sign.

Symmetries and Structure. The symmetries and structure derived for the DFS can be rewritten specifically for the DFT as

$$x[-n \ \mathrm{mod}\ N] \stackrel{\mathrm{DFT}}{\longleftrightarrow} X[-k \ \mathrm{mod}\ N] \qquad (4.75)$$

$$x^*[n] \stackrel{\mathrm{DFT}}{\longleftrightarrow} X^*[-k \ \mathrm{mod}\ N] \qquad (4.76)$$

The following symmetries hold only for *real* signals:

$$X[k] = X^*[-k \ \mathrm{mod}\ N] \qquad (4.77)$$

$$|X[k]| = |X[-k \ \mathrm{mod}\ N]| \qquad (4.78)$$

$$\measuredangle X[k] = -\measuredangle X[-k \ \mathrm{mod}\ N] \qquad (4.79)$$

$$\mathrm{Re}\{X[k]\} = \mathrm{Re}\{X[-k \ \mathrm{mod}\ N]\} \qquad (4.80)$$

$$\mathrm{Im}\{X[k]\} = -\mathrm{Im}\{X[-k \ \mathrm{mod}\ N]\} \qquad (4.81)$$

Finally, if $x[n]$ is real and symmetric (using the symmetry definition in (4.74), then the DFT is real:

$$x[k] = x[N-k], \qquad k = 1, 2, \ldots, \lfloor (N-1)/2 \rfloor \Longleftrightarrow X[k] \in \mathbb{R} \qquad (4.82)$$

while, for real antisymmetric signals we have that the DFT is purely imaginary.

Linearity and Shifts. The DFT is obviously a linear operator. A circular shift in the discrete-time domain leads to multiplication by a phase term in the frequency domain:

$$x[(n-n_0) \ \mathrm{mod}\ N] \stackrel{\mathrm{DFT}}{\longleftrightarrow} W_N^{kn_0} X[k] \qquad (4.83)$$

while the finite-length equivalent of the modulation theorem states

$$W_N^{-nk_0} x[n] \stackrel{\mathrm{DFT}}{\longleftrightarrow} X[(k-k_0) \ \mathrm{mod}\ N] \qquad (4.84)$$

Discrete-Time Fourier Transform (DTFT)

used for:	infinite, two sided signals ($x[n] \in \ell_2(\mathbb{Z})$)
analysis formula:	$X(e^{j\omega}) = \sum_{n=-\infty}^{\infty} x[n] e^{-j\omega n}$
synthesis formula:	$x[n] = \frac{1}{2\pi} \int_{-\pi}^{\pi} X(e^{j\omega}) e^{j\omega n} d\omega$
symmetries:	$x[-n] \stackrel{\text{DTFT}}{\longleftrightarrow} X(e^{-j\omega})$
	$x^*[n] \stackrel{\text{DTFT}}{\longleftrightarrow} X^*(e^{-j\omega})$
shifts:	$x[n-n_0] \stackrel{\text{DTFT}}{\longleftrightarrow} e^{-j\omega n_0} X(e^{j\omega})$
	$e^{j\omega_0 n} x[n] \stackrel{\text{DTFT}}{\longleftrightarrow} X(e^{j(\omega-\omega_0)})$
Parseval:	$\sum_{n=-\infty}^{\infty} |x[n]|^2 = \frac{1}{2\pi} \int_{-\pi}^{\pi} |X(e^{j\omega})|^2 d\omega$

Some DTFT pairs

$x[n] = \delta[n-k]$	$X(e^{j\omega}) = e^{-j\omega k}$
$x[n] = 1$	$X(e^{j\omega}) = \tilde{\delta}(\omega)$
$x[n] = u[n]$	$X(e^{j\omega}) = \frac{1}{1-e^{-j\omega}} + \frac{1}{2}\tilde{\delta}(\omega)$
$x[n] = a^n u[n], \quad |a| < 1$	$X(e^{j\omega}) = \frac{1}{1-ae^{-j\omega}}$
$x[n] = e^{j\omega_0 n}$	$X(e^{j\omega}) = \tilde{\delta}(\omega - \omega_0)$
$x[n] = \cos(\omega_0 n + \phi)$	$X(e^{j\omega}) = \frac{1}{2}\left[e^{j\phi}\tilde{\delta}(\omega - \omega_0) + e^{-j\phi}\tilde{\delta}(\omega + \omega_0)\right]$
$x[n] = \sin(\omega_0 n + \phi)$	$X(e^{j\omega}) = \frac{-j}{2}\left[e^{j\phi}\tilde{\delta}(\omega - \omega_0) - e^{-j\phi}\tilde{\delta}(\omega + \omega_0)\right]$
$x[n] = \begin{cases} 1 & \text{for } 0 \leq n \leq N-1 \\ 0 & \text{otherwise} \end{cases}$	$X(e^{j\omega}) = \frac{\sin((N/2)\omega)}{\sin(\omega/2)} e^{-j\frac{N-1}{2}\omega}$

Discrete Fourier Series (DFS)

used for:	periodic signals ($\tilde{x}[n] \in \tilde{\mathbb{C}}^N$)
analysis formula:	$\tilde{X}[k] = \sum_{n=0}^{N-1} \tilde{x}[n] W_N^{nk}, \quad k = 0, \ldots, N-1$
synthesis formula:	$\tilde{x}[n] = \frac{1}{N} \sum_{k=0}^{N-1} \tilde{X}[k] W_N^{-nk}, \quad n = 0, \ldots, N-1$

symmetries:	$\tilde{x}[-n] \overset{\text{DFS}}{\longleftrightarrow} \tilde{X}[-k]$
	$\tilde{x}^*[n] \overset{\text{DFS}}{\longleftrightarrow} \tilde{X}^*[-k]$
shifts:	$\tilde{x}[(n-n_0)] \overset{\text{DFS}}{\longleftrightarrow} W_N^{kn_0} \tilde{X}[k]$
	$W_N^{-nk_0} \tilde{x}[n] \overset{\text{DFS}}{\longleftrightarrow} \tilde{X}[(k-k_0)]$
Parseval:	$\sum_{n=0}^{N-1} |\tilde{x}[n]|^2 = \frac{1}{N} \sum_{k=0}^{N-1} |\tilde{X}[k]|^2$

Discrete Fourier Transform (DFT)

used for:	finite support signals $(x[n] \in \mathbb{C}^N)$
analysis formula:	$X[k] = \sum_{n=0}^{N-1} x[n] W_N^{nk}, \quad k = 0, \ldots, N-1$
synthesis formula:	$x[n] = \frac{1}{N} \sum_{k=0}^{N-1} X[k] W_N^{-nk}, \quad n = 0, \ldots, N-1$
symmetries:	$x[-n \bmod N] \overset{\text{DFT}}{\longleftrightarrow} X[-k \bmod N]$
	$x^*[n] \overset{\text{DFT}}{\longleftrightarrow} X^*[-k \bmod N]$
shifts:	$x[(n-n_0) \bmod N] \overset{\text{DFT}}{\longleftrightarrow} W_N^{kn_0} X[k]$
	$W_N^{-nk_0} x[n] \overset{\text{DFT}}{\longleftrightarrow} X[(k-k_0) \bmod N]$
Parseval:	$\sum_{n=0}^{N-1} |x[n]|^2 = \frac{1}{N} \sum_{k=0}^{N-1} |X[k]|^2$

Some DFT pairs for length-N signals $\qquad (n, k = 0, 1, \ldots, N-1)$

$x[n]$	$X[k]$
$x[n] = \delta[n-k]$	$X[k] = e^{-j\frac{2\pi}{N}k}$
$x[n] = 1$	$X[k] = N\delta[k]$
$x[n] = e^{j\frac{2\pi}{N}Ln}$	$X[k] = N\delta[k-L]$
$x[n] = \cos\left(\frac{2\pi}{N}Ln + \phi\right)$	$X[k] = \frac{N}{2}\left[e^{j\phi}\delta[k-L] + e^{-j\phi}\delta[k-N+L]\right]$
$x[n] = \sin\left(\frac{2\pi}{N}Ln + \phi\right)$	$X[k] = \frac{-jN}{2}\left[e^{j\phi}\delta[k-L] - e^{-j\phi}\delta[k-N+L]\right]$
$x[n] = \begin{cases} 1 & \text{for } n \leq M-1 \\ 0 & \text{for } M \leq n \leq N-1 \end{cases}$	$X[k] = \frac{\sin((\pi/N)Mk)}{\sin((\pi/N)k)} e^{-j\frac{\pi}{N}(M-1)k}$

Energy Conservation. Parseval's theorem for the DFT is (obviously) identical to (4.70):

$$\sum_{n=0}^{N-1}|x[n]|^2 = \frac{1}{N}\sum_{k=0}^{N-1}|X[k]|^2 \qquad (4.85)$$

4.7 Fourier Analysis in Practice

In the previous Sections, we have developed three frequency representations for the three main types of discrete-time signals; the derivation was eminently theoretical and concentrated mostly upon the mathematical properties of the transforms seen as a change of basis in Hilbert space. In the following Sections we will see how to put the Fourier machinery to practical use.

We have seen two fundamental ways to look at a signal: its time-domain representation, in which we consider the values of the signal as a function of discrete time, and its frequency-domain representation, in which we consider its energy and phase content as a function of digital frequency. The information contained in each of the two representations is exactly the same, as guaranteed by the invertibility of the Fourier transform; yet, from an analytical point of view, we can choose to concentrate on one domain or the other according to what we are specifically seeking. Consider for instance a piece of music; such a signal contains two coexisting perceptual features, *meter* and *key*. Meter can be determined by looking at the duration patterns of the played notes: its "natural" domain is therefore the time domain. The key, on the other hand, can be determined by looking at the pitch patterns of the played notes: since pitch is related to the frequency content of the sound, the natural domain of this feature is the frequency domain.

We can recall that the DTFT is mostly a theoretical analysis tool; the DTFTs which can be computed exactly (i.e. those in which the sum in (4.13) can be solved in closed form) represent only a small set of sequences; yet, these sequences are highly representative and they will be used over and over to illustrate a prototypical behavior. The DFT,[8] on the other hand, is fundamentally a *numerical* tool in that it defines a finite set of operations which can be computed in a finite amount of time; in fact, a very efficient algorithmic implementation of the DFT exists under the name of Fast Fourier

---

[8] This also applies to the DFS, of course, which is formally identical. As a general remark, whenever we talk about the DFT of a length-N signal, the same holds for the DFS of an N-periodic signal; for simplicity, from now on we will just concentrate on the DFT.

Transform (FFT) which only requires a number of operations on the order of $N(\log N)$ for an N-point data vector. The DFT, as we know, only applies to finite-length signals but this is actually acceptable since, in practice, all measured signals have finite support; in principle, therefore, the DFT suffices for the spectral characterization of real-world sequences. Since the transform of a finite-length signal and its DTFT are related by (4.43) or by (4.44) according to the underlying model for the infinite-length extension, we can always use the DTFT to illustrate the fundamental concepts of spectral analysis for the general case and then particularize the results for finite-length sequences.

4.7.1 Plotting Spectral Data

The first question that we ask ourselves is how to represent spectral data. Since the transform values are complex numbers, it is customary to separately plot their magnitude and their phase; more often than not, we will concentrate on the magnitude only, which is related to the energy distribution of the signal in the frequency domain.[9] For infinite sequences whose DTFT can be computed exactly, the graphical representation of the transform is akin to a standard function graph – again, the interest here is mostly theoretical. Consider now a finite-length signal of length N; its DFT can be computed numerically, and it yields a length-N vector of complex spectral values. These values can be displayed as such (and we obtain a plain DFT plot) or they can be used to obtain the DTFT of the periodic or finite-support extension of the original signal.

Consider for example the length-16 triangular signal $x[n]$ in Figure 4.15; note in passing that the signal is symmetric according to our definition in (4.74) so that its DFT is real. The DFT coefficients $|X[k]|$ are plotted in Figure 4.16; according to the fact that $x[n]$ is a real sequence, the set of DFT coefficients is symmetric (again according to (4.74)). The k-th DFT coefficient corresponds to the frequency $(2\pi/N)k$ and, therefore, the plot's abscissa extends implicitly from 0 to 2π; this is a little different than what we are used to in the case of the DTFT, where we usually consider the $[-\pi, \pi]$ interval, but it is customary. Furthermore, the difference is easily eliminated if we consider the sequence of $X[k]$ as being N-periodic (which it is, as we showed in Section 4.3) and plot the values from $-k/2$ to $k/2$ for k even, or from $-(k-1)/2$ to $(k-1)/2$ for k odd.

---

[9] A notable exception is the case of transfer function for digital filters, in which phase information is extremely important; we will study this in the next Chapter.

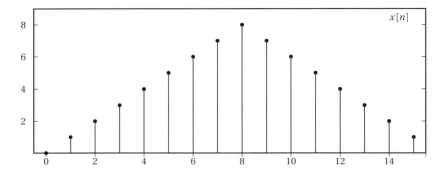

Figure 4.15 Length-16 signal.

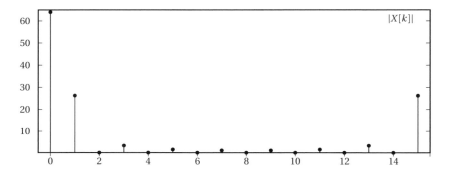

Figure 4.16 Magnitude DFT (or, equivalently, one period of the DFS) of the signal in Figure 4.15.

This can be made explicit by considering the N-periodic extension of $x[n]$ and by using the DFS-DTFT relationship (4.23); the standard way to plot this is as in Figure 4.17. Here the N pulse trains $\tilde{\delta}(\omega - (2\pi/N)k)$ are represented as lines (or arrows) scaled by the magnitude of the corresponding DFT coefficients. By plotting the representative $[-\pi, \pi]$ interval, we can appreciate, in full, the symmetry of the transform's magnitude.

By considering the finite-support extension of $x[n]$ instead, and by plotting the magnitude of its DTFT, we obtain Figure 4.18. The points in the plot can be computed directly from the summation defining the DTFT (which, for finite-support signals only contains a finite number of terms) and by evaluating the sum over a sufficiently fine grid of values for ω in the $[-\pi, \pi]$ interval; alternatively, the whole set of points can be obtained in one shot from an FFT with a sufficient amount of zero-padding (this method will be precised later). Again, the DTFT of a finite-support extension is just a smooth interpolation of the original DFT points and no new information is added.

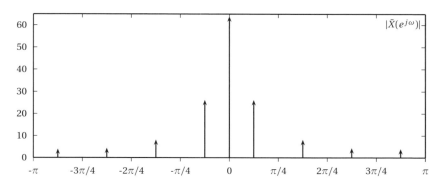

Figure 4.17 Magnitude DTFT of the periodic extension of the signal in Figure 4.15.

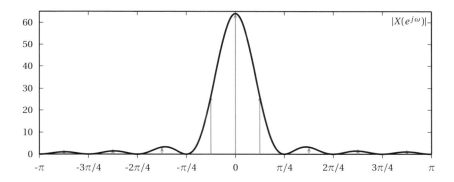

Figure 4.18 Magnitude DTFT of the finite-support extension of the signal in Figure 4.15. The Lagrange interpolation between DFT values is made apparent by the lines in gray.

4.7.2 Computing the Transform: the FFT

The Fast Fourier Transform, or FFT, is *not* another type of transform but simply the name of an efficient algorithm to compute the DFT. The algorithm, in its different flavors, is so ubiquitous and so important that the acronym FFT is often used liberally to indicate the DFT (or the DFS, which would be more appropriate since the underlying model is that of a periodic signal).

We have already seen in (4.6) that the DFT can be expressed in terms of a matrix vector multiplication:

$$\mathbf{X} = \mathbf{W}\mathbf{x}$$

as such, the computation of the DFT requires a number of operations on the order of N^2. The FFT algorithm exploits the highly structured nature of

W to reduce the number of operations to $N \log(N)$. In matrix form this is equivalent to decomposing **W** into the product of a series of matrices with mostly zero or unity elements. The algorithmic details of the FFT can be found in the bibliography; we can mention, however, that the FFT algorithm is particularly efficient for data lengths which are a power of 2 and that, in general, the more prime factors the data length can be decomposed into, the more efficient the FFT implementation.

4.7.3 Cosmetics: Zero-Padding

FFT algorithms are tailored to the specific length of the input signal. When the input signal's length is a large prime number or when only a subset of FFT algorithms is available (when, for instance, all we have is the radix-2 algorithm, which processes input vectors with lengths of a power of 2) it is customary to extend the length of the signal to match the algorithmic requirements. This is usually achieved by *zero padding*, i.e. the length-N data vector is extended to a chosen length M by appending $(M-N)$ zeros to it. Now, the maximum resolution of an N-point DFT, i.e. the separation between frequency components, is $2\pi/N$. By extending the signal to a longer length M, we are indeed reducing the separation between frequency components. One may think that this artificial increase in resolution allows the DFT to show finer details of the input signal's spectrum. It is not so.

The M-point DFT $\mathbf{X}^{(M)}$ of an N-point data vector **x**, obtained via zero-padding, can be obtained directly from the "canonical" N-point DFT of the vector $\mathbf{X}^{(N)}$ via a simple matrix multiplication:

$$\mathbf{X}^{(M)} = \mathbf{M}_{M,N} \mathbf{X}^{(N)} \tag{4.86}$$

where the $M \times N$ matrix $\mathbf{M}_{M,N}$ is given by

$$\mathbf{M}_{M,N} = \mathbf{W}'_M \mathbf{W}_N^H$$

where $\mathbf{W}_N$ is the standard DFT matrix and $\mathbf{W}'_M$ is the $M \times N$ matrix obtained by keeping just the first N columns of the standard DFT matrix $\mathbf{W}_M$. The fundamental meaning of (4.86) is that, by zero padding, we are adding no information to the spectral representation of a finite-length signal. Details of the spectrum which were not apparent in an N-point DFT are still not apparent in a zero-padded version of the same. It can be shown that (4.86) is a form of Lagrangian interpolation of the original DFT samples; therefore the zero-padded DFT is more attractive in a "cosmetic" fashion since the new points, when plotted, show a smooth curve between the original DFT points (and this is how plots such as the one in Figure 4.18 are obtained).

4.7.4 Spectral Analysis

The spectrum is a complete, alternative representation of a signal; by analyzing the spectrum, one can obtain, at a glance, the fundamental information, reguired to characterize and classify a signal in the frequency domain.

Magnitude The magnitude of a signal's spectrum, obtained by the Fourier transform, represents the energy distribution in frequency for the signal. It is customary to broadly classify discrete-time signals into three classes:

- **Lowpass** (or *baseband*) signals, for which the magnitude spectrum is concentrated around $\omega = 0$ and negligible elsewhere (Fig. 4.19).

- **Highpass** signals, for which the spectrum is concentrated around $\omega = \pi$ and negligible elsewhere, notably around $\omega = 0$ (Fig. 4.20).

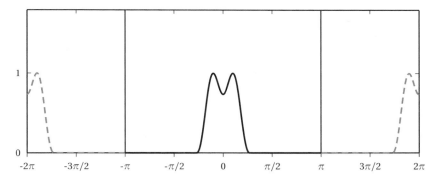

Figure 4.19 Lowpass spectrum. Note in this and the following figures, the 2π-periodicity of the spectrum is made explicit (spectral replicas are plotted in gray).

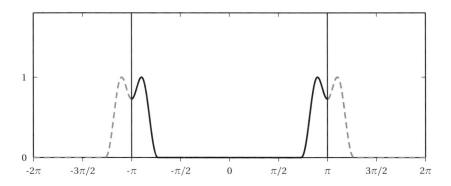

Figure 4.20 Highpass spectrum.

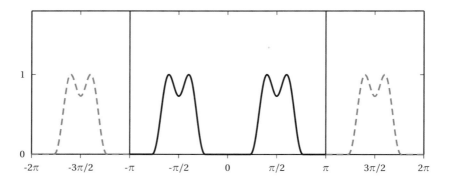

Figure 4.21 Bandpass spectrum.

- **Bandpass** signals, signals, for which the spectrum is concentrated around $\omega = \pm\omega_p$ and negligible elsewhere, notably around $\omega = 0$ and $\omega = \pi$ (Fig. 4.21).

For real-valued signals, the magnitude spectrum is a symmetric function and the above classifications take this symmetry into account. Remember also, that all spectra of discrete-time signals are 2π-periodic functions so that the above definitions are to be interpreted in a 2π-periodic fashion. For once, this is made explicit in Figures 4.19 to 4.21 where the plotting range, instead of the customary $[-\pi, \pi]$ interval, is extended from -2π to 2π.

Phase As we have stated before, the Fourier representation allows us to think of any signal as the sum of the outputs of a (potentially infinite) number of sinusoidal generators. While the magnitude of the spectrum defines the inherent power produced by each of the generators, its phase defines the *relative alignment* of the generated sinusoids. This alignment determines the *shape* of the signal in the discrete-time domain. To illustrate this with an example, consider the following 64-periodic signal:[10]

$$\tilde{x}[n] = \sum_{i=0}^{3} \frac{1}{2i+1} \sin\left(\frac{2\pi}{64}(2i+1)n + \phi_i\right) \qquad (4.87)$$

---

[10] The signal is the sum of the first four terms of the canonical trigonometric expansion of a square wave of period 64.

The magnitude of its DFS $\tilde{X}[k]$ is independent of the values of $\phi_i = 0$, $i = 0, 1, 2, 3$, and it is plotted in Figure 4.22. If the phase terms are uniformly zero, i.e. $\phi_i = 0$, $i = 0, 1, 2, 3$, $\tilde{x}[n]$ is the discrete-time periodic signal plotted in Figure 4.23; the alignment of the constituent sinusoids is such that the "square wave" exhibits a rather sharp transition between half-periods and a rather flat behavior over the half-period intervals. In addition, it should be noted with a zero phase term, the periodic signal is symmetric and that therefore the DFS coefficients are real. Now consider modifying the individual phases so that $\phi_i = 2\pi i/3$; in other words, we introduce a linear phase term in the constituent sinusoids. While the DFS magnitude remains exactly the same, the resulting time-domain signal is the one depicted in Figure 4.24; lack of alignment between sinusoids creates a "smearing" of the signal which no longer resembles a square wave.

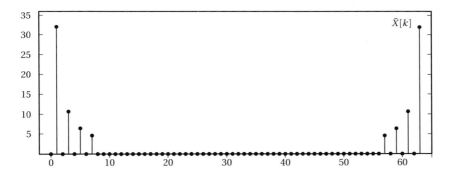

Figure 4.22 Magnitude DFS of the signal in (4.87).

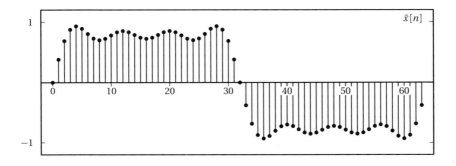

Figure 4.23 The signal in (4.87) with $\phi_i = 0$.

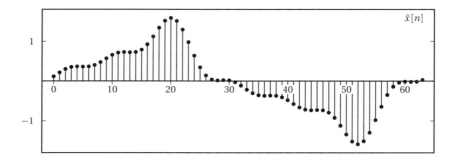

Figure 4.24 The signal in (4.87) with $\phi_i = 2\pi i/3$.

4.8 Time-Frequency Analysis

Recall our example at the beginning of this Chapter, when we considered the time and frequency information contained in a piece of music. We stated that the melodic information is related to the frequency content of the signal; obviously this is only partially true, since the melody is determined not only by the pitch values but also by their duration and order. Now, if we take a global Fourier Transform of the entire musical piece we have a *comprehensive* representation of the frequency content of the piece: in the resulting spectrum there is information about the frequency of each played note.[11] The time information, however, that is the information pertaining to the order in which the notes are played, is completely hidden by the spectral representation. This makes us wonder whether there exists a *time-frequency* representation of a signal, in which both time and frequency information are readily apparent.

4.8.1 The Spectrogram

The simplest time-frequency transformation is called the spectrogram. The recipe involves splitting the signal into small consecutive (and possibly overlapping) length-N pieces and computing the DFT of each. What we obtain is the following function of discrete-time and of a dicrete frequency index:

$$S[k,m] = \sum_{i=0}^{N-1} x[mM+i] W_N^{ik} \tag{4.88}$$

---

[11] Of course, even with the efficiency of the FFT algorithm, the computation of the DFT of an hour-long signal is beyond practical means.

where M, $1 \leq M \leq N$ controls the overlap between segments. In matrix notation we have

$$\mathbf{S} = \mathbf{W}_N \begin{bmatrix} x[0] & x[M] & x[2M] & \cdots \\ x[1] & x[M+1] & x[2M+1] & \cdots \\ \vdots & \vdots & \vdots & \cdots \\ x[N-1] & x[M+N-1] & x[L] & \cdots \end{bmatrix} \quad (4.89)$$

The resulting spectrogram is therefore an $N \times \lfloor L/M \rfloor$ matrix, where L is the total length of the signal $x[n]$. It is usually represented graphically as a plot in which the x-axis is the discrete-time index m, the y-axis is the discrete frequency index k and a color is the magnitude of $S[k, m]$, with darker colors for larger values.

As an example of the insight we can gain from the spectrogram, consider analyzing the well-known *Bolero* by Ravel. Figure 4.25 shows the spectrogram of the initial 37 seconds of the piece. In the first 13 seconds the only instrument playing is the snare drum, and the vertical line in the spectrogram represents, at the same time, the wide frequency content of a percussive instrument and its rhythmic pattern: if we look at the spacing between lines, we can identify the "trademark" drum pattern of Ravel's *Bolero*. After 13 seconds, the flute starts playing the theme; this is identifiable in the dark horizontal stripes which denote a high energy content around the frequencies which correspond to the pitches of the melody; with further analysis we could even try to identify the exact notes. The clarity of this plot is due to

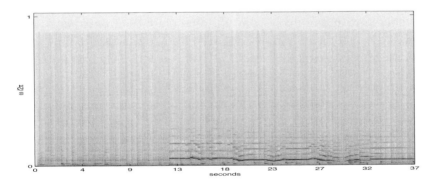

Figure 4.25 Spectrogram representation of the beginning of Ravel's *Bolero*. DFT size is 1024 samples, overlap is 512 samples.

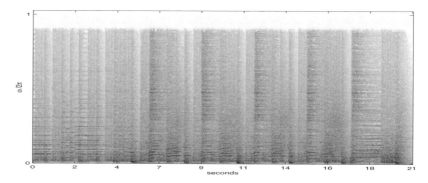

Figure 4.26 Spectrogram representation of the end of Ravel's *Bolero*.

the simple nature of the signal; if we now plot the spectrogram of the last 20 seconds of the piece, we obtain Figure 4.26. Here the orchestra is playing full blast, as indicated by the high energy activity across the whole spectrum; we can only detect the onset of the rhythmic shouts that precede the final chord.

4.8.2 The Uncertainty Principle

Each of the columns of **S** represents the "local" spectrum of the signal for a time interval of length N. We can therefore say that the *time resolution* of the spectrogram is N samples since the value of the signal at time n_0 influences the DFT of the N-point window around n_0. Seen from another point of view, the time information is "smeared" over an N-point interval. At the same time, the *frequency resolution* of the spectrogram is $2\pi/N$ (and we cannot increase it by zero-padding, as we have just shown). The conflict is therefore apparent: if we want to increase the frequency resolution we need to take longer windows but in so doing, we lose the time localization of the spectrogram; likewise, if we want to achieve a fine resolution in time, the corresponding spectral information for each "time slice" will be very coarse. It is rather easy to show that the amount of overlap does not change the situation. In practice, we need to choose an optimal tradeoff taking the characteristics of the signal into consideration.

The above problem, described for the case of the spectrogram, is actually a particular instance of a general uncertainty principle for time-frequency analysis. The principle states that, independently of the analysis tools that we put in place, we can never hope to achieve arbitrarily good resolution in both time and frequency since there exists a lower bound greater than zero for the product of the localization measure in time and frequency.

4.9 Digital Frequency vs. Real Frequency

The conceptual representation of discrete-time signals relies on the notion of a dimensionless "time", indicated by the integer index n. The absence of a physical dimension for time has the happy consequence that all discrete-time signal processing tools become indifferent to the underlying physical nature of the actual signals: stock exchange values or sampled orchestral music are just sequences of numbers. Similarly, we have just derived a frequency representation for signals which is based on the notion of a dimensionless frequency; because of the periodicity of the Fourier basis, all we know is that π is the highest digital frequency that we can represent in this model. Again, the power of generality is (or will soon be) apparent: a digital filter which is designed to remove the upper half of a signal's spectrum can be used with any type of input sequence, with the same results. This is in stark contrast with the practice of analog signal processing in which a half-band filter (made of capacitors, resistors and other electronic components) must be redesigned for any new class of input signals.

This dimensionless abstraction, however, is not without its drawbacks from the point of view of hands-on intuition; after all, we are all very familiar with signals in the real world for which time is expressed in seconds and frequency is expressed in hertz. We say, for instance, that speech has a bandwidth up to 4 KHz, that the human ear is sensitive to frequencies up to 20 KHz, that a cell phone transmits in the GHz band, and so on. What does "π" mean in these cases? The precise, formal link between real-world signal and discrete-time signal processing is given by the Sampling Theorem, which we will study later. The fundamental idea, however, is that we can remove the abstract nature of a discrete-time signal (and, correspondingly, of a dimensionless frequency) *by associating a time duration to the interval between successive discrete-time indices in the sequence.*

Let us say that the "real-world" time between indices n and $n+1$ in a discrete-time sequence is T_s seconds (where T_s is generally very small); this can correspond to sampling a signal every T_s seconds or to generating a synthetic sequence with a DSP chip whose clock cycle is T_s seconds. Now, recall that the phase increment between successive samples of a generic complex exponential $e^{j\omega_0 n}$ is ω_0 radians. The oscillation, therefore, completes a full cycle in $n_0 = (2\pi/\omega_0)$ samples. If T_s is the real-world time between samples, the full cycle is completed in $n_0 T_s$ seconds and so its "real-world" frequency is $f_0 = 1/(n_0 T_s)$ hertz. The relationship between the digital frequency ω_0 and the "real" frequency f_0 in Hertz as determined by the "clock" period T_s is therefore

$$f_0 \xrightarrow{T_s} \frac{1}{2\pi} \frac{\omega_0}{T_s} \qquad (4.90)$$

In particular, the highest real frequency which can be represented in the discrete-time system (which corresponds to $\omega = \pi$) is

$$F_{\max} = \frac{F_s}{2}$$

where we have used $F_s = (1/T_s)$; F_s is just the operating frequency of the discrete time system (also called the *sampling frequency* or clock frequency). With this notation, the digital frequency ω_0 corresponding to a real frequency f_0 is

$$\omega_0 = 2\pi \frac{f_0}{F_s}$$

The compact disk system, for instance, operates at a frequency $F_s = 44.1$ KHz; the maximum representable frequency for the system is 22.05 KHz (which constitutes the highest-pitched sound which can be encoded on, and reproduced by, a CD).

Examples

Example 4.1: The structure of DFT formulas

The DFT and inverse DFT (IDFT) formulas have a high degree of symmetry. Indeed, we can use the DFT algorithm to compute the IDFT with just a little manipulation: this can be useful if we have a "black box" FFT routine and we want to compute an inverse transform.

In the space of length-N signals, indicate the DFT of a signal $\mathbf{x}$ as

$$\mathbf{X} = \begin{bmatrix} X[0] & X[1] & \ldots & X[N-1] \end{bmatrix}^T = \text{DFT}\{\mathbf{x}\}$$

so that we can also write

$$\text{DFT}\{\mathbf{x}\}[n] = X[n]$$

Now consider the time-reversed signal

$$\mathbf{X}_r = \begin{bmatrix} X[N-1] & X[N-2] & \ldots & X[1] & X[0] \end{bmatrix}^T$$

we can show that

$$x[n] = \frac{1}{N} W_N^n \cdot \text{DFT}\{\mathbf{X}_r\}[n]$$

so that the inverse DFT can be obtained as the DFT of a time-reversed and scaled version of the original DFT. Indeed, with the change of variable $m = (N-1) - k$, we have

$$\begin{aligned}
\text{DFT}\{\mathbf{X}_r\}[n] &= \sum_{k=0}^{N-1} X[(N-1)-k] \, e^{-j\frac{2\pi}{N}kn} \\
&= \sum_{m=0}^{N-1} X[m] \, e^{-j\frac{2\pi}{N}(N-1-m)n} \\
&= \sum_{m=0}^{N-1} X[m] \, e^{j\frac{2\pi}{N}mn} \, e^{j\frac{2\pi}{N}n} \, e^{-j\frac{2\pi}{N}Nn} \\
&= e^{j\frac{2\pi}{N}n} \sum_{m=0}^{N-1} X[m] \, e^{j\frac{2\pi}{N}mn} = e^{j\frac{2\pi}{N}n} \cdot N \cdot x[n]
\end{aligned}$$

Example 4.2: The DTFT of the step function

In the delta-function formalism, the Fourier transform of the unit signal $x[n] = 1$ is the pulse train $\tilde{\delta}(\omega)$. Intuitively, the reasoning goes as follows: the unit signal has the same value over its entire infinite, two-sided support; nothing *ever* changes, there is not even the minutest glimpse of movement in the signal, ergo its spectrum can only have a nonzero value at the zero frequency. Recall that a frequency of zero is the frequency of dead quiet; the spectral value at $\omega = 0$ is also known as the DC component (for Direct Current), as opposed to a livelier AC (Alternating Current). At the same time, the unit signal has a very large energy, an infinite energy to be precise; imagine it as the voltage at the poles of a battery connected to a light bulb: to keep the light on for all eternity (i.e. over $\mathbb{Z}$) the energy must indeed be infinite. Our delta function captures this duality very effectively, if not rigorously.

Now consider the unit step $u[n]$; this is a somewhat stranger entity since it still possesses infinite energy and it still is a *very* quiet signal – except in $n = 0$. The transition in the origin is akin to flipping a switch in the battery/light bulb circuit above with the switch remaining on for the rest of (positive) eternity. As for the Fourier transform, intuitively we will still have a delta in zero (because of the infinite energy) but also some nonzero values over the entire frequency range because of the "movement" in $n = 0$. We know that for $|a| < 1$ it is

$$a^n \, u[n] \xleftrightarrow{\text{DTFT}} \frac{1}{1 - a\,e^{-j\omega}}$$

so that it is tempting to let $a \to 1$ and just say

$$u[n] \xleftrightarrow{\text{DTFT ??}} \frac{1}{1 - e^{-j\omega}}$$

This is not quite correct; even intuitively, the infinite energy delta is missing. To see what's wrong, let us try to find the inverse Fourier transform of the above expression; by using the substitution $e^{j\omega} = z$ and contour integration on the unit circle we have

$$\hat{u}[n] = \frac{1}{2\pi} \int_{-\pi}^{\pi} \frac{e^{j\omega n}}{1 - e^{-j\omega}} d\omega = \frac{1}{2\pi} \oint_C \frac{z^n}{1 - z^{-1}} \frac{dz}{jz}$$

Since there is a pole on the contour, we need have to use Cauchy's principal value theorem for the indented integration contour shown in Figure 4.27.

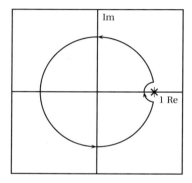

Figure 4.27 Indented integration contour.

For $n \geq 0$ there are no poles other than in $z = 1$ and we can use the "half-residue" theorem to obtain

$$\oint_{C'} \frac{z^n}{z-1} dz = j\pi [\text{Residue at } z = 1] = j\pi$$

so that

$$\hat{u}[n] = \frac{1}{2} \qquad \text{for } n \geq 0$$

For $n < 0$ there is a (multiple) pole in the origin; with the change of variable $v = z^{-1}$ we have

$$\oint_{C'} \frac{z^n}{z-1} dz = \oint_{C''} \frac{v^{-(n+1)}}{1-v} dv$$

where C'' is the same contour as C' but oriented clockwise. Because of this inversion it is

$$\hat{u}[n] = -\frac{1}{2} \qquad \text{for } n < 0$$

In conclusion

$$\text{DTFT}^{-1}\left\{\frac{1}{1-e^{-j\omega}}\right\} = \hat{u}[n] = \begin{cases} +1/2 & \text{for } n \geq 0 \\ -1/2 & \text{for } n < 0 \end{cases}$$

But this is almost good! Indeed,

$$u[n] = \hat{u}[n] + \frac{1}{2}$$

so that finally the DTFT of the unit step is

$$U(e^{j\omega}) = \frac{1}{1-e^{-j\omega}} + \frac{1}{2}\tilde{\delta}(\omega) \tag{4.91}$$

and its magnitude is sketched in Figure 4.28.

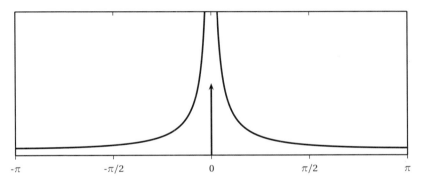

Figure 4.28 Magnitude spectrum for the unit step.

Further Reading

A nice engineering book on Fourier theory is *The Fourier Transform and Its Applications*, by R. Bracewell (McGraw-Hill, 1999). A more mathematically oriented textbook is *Fourier Analysis*, by T. W. Korner (Cambridge University Press, 1989), as is P. Bremaud's book, *Mathematical Principles of Signal Processing* (Springer, 2002).

Exercises

Exercise 4.1: DFT of elementary functions. Derive the formula for the DFT of the length-N signal $x[n] = \cos((2\pi/N)Ln + \phi)$.

Exercise 4.2: Real DFT. Compute the DFT of the length-4 signal $x[n] = \{a,b,c,d\}$. For which values of a,b,c,d is the DFT real?

Exercise 4.3: Limits. What is the value of the limit

$$\lim_{N \to \infty} \sum_{n=-N}^{N} \cos(\omega_0 n)$$

(in a signal processing sense)?

Exercise 4.4: Estimating the DFT graphically. Consider a length-64 signal $x[n]$ which is the sum of the three sinusoidal signals plotted in the following Figure (the signals are plotted as continuous lines just for clarity). Compute the DFT coefficients $X[k]$, $k = 0, 1, \ldots, 63$.

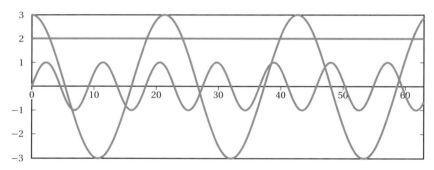

Exercise 4.5: The structure of DFT formulas. Consider a length-N signal $x[n]$, $N = 0, \ldots, N-1$; what is the length-N signal $y[n]$ obtained as

$$y[n] = \text{DFT}\{\text{DFT}\{x[n]\}\}$$

(i.e. by applying the DFT algorithm twice in a row)?

Exercise 4.6: Two DFTs for the price of one. When you compute the DFT of an length-N *real* signal, your data contains N real values while the DFT vector contains N *complex* values: there is clearly a redundancy of a factor of two in the DFT vector, which is apparent when you consider its

Hermitian symmetry (i.e. $X[k] = X^*[N-k]$). You can exploit this fact to compute the DFT of two real length-N signals for the price of one. This is useful if you have a pre-packaged FFT algorithm which you want to apply to real data.

Assume $x[n]$ and $y[n]$ are two length-N real signals. Build a complex signal $c[n] = x[n] + jy[n]$ and compute its DFT $C[k]$, $k = 0, 1, \ldots, N-1$. Show that

$$X[k] = \frac{1}{2}(C[k] + C^*[N-k])$$
$$Y[k] = \frac{1}{2j}(C[k] - C^*[N-k])$$

$k = 0, 1, \ldots, N-1$

where $X[k]$ and $Y[k]$ are the DFTs of $x[n]$ and $y[n]$, respectively.

Exercise 4.7: The Plancherel-Parseval equality. Let $x[n]$ and $y[n]$ be two complex valued sequences and $X(e^{jw})$ and $Y(e^{jw})$, their corresponding DTFTs.

(a) Show that

$$\langle x[n], y[n] \rangle = \frac{1}{2\pi} \langle X(e^{jw}), Y(e^{jw}) \rangle$$

where we use the inner products for $\ell_2(\mathbb{Z})$ and $L_2([-\pi, \pi])$, respectively.

(b) What is the physical meaning of the above formula when $x[n] = y[n]$?

Exercise 4.8: Numerical computation of the DFT. Consider the signal $x(n) = \cos(2\pi f_0 n)$. Compute and draw the DFT of the signal in $N = 128$ points, for

(a) $f_0 = 21/128$

(b) $f_0 = 21/127$

You can use any numerical package to do this. Explain the differences that we can see in these two spectra.

Exercise 4.9: DTFT vs. DFT. Consider the following infinite non-periodic discrete time signal:

$$x[n] = \begin{cases} 0 & n < 0 \\ 1 & 0 \leq n < a \\ 0 & n \geq a \end{cases}$$

(a) Compute its DTFT $X(e^{j\omega})$.

We want to visualize the magnitude of $X(e^{j\omega})$ using a numerical package (for instance, Matlab). Most numeric packages cannot handle continuous sequences such as $X(e^{j\omega})$; therefore we need to consider only a finite number of points for the spectrum.

(b) Plot 10,000 points of one period of $|X(e^{j\omega})|$ (from 0 to 2π) for $a = 20$.

The DTFT is mostly a theoretical analysis tool, and in many cases, we will compute the DFT. Moreover, for obvious reasons, numeric computation programs as, Matlab, only compute the DFT. Recall that in Matlab we use the Fast Fourier Transform (FFT), an efficient algorithm to compute the DFT.

(c) Generate a finite sequence $x_1[n]$ of length $N = 30$ such that $x_1[n] = x[n]$ for $n = 1, \ldots, N$. Compute its DFT and plot its magnitude. Compare it with the plot obtained in (b).

(d) Repeat now for different values of $N = 50, 100, 1000$. What can you conclude?

Chapter 5

Discrete-Time Filters

The previous Chapters gave us a thorough overview on both the nature of discrete-time signals and on the tools used in analyzing their properties. In the next few Chapters, we will study the fundamental building block of any digital signal processing system, that is, the linear filter. In the discrete-time world, filters are nothing but procedures which store and manipulate mathematically the numerical samples appearing at their input and their output; in other words, any discrete-time filter can be described procedurally in the form of an algorithm. In the special case of linear and time-invariant filters, such an algorithm can be concisely described mathematically by a constant-coefficient difference equation.

5.1 Linear Time-Invariant Systems

In its most general form, a *discrete-time system* can be described as a black box accepting a number of discrete-time sequences as inputs and producing another number of discrete-time sequences at its output.

In this book we are interested in studying the class of *linear time-invariant* (LTI) discrete-time systems with a single input and a single output; a system of this type is referred to as a *filter*. A linear time-invariant system $\mathcal{H}$ can thus be viewed as an operator which transforms an input sequence into an output sequence:

$$y[n] = \mathcal{H}\{x[n]\}$$

Figure 5.1 A single-input, single-output discrete-time system (black-box view).

Linearity is expressed by the equivalence

$$\mathcal{H}\{\alpha x_1[n] + \beta x_2[n]\} = \alpha \mathcal{H}\{x_1[n]\} + \beta \mathcal{H}\{x_2[n]\} \quad (5.1)$$

for any two sequences $x_1[n]$ and $x_2[n]$ and any two scalars $\alpha, \beta \in \mathbb{C}$. Time-invariance is expressed by

$$y[n] = \mathcal{H}\{x[n]\} \iff \mathcal{H}\{x[n-n_0]\} = y[n-n_0] \quad (5.2)$$

Linearity and time-invariance are very reasonable and "natural" requirements for a signal processing system. Imagine a recording system: linearity implies that a signal obtained by recording a violin and a piano playing together is the same as the sum of the signals obtained recording the violin and the piano separately (but in the same recording room). Multi-track recordings in music production are an application of this concept. Time invariance basically means that the system's behavior is independent of the time the system is turned on. Again, to use a musical example, this means that a given digital recording played back by a digital player will sound the same, regardless of when it is played.

Yet, simple as these properties, linearity and time-invariance taken together have an incredibly powerful consequence on a system's behavior. Indeed, a linear time-invariant system turns out to be *completely* characterized by its response to the input $x[n] = \delta[n]$. The sequence $h[n] = \mathcal{H}\{\delta[n]\}$ is called the *impulse response* of the system and $h[n]$ is all we need to know to determine the system's output for *any* other input sequence. To see this, we know that for any sequence we can always write the canonical orthonormal expansion (i.e. the reproducing formula in (2.18))

$$x[n] = \sum_{k=-\infty}^{\infty} x[k]\delta[n-k]$$

and therefore, if we let $\mathcal{H}\{\delta[n]\} = h[n]$, we can apply (5.1) and (5.2) to obtain

$$y[n] = \mathcal{H}\{x[n]\} = \sum_{k=-\infty}^{\infty} x[k]h[n-k] \quad (5.3)$$

5.2 Filtering in the Time Domain

The summation in (5.3) is called the *convolution* of sequences $x[n]$ and $h[n]$ and is denoted by the operator "$*$" so that (5.3) can be shorthanded to

$$y[n] = x[n] * h[n]$$

This is the general expression for a filtering operation in the discrete-time domain. To indicate a specific value of the convolution at a given time index n_0, we may use the notation $y[n_0] = (x * h)[n_0]$

5.2.1 The Convolution Operator

Clearly, for the convolution of two sequences to exist, the sum in (5.3) must be finite and this is always the case if both sequences are absolutely summable. As in the case of the DTFT, absolute summability is just a sufficient condition and the sum (5.3) can be well defined in certain other cases as well.

Basic Properties. The convolution operator is easily shown to be linear and time-invariant (which is rather intuitive seeing as it describes the behavior of an LTI system):

$$x[n] * (\alpha \cdot y[n] + \beta \cdot w[n]) = \alpha \cdot x[n] * y[n] + \beta \cdot x[n] * w[n] \tag{5.4}$$

$$w[n] = x[n] * y[n] \iff x[n] * y[n-k] = w[n-k] \tag{5.5}$$

The convolution is also commutative:

$$x[n] * y[n] = y[n] * x[n] \tag{5.6}$$

which is easily shown via a change of variable in (5.3). Finally, in the case of square summable sequences, it can be shown that the convolution is associative:

$$(x[n] * h[n]) * w[n] = x[n] * (h[n] * w[n]) \tag{5.7}$$

This last property describes the effect of connecting two filters $\mathscr{H}$ and $\mathscr{W}$ in cascade and it states that the resulting effect is that of a single filter whose impulse response is the convolution of the two original impulse responses. As a corollary, because of the commutative property, the order of the two filters in the cascade is completely irrelevant. More generally, a sequence of filtering operations can be performed in any order.

Please note that associativity does not necessarily hold for sequences which are not square-summable. A classic counterexample is the following: consider the three sequences

$x[n] = u[n]$ *the unit step*
$y[n] = \delta[n] - \delta[n-1]$ *the first-difference operator*
$w[n] = 1$ *a constant signal*

where clearly $x[n], w[n] \notin \ell_2(\mathbb{Z})$. It is easy to verify that

$$x[n] * (y[n] * w[n]) = 0$$
$$(x[n] * y[n]) * w[n] = 1$$

Convolution and Inner Product. It is immediately obvious that, for two sequences $x[n]$ and $h[n]$, we can write:

$$x[n] * h[n] = \langle h^*[n-k], x[k] \rangle$$

that is, the value at index n of the convolution of two sequences is the inner product (in $\ell_2(\mathbb{Z})$) of the first sequence – conjugated,[1] time-reversed and re-centered at n – with the input sequence. The above expression describes the output of a filtering operation as a series of "localized" inner products; filtering, therefore, measures the time-localized similarity (in the inner product sense, i.e. in the sense of the correlation) between the input sequence and a prototype sequence (the time-reversed impulse response).

In general, the convolution operator for a signal is defined with respect to the inner product of its underlying Hilbert space. For the space of N-periodic sequences, for instance, the convolution is defined as

$$\tilde{x}[n] * \tilde{y}[n] = \sum_{k=0}^{N-1} \tilde{x}[k]\tilde{y}[n-k] \tag{5.8}$$

$$= \sum_{k=0}^{N-1} \tilde{x}[n-k]\tilde{y}[k] \tag{5.9}$$

which is consistent with the inner product definition in (3.55). We will also consider the convolution of DTFTs. In this case, since we are in the space of 2π-periodic functions of a real variable, the convolution is defined as

---

[1] Since we consider only real impulse responses, the conjugation operator is in this case redundant.

$$X(e^{j\omega}) * Y(e^{j\omega}) = \frac{1}{2\pi} \langle X^*(e^{j(\omega-\sigma)}), Y(e^{j\sigma}) \rangle \qquad (5.10)$$

$$= \frac{1}{2\pi} \int_{-\pi}^{\pi} X(e^{j(\omega-\sigma)}) Y(e^{j\sigma}) d\sigma \qquad (5.11)$$

$$= \frac{1}{2\pi} \int_{-\pi}^{\pi} X(e^{j\sigma}) Y(e^{j(\omega-\sigma)}) d\sigma \qquad (5.12)$$

which is consistent with the inner product definition in (3.30).

5.2.2 Properties of the Impulse Response

As we said, an LTI system is completely described by its impulse response, i.e. by $h[n] = \mathcal{H}\{x[n]\}$.

FIR vs IIR. Since the impulse response is defined as the transformation of the discrete-time delta and since the delta is an infinite-length signal, *the impulse response is always an infinite-length signal*, i.e. a sequence. The nonzero values of the impulse response are usually called *taps*. Two distinct cases are possibles:

- **IIR filters:** when the number of taps is infinite.

- **FIR filters:** when the number of taps is finite (i.e. the impulse response is a finite-support sequence).

Note that in the case of FIR filters, the convolution operator entails only a finite number of sums and products; if $h[n] = 0$ for $n < N$ and $n \geq M$, we can invoke commutativity and rewrite (5.3) as

$$y[n] = \sum_{k=N}^{M-1} h[k] x[n-k]$$

Thus, convolution sums involving a finite-support impulse response are always well defined.

Causality. A system is called *causal* if its output does not depend on future values of the input. In practice, a causal system is the only type of "real-time" system we can actually implement, since knowledge of the future is normally not an option in real life. Yet, noncausal filters maintain a practical interest since in some application (usually called "batch processing") we may have access to the entirety of a discrete-time signal, which has been previously stored on some form of memory support.[2] A filter whose output depends exclusively on future values of the input is called *anticausal*.

For an LTI system, causality implies that the associated impulse response is zero for negative indices; this is the only way to remove all "future" terms in the convolution sum (5.3). Similarly, for anticausal systems, the impulse response must be zero for all positive indices. Clearly, between the strict causal and anticausal extremes, we can have intermediate cases: consider for example a filter $\mathscr{F}$ whose impulse response is zero for $n < -M$ with $M \in \mathbb{N}^+$. This filter is technically noncausal, but only in a "finite" way. If we consider the *pure delay* filter $\mathscr{D}$, whose impulse response is

$$d[n] = \delta[n-1]$$

we can easily see that $\mathscr{F}$ can be made strictly causal by cascading M delays in front of it. Clearly, an FIR filter is always causal up to a delay.

Stability. A system is called bounded-input bounded-output stable (BIBO stable) if its output is bounded for all bounded input sequences. Again, stability is a very natural requirement for a filter, since it states that the output will not "blow up" when the input is reasonable. Linearity and time-invariance do not guarantee stability (as anyone who has ever used a hands-free phone has certainly experienced).

A bounded sequence $x[n]$ is one for which it is possible to find a finite value $L \in \mathbb{R}^+$ so that $|x[n]| < L$ for all n. A necessary and sufficient condition for an LTI system $\mathscr{H}$ to be BIBO stable is that its impulse response $h[n]$ be absolutely summable. The sufficiency of the condition is proved as follows: if $x[n] < L$ for all n, then we have

$$|y[n]| = |h[n] * x[n]| = \left| \sum_{k=-\infty}^{\infty} h[k] x[n-k] \right|$$

$$\leq \sum_{k=-\infty}^{\infty} |h[k] x[n-k]| \leq L \sum_{k=-\infty}^{\infty} |h[k]|$$

and the last term is finite if $h[n]$ is absolutely summable. Conversely, assume that $h[n]$ is not absolutely summable and consider the signal

$$x[n] = \operatorname{sign}(h[-n])$$

$x[n]$ is clearly bounded, since it takes values only in $\{-1, 0, +1\}$, and yet

$$y[0] = (h * x)[0] = \sum_{k=-\infty}^{\infty} h[k] x[-k] = \sum_{k=-\infty}^{\infty} |h[k]| = \infty$$

---

[2] Clearly, to have a discrete-time signal stored in memory, the signal must be a finite-support sequence. If the support is sufficiently large, however, we can consider the signal as a full-fledged sequence.

Note that in the case of FIR filters, the convolution sum only involves a finite number of terms. As a consequence, *FIR filters are always stable.*

5.3 Filtering by Example – Time Domain

So far, we have described a filter from a very abstract point of view, and we have shown that a filtering operation corresponds to a convolution with a defining sequence called the impulse response. We now take a diametrically opposite standpoint: we introduce a very practical problem and arrive at a solution which defines an LTI system. Once we recognize that the solution is indeed a discrete-time filter, we will be able to make use of the theoretical results of the previous Sections in order, to analyze its properties.

Consider a sequence like the one in Figure 5.2; we are clearly in the presence of a "smooth" signal corrupted by noise, which appears as little wiggles in the plot. Our goal is the removal of the noise, i.e. to smooth out the signal, in order to improve its readability.

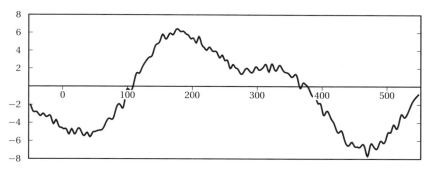

Figure 5.2 Noisy signal.

5.3.1 FIR Filtering

An intuitive and basic approach to remove noise from data is to replace each point of the sequence $x[n]$ by a *local average*, which can be obtained by taking the average of the sample at n and its $N-1$ predecessors. Each point of the "de-noised" sequence can therefore be computed as

$$y[n] = \frac{1}{N} \sum_{k=0}^{N-1} x[n-k] \qquad (5.13)$$

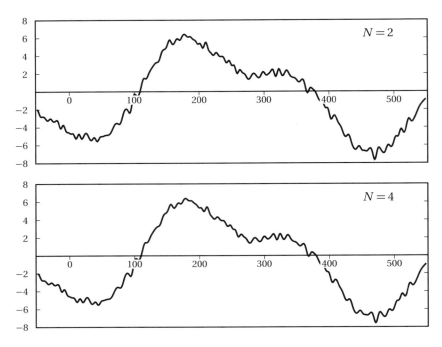

Figure 5.3 Moving averages for small values of N.

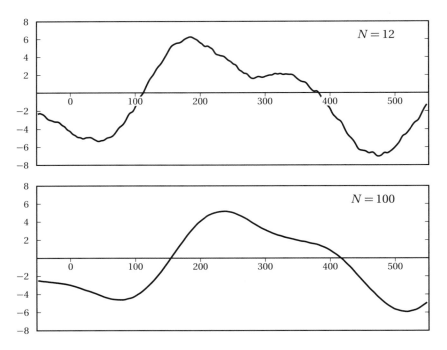

Figure 5.4 Moving averages for large values of N.

This is easily recognized as a convolution sum, and we can obtain the impulse response of the associated filter by letting $x[n] = \delta[n]$; it is easy to see that

$$h[n] = \frac{1}{N} \sum_{k=0}^{N-1} \delta[n-k] = \begin{cases} \frac{1}{N} & \text{for } 0 \leq n < N \\ 0 & \text{for } n < 0 \text{ and } n \geq N \end{cases} \quad (5.14)$$

The impulse response, as it turns out, is a finite-support sequence so the filter that we have just built, is an FIR filter; this particular filter goes under the name of *Moving Average* (MA) filter. The "smoothing power" of this filter is dependent on the number of samples we take into account in the average or, in other words, on the length N of its impulse response. The filtered version of the original sequence for small and large values of N is plotted in Figures 5.3 and 5.4 respectively. Intuitively we can see that as N grows, more and more wiggles are removed. We will soon see how to handle the "smoothing power" of a filter in a precise, quantitative way. A general characteristic of FIR filters, that should be immediately noticed is that the value of the output does not depend on values of the input which are more than N steps away; FIR filters are therefore *finite memory* filters. Another aspect that we can mention at this point concerns the *delay* introduced by the filter: each output value is the average of a window of N input values whose representative sample is the one falling in the middle; thus, there is a delay of $N/2$ samples between input and output, and the delay grows with N.

5.3.2 IIR Filtering

The moving average filter that we built in the previous Section has an obvious drawback; the more we want to smooth the signal, the more points we need to consider and, therefore, the more computations we have to perform to obtain the filtered value. Consider now the formula for the output of a length-M moving average filter:

$$y_M[n] = \frac{1}{M} \sum_{k=0}^{M-1} x[n-k] \quad (5.15)$$

We can easily see that

$$\begin{aligned} y_M[n] &= \frac{M-1}{M} y_{M-1}[n-1] + \frac{1}{M} x[n] \\ &= \lambda y_{M-1}[n-1] + (1-\lambda) x[n] \end{aligned}$$

where we have defined $\lambda = (M-1)/M$. Now, as M grows larger, we can safely assume that if we compute the average over $M-1$ or over M points,

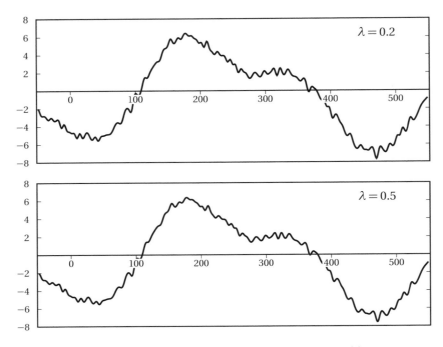

Figure 5.5 Moving averages for different values of λ.

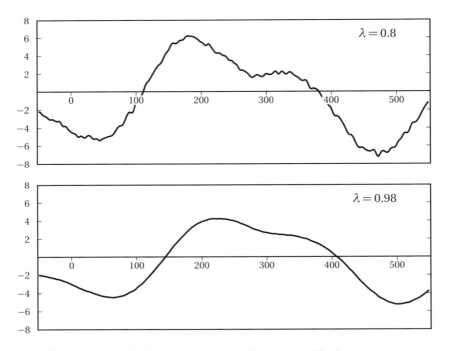

Figure 5.6 Leaky integrator outputs for values of λ close to one.

the result is basically the same: in other words, for M large, we can say that $y_{M-1}[n] \approx y_M[n]$. This suggests a new way to compute the smoothed version of a sequence in a *recursive* fashion:

$$y[n] = \lambda y[n-1] + (1-\lambda)x[n] \tag{5.16}$$

This no longer looks like a convolution sum; it is, instead, an instance of a *constant coefficient difference equation*. We might wonder whether the transformation realized by (5.16) is still linear and time-invariant and, in this case, what its impulse response is. The first problem that we face in addressing this question stems from the recursive nature of (5.16): each new output value depends on the previous output value. We need to somehow define a starting value for $y[n]$ or, in system theory parlance, we need to set the *initial conditions*. The choice which guarantees that the system defined by (5.16) is linear and time-invariant corresponds to the requirement that the system response to a sequence identically zero, be zero for all n; this requirement is also known as *zero initial conditions*, since it corresponds to setting $y[n] = 0$ for $n < N_0$ where N_0 is some time in the past.

The linearity of (5.16) can now be proved in the following way: assume that the output sequence for the system defined by (5.16) is $y[n]$ when the input is $x[n]$. It is immediately obvious that $y_1[n] = \alpha y[n]$ satisfies (5.16) for an input equal to $\alpha x[n]$. All we need to prove is that this is the only solution. Assume this is not the case and call $y_2[n]$ the other solution; we have

$$y_1[n] = \lambda y_1[n-1] + (1-\lambda)(\alpha x[n])$$
$$y_2[n] = \lambda y_2[n-1] + (1-\lambda)(\alpha x[n])$$

We can now subtract the second equation from the first. What we find is that the sequence $y_1[n] - y_2[n]$ is the system's response to the zero sequence, and therefore is zero for all n. Linearity with respect to the sum and time invariance can be proven in exactly the same way.

Now that we know that (5.16) defines an LTI system, we can try to compute its impulse response. Assuming zero initial conditions and $x[n] = \delta[n]$, we have

$$\begin{aligned} y[n] &= 0 \quad &&\text{for } n < 0 \\ y[0] &= 1 - \lambda \\ y[1] &= (1-\lambda)\lambda \\ y[2] &= (1-\lambda)\lambda^2 \\ &\vdots \\ y[n] &= (1-\lambda)\lambda^n \end{aligned} \tag{5.17}$$

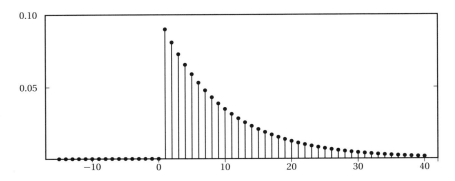

Figure 5.7 Impulse response (portion) of the leaky integrator for $\lambda = 0.9$.

so that the impulse response (shown in Figure 5.7) is

$$h[n] = (1-\lambda)\lambda^n u[n] \tag{5.18}$$

The impulse response clearly defines an IIR filter and therefore the immediate question is whether the filter is stable. Since a sufficient condition for stability is that the impulse response is absolutely summable, we have

$$\sum_{n=-\infty}^{\infty} |h[n]| = \lim_{n \to \infty} |1-\lambda| \frac{1-|\lambda|^{n+1}}{1-|\lambda|} \tag{5.19}$$

We can see that the above limit is finite for $|\lambda| < 1$ and so the system is BIBO stable for these values. The value of λ (which is, as we will see, the *pole* of the system) determines the smoothing power of the filter (Fig. 5.5). As $\lambda \to 1$, the input is smoothed more and more as can be seen in Figure 5.6, at a constant computational cost. The system implemented by (5.16) is often called a *leaky integrator*, in the sense that it approximates the behavior of an integrator with a leakage (or forgetting) factor λ. The delay introduced by the leaky integrator is more difficult to analyze than for the moving average but, again, it grows with the smoothing power of the filter; we will soon see how to proceed in order to quantify the delay introduced by IIR filters.

As we can infer from this simple analysis, IIR filters are much more delicate entities than FIR filters; in the next Chapters we will also discover that their design is also much less straightforward and offers less flexibility. This is why, in practice, FIR filters are the filters of choice. IIR filters, however, and especially the simplest ones such as the leaky integrator, are extremely attractive when computational power is a scarce resource.

5.4 Filtering in the Frequency Domain

The above examples have introduced the notion of filtering in an operational and intuitive way. In order to make more precise statements on the characteristics of a discrete-time filter we need to move to the frequency domain. What does a filtering operation translate to in the frequency domain? The fundamental result of this Section is the convolution theorem for discrete-time signals: a convolution in the discrete-time domain is equivalent to a multiplication of Fourier transforms in the frequency domain. This result opens up a very fruitful perspective on filtering and filter design, together with alternative approaches to the implementation of filtering devices, as we will see momentarily.

5.4.1 LTI "Eigenfunctions"

Consider the case of a complex exponential sequence of frequency ω_0 as the input to a linear time-invariant system $\mathcal{H}$; we have

$$\begin{aligned}
\mathcal{H}\{e^{j\omega_0 n}\} &= \sum_{k=-\infty}^{\infty} e^{j\omega_0 k} h[n-k] \\
&= \sum_{k=-\infty}^{\infty} h[k] e^{j\omega_0(n-k)} \\
&= e^{j\omega_0 n} \sum_{k=-\infty}^{\infty} h[k] e^{-j\omega_0 k} \\
&= H(e^{j\omega_0}) e^{j\omega_0 n}
\end{aligned} \qquad (5.20)$$

where $H(e^{j\omega_0})$ (i.e. the DTFT of $h[n]$ at $\omega = \omega_0$) is called the *frequency response* of the filter at frequency ω_0. The above result states the fundamental fact that *complex exponentials are eigensequences*[3] *of linear-time invariant systems*. We notice the following two properties:

- Using the polar form, $H(e^{j\omega_0}) = A_0 \, e^{j\theta_0}$, and we can write:

$$\mathcal{H}\{e^{j\omega_0 n}\} = A_0 \, e^{j(\omega_0 n + \theta_0)}$$

 i.e. the output oscillation is scaled in amplitude by $A_0 = |H(e^{j\omega_0})|$, the magnitude of the DTFT, and it is shifted in phase by $\theta_0 = \angle H(e^{j\omega_0})$, the phase of the DTFT.

---

[3] In continuous time, complex exponential functions are *eigenfunctions* of LTI system. In discrete time we use the slightly less standard term *eigensequences* to indicate input signal whose shape is not changed by a filtering operation.

- If the input to a linear time-invariant system is a sinusoidal oscillation, the output is always be a sinusoidal oscillation at the same frequency (or zero if $H(e^{j\omega_0}) = 0$). In other words, linear time-invariant systems cannot shift or duplicate frequencies.[4]

5.4.2 The Convolution and Modulation Theorems

Consider two sequences $x[n]$ and $h[n]$, both absolutely summable. The discrete-time Fourier transform of the convolution $y[n] = x[n] * h[n]$ is

$$Y(e^{j\omega}) = X(e^{j\omega})H(e^{j\omega}) \qquad (5.21)$$

The proof is as follows: if we take the DTFT of the convolution sum, we have

$$Y(e^{j\omega}) = \sum_{n=-\infty}^{\infty} \sum_{k=-\infty}^{\infty} x[k]h[n-k]e^{-j\omega n}$$

and by interchanging the order of summation (which can be done because of the absolute summability of both sequences) and by splitting the complex exponential, we obtain

$$Y(e^{j\omega}) = \sum_{k=-\infty}^{\infty} x[k]e^{-j\omega k} \sum_{n=-\infty}^{\infty} h[n-k]e^{-j\omega(n-k)}$$

from which the result immediately follows after a change of variable. Before discussing the implications of the theorem, we to state and prove its dual, which is called the modulation theorem.

Consider now the discrete-time sequences $x[n]$ and $w[n]$, both absolutely summable, with discrete-time Fourier transforms $X(e^{j\omega})$ and $W(e^{j\omega})$. The discrete-time Fourier transform of the product $y[n] = x[n]w[n]$ is

$$Y(e^{j\omega}) = X(e^{j\omega}) * W(e^{j\omega}) \qquad (5.22)$$

where the DTFT convolution is via the convolution operator for 2π-periodic functions, defined in (5.12). This is easily proven as follows: we begin with the DTFT inversion formula of the DTFT convolution:

$$\frac{1}{2\pi} \int_{-\pi}^{\pi} (X*Y)(e^{j\omega}) e^{j\omega n} d\omega$$

$$= \frac{1}{2\pi} \int_{-\pi}^{\pi} \frac{1}{2\pi} \int_{-\pi}^{\pi} X(e^{j(\omega-\sigma)}) Y(e^{j\sigma}) e^{j\omega n} d\sigma d\omega$$

---

[4] This strength is also a weakness in some applications and that is why sometimes *nonlinear transformations* are necessary.

and we split the last integral to obtain

$$\left(\frac{1}{2\pi}\int_{-\pi}^{\pi}X(e^{j(\omega-\sigma)})e^{j(\omega-\sigma)n}\,d\omega\right)\left(\frac{1}{2\pi}\int_{-\pi}^{\pi}Y(e^{j\sigma})e^{j\sigma n}\,d\sigma\right)$$
$$=x[n]y[n]$$

These fundamental results are summarized in Table 5.1.

Table 5.1 The convolution and modulation theorems.

Time Domain	Frequency Domain
$x[n] * y[n]$	$X(e^{j\omega})Y(e^{j\omega})$
$x[n]y[n]$	$X(e^{j\omega}) * Y(e^{j\omega})$

5.4.3 Properties of the Frequency Response

Since an LTI system is completely characterized by its impulse response, it is also uniquely characterized by its frequency response. The frequency response provides us with a different perspective on the properties of a given filter, which are embedded in the magnitude and the phase of the response.

Just as the impulse response completely characterizes a filter in the discrete-time domain, its Fourier transform, called the filter's *frequency response*, completely characterizes the filter in the frequency domain. The properties of LTI systems are described in terms of their DTFTs magnitude and phase, each of which controls different features of the system's behavior.

Magnitude. The most powerful intuition arising from the convolution theorem is obtained by considering the magnitude of the spectra involved in a filtering operation. Recall that a Fourier spectrum represents the energy distribution of a signal in frequency; by appropriately "shaping" the magnitude spectrum of a filter's impulse response we can easily boost, attenuate, and even completely eliminate, a given part of the frequency content in the filtered input sequence. According to the way the magnitude spectrum is affected by the filter, we can classify filters into three broad categories (here as before we assume that the impulse response is real, and therefore the associated magnitude spectrum is symmetric; in addition, the 2π periodicity of the spectrum is implicitly understood):

- **Lowpass filters**, for which the magnitude of the transform is concentrated around $\omega = 0$; these filters preserve the low- frequency energy of the input signals and attenuate or eliminate the high-frequency components.

- **Highpass filters**, for which the magnitude of the transform is concentrated around $\omega = \pm\pi$; these filters preserve the high-frequency energy of the input signals and attenuate or eliminate the low-frequency components.

- **Bandpass filters**, for which the magnitude of the transform is concentrated around $\omega = \pm\omega_p$; these filters preserve the energy of the input signals around the frequency ω_p and attenuate the signals elsewhere, notably around $\omega = 0$ and $\omega = \pm\pi$.

- **Allpass filters**, for which the magnitude of the transform is a *constant* over the entire $[-\pi, \pi]$ interval. These filters do not affect their input's spectral magnitude (except for a constant gain factor) and they are designed entirely in terms of their phase response (typically, to introduce, or compensate for, a delay).

The frequency interval (or intervals) for which the magnitude of the frequency response is zero (or practically negligible) is called the *stopband*. Conversely, the frequency interval (or intervals) for which the magnitude is non-negligible is called the *passband*.

Phase. The phase response of a filter has an equally important effect on the output signal, even though its impact is less intuitive.

By and large, the phase response acts as a generalized delay. Consider Equation (5.20) once more; we can see that a single sinusoidal oscillation undergoes a phase shift equal to the phase of the impulse response's Fourier transform. A phase offset for a sinusoid is equivalent to a delay in the time domain. This is immediately obvious in the case of a trigonometric function defined on the real line since we can always write

$$\cos(\omega t + \phi) = \cos(\omega(t - t_0)), \qquad t_0 = -\frac{\phi}{\omega}$$

For discrete-time sinusoids, it is not always possible to express the phase offset in terms of an integer number of samples (exactly for the same reasons for which a discrete- time sinusoid is not always periodic in its index n); yet the effect is the same, in that a phase offset corresponds to an implicit delay of the sinusoid. When the phase offset for a complex exponential is not an integer multiple of its frequency, we say that we are in the presence of a *fractional delay*. Now, since each sinusoidal component of the input signal may be delayed by an arbitrary amount, the output signal will be composed of sinusoids whose relative alignment may be very different from the original. Phase alignment determines the shape of the signal in the time domain, as we have seen in Section 4.7.4. A filter with unit magnitude across the spectrum, which does not affect the amplitude of the sinusoidal components, but whose phase response is not linear, can completely change the shape of a filtered signal.[5]

Linear Phase. A very important type of phase response is *linear phase*:

$$\angle H(e^{j\omega}) = e^{-j\omega d} \tag{5.23}$$

Consider a simple system which just delays its input, i.e. $y[n] = x[n - D]$ with $D \in \mathbb{Z}$; this is obviously an LTI system with impulse response $h[n] = \delta[n - D]$ and frequency response $H(e^{j\omega}) = e^{-j\omega D}$. This means that, if the value d in (5.23) is an integer, (5.23) defines a pure delay system; since the magnitude is constant and equal to one, this is an example of an allpass filter. If d is not an integer, (5.23) still defines an allpass delay system for which the delay is fractional, and we should interpret its effect as explained in the previous Section. In particular, if we think of the original signal in terms of its Fourier reconstruction formula, the fractionally delayed output is obtained by stepping forward the initial phase of *all* oscillators by a non-integer multiple of the frequency. In the discrete-time domains, we have a signal which takes values "between" the original samples but, since the relative phase of any one oscillator, with respect to the others, has remained the same as in the original signal, the shape of the signal in the time domain is unchanged.

For a general filter with linear phase we can always write

$$H(e^{j\omega}) = |H(e^{j\omega})| e^{-j\omega d}$$

In other words, the net effect of the linear phase filter is that of a cascade of two systems: a zero-phase filter which affects only the spectral magnitude of the input and therefore introduces no phase distortion, followed by a (possibly fractional) delay system (which, again, introduces just a delay but no phase distortion).

Group Delay. When a filter does not have linear phase, it is important to quantify the amount of phase distortion both in amount and in location.

---

[5] In all fairness, the phase response of a system is not very important in most audio applications, since the human ear is largely insensitive to phase. Phase is however extremely important in data transmission applications.

Nonlinear phase is not always a problem; if a filter's phase is nonlinear just in the stopband, for instance, the actual phase distortion is negligible. The concept of group delay is a measure of nonlinearity in the phase; the idea is to express the phase response around any given frequency ω_0 using a first order Taylor approximation. Define $\varphi(\omega) = \measuredangle H(e^{j\omega})$ and approximate $\varphi(\omega)$ around ω_0 as $\varphi(\omega_0 + \tau) = \varphi(\omega_0) + \tau \varphi'(\omega_0)$; we can write

$$\begin{aligned} H(e^{j(\omega_0+\tau)}) &= \left|H(e^{j(\omega_0+\tau)})\right| e^{j\varphi(\omega_0+\tau)} \\ &\approx \left(\left|H(e^{j(\omega_0+\tau)})\right| e^{j\varphi(\omega_0)}\right) e^{j\varphi'(\omega_0)\tau} \end{aligned} \quad (5.24)$$

so that, approximately, the frequency response of the filter is linear phase for at least a *group* of frequencies around a given ω_0. The delay for this group of frequencies is the negative of the derivative of the phase, from which the definition of group delay is

$$\mathrm{grd}\{H(e^{j\omega})\} = -\varphi'(\omega) = -\frac{d\measuredangle H(e^{j\omega})}{d\omega} \quad (5.25)$$

For truly linear phase systems, the group delay is a constant. Deviations from a constant value quantify the amount of phase distortion introduced by a filter in terms of the (possibly non-integer) number of samples by wich a frequency component is delayed.

5.5 Filtering by Example – Frequency Domain

Now that we know what to look for in a filter, we can revisit the "empirical" de-noising filters introduced in Section 5.3. Both filters are realizable, in the sense that they can be implemented with practical and efficient algorithms, as we will study in the next Chapters. Their frequency responses allow us to qualify and quantify precisely their smoothing properties, which we previously described, in an intuitive fashion.

Moving Average. The frequency response of the moving average filter (Sect. 5.3.1) can be shown to be

$$H(e^{j\omega}) = \frac{1}{N} \frac{\sin(\omega N/2)}{\sin(\omega/2)} e^{-j\frac{N-1}{2}\omega} \quad (5.26)$$

In the above expression, it is easy to separate the magnitude and the phase, which are plotted in Figure 5.8. The group delay for the filter is the constant $(N-1)/2$, which means that the filter delays its output by $(N-1)/2$ samples (i.e. there is a fractional delay for N even). This formalizes the intuition that the "representative sample" for an averaging window of N samples is the

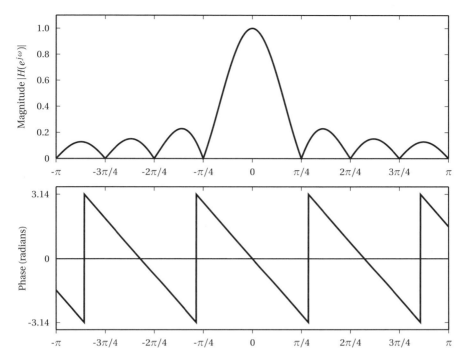

Figure 5.8 Magnitude and phase response of the moving average filter for $N=8$.

sample in the middle. If N is even, this does not correspond to a real sample but to a "ghost" sample in the middle.

Leaky Integrator. The frequency response of the leaky integrator in Section 5.3.2 is

$$H(e^{j\omega}) = \frac{1-\lambda}{1-\lambda e^{-j\omega}} \qquad (5.27)$$

Magnitude and phase are, respectively,

$$\left|H(e^{j\omega})\right|^2 = \frac{(1-\lambda)^2}{1+\lambda^2 - 2\lambda\cos(\omega)} \qquad (5.28)$$

$$\angle H(e^{j\omega}) = \arctan\left[-\frac{\lambda \sin(\omega)}{1-\lambda\cos(\omega)}\right] \qquad (5.29)$$

and they are plotted in Figure 5.9. The group delay, also plotted in Figure 5.9, is obtained by differentiating the phase response:

$$\mathrm{grd}\{H(e^{j\omega})\} = \frac{\lambda\cos(\omega) - \lambda^2}{1+\lambda^2 - 2\lambda\cos(\omega)} \qquad (5.30)$$

The group delay indicates that, for the frequencies for which the magnitude is not very small, the delay increases with the smoothing power of the filter.

Note that, according to the classification in Section 5.4.3, both the moving average and the leaky integrator are lowpass filters.

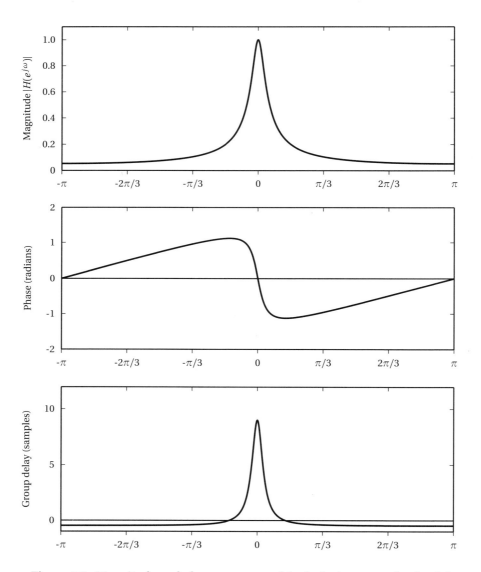

Figure 5.9 Magnitude and phase response of the leaky integrator for $\lambda = 0.9$.

5.6 Ideal Filters

The frequency characterization introduced in Section 5.4.3 immediately leads to questions such as "What is the best lowpass filter?" or "Can I have a highpass filter with zero delay?" It turns out that the answers to such questions are given by *ideal filters*. Ideal filters are what the (Platonic) name suggests: theoretical abstractions which capture the essence of the basic filtering operation but which are not realizable in practice. In a way, they are the "gold standard" of filter design.

Ideal Lowpass. The ideal lowpass filter is a filter which "kills" all frequency content above a *cutoff frequency* ω_c and leaves all frequency content below ω_c untouched; it is defined in the frequency domain as

$$H_{lp}(e^{j\omega}) = \begin{cases} 1 & |\omega| \leq \omega_c \\ 0 & \omega_c < |\omega| \leq \pi \end{cases} \quad (5.31)$$

and clearly, the filter has zero phase delay. The ideal lowpass can also be defined in terms of its *bandwidth* $\omega_b = 2\omega_c$. The DTFT inversion formula gives the corresponding impulse response:

$$h_{lp}[n] = \frac{\sin(\omega_c n)}{\pi n} \quad (5.32)$$

The impulse response is a symmetric infinite sequence and the filter is therefore IIR; unfortunately, however, it can be proved that no realizable system (i.e. no algorithm with a finite number of operations per output sample) can exactly implement the above impulse response. More bad news: the decay of the impulse response is slow, going to zero only as $1/n$, and it is not

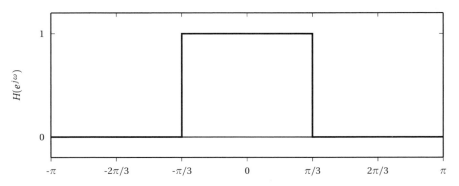

Figure 5.10 Frequency response of the ideal lowpass filter, $\omega_c = \pi/3$.

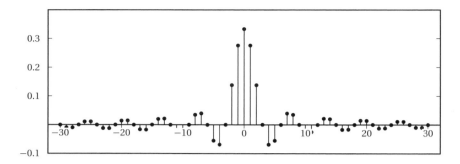

Figure 5.11 Impulse response (portion) of the ideal lowpass filter, $\omega_c = \pi/3$.

absolutely summable; this means that any FIR approximation of the ideal lowpass obtained by truncating $h[n]$ needs a lot of samples to achieve some accuracy and that, in any case, convergence to the ideal frequency response is only be in the mean square sense. An immediate consequence of these facts is that, when designing realizable filters, we will take an entirely different approach.

Despite these practical difficulties, the ideal lowpass and its associated DTFT pair are so important as a theoretical paradigm, that two special function names are used to denote the above expressions. These are defined as follows:

$$\text{rect}(x) = \begin{cases} 1 & |x| \leq 1/2 \\ 0 & |x| > 1/2 \end{cases} \tag{5.33}$$

$$\text{sinc}(x) = \begin{cases} \dfrac{\sin(\pi x)}{\pi x} & x \neq 0 \\ 1 & x = 0 \end{cases} \tag{5.34}$$

Note that the sinc function is zero for all integer values of the argument except zero. With this notation, and with respect to the bandwidth of the filter, the ideal lowpass filter's frequency response between $-\pi$ and π becomes

$$H_{lp}(e^{j\omega}) = \text{rect}\left(\frac{\omega}{\omega_b}\right) \tag{5.35}$$

(obviously 2π-periodized over all $\mathbb{R}$). Its impulse response in terms of bandwidth becomes

$$h_{lp}[n] = \frac{\omega_b}{2\pi} \text{sinc}\left(\frac{\omega_b}{2\pi} n\right) \tag{5.36}$$

or, in terms of cutoff frequency,

$$h_{lp}[n] = \frac{\omega_c}{\pi} \text{sinc}\left(\frac{\omega_c}{\pi} n\right) \tag{5.37}$$

The DTFT pair:

$$\frac{\omega_b}{2\pi} \text{sinc}\left(\frac{\omega_b}{2\pi} n\right) \xleftrightarrow{\text{DTFT}} \text{rect}\left(\frac{\omega}{\omega_b}\right) \tag{5.38}$$

constitutes one of the fundamental relationships of digital signal processing. Note that as $\omega_b \to 2\pi$, we re-obtain the well-known DTFT pair $\delta[n] \longleftrightarrow 1$, while as $\omega_b \to 0$ we can re-normalize by $(2\pi/\omega_b)$ to obtain $1 \longleftrightarrow \tilde{\delta}(\omega)$.

Ideal Highpass. The ideal highpass filter with cutoff frequency ω_c is the complementary filter to the ideal lowpass, in the sense that it eliminates all frequency content below the cutoff frequency. Its frequency response is

$$H_{hp}(e^{j\omega}) = \begin{cases} 0 & |\omega| \leq \omega_c \\ 1 & \omega_c < |\omega| \leq \pi \end{cases} \tag{5.39}$$

where the 2π-periodicity is as usual implicitly assumed. From the relation $H_h(e^{j\omega}) = 1 - \text{rect}(\omega/2\omega_c)$ the impulse response is easily obtained as

$$h_{hp}[n] = \delta[n] - \frac{\omega_c}{\pi} \text{sinc}\left(\frac{\omega_c}{\pi} n\right)$$

Ideal Bandpass. The ideal bandpass filter with center frequency ω_0 and bandwidth ω_b, $\omega_b/2 < \omega_0$ is defined in the frequency domain between $-\pi$ and π as

$$H_{bp}(e^{j\omega}) = \begin{cases} 1 & \omega_0 - \omega_b/2 \leq \omega \leq \omega_0 + \omega_b/2 \\ 1 & -\omega_0 - \omega_b/2 \geq \omega \geq -\omega_0 + \omega_b/2 \\ 0 & \text{elsewhere} \end{cases} \tag{5.40}$$

where the 2π-periodicity is, as usual, implicitly assumed. It is left as an exercise to prove that the impulse response is

$$h_{bp}[n] = 2\cos(\omega_0 n) \frac{\omega_b}{2\pi} \text{sinc}\left(\frac{\omega_b}{2\pi} n\right) \tag{5.41}$$

Hilbert Filter. The Hilbert filter is defined in the frequency domain as

$$H(e^{j\omega}) = \begin{cases} -j & 0 \leq \omega < \pi \\ +j & -\pi \leq \omega < 0 \end{cases} \tag{5.42}$$

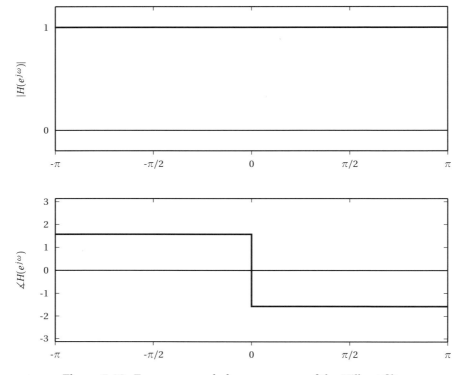

Figure 5.12 Frequency and phase response of the Hilbert filter.

where the 2π-periodicity is, as usual, implicitly assumed. Its impulse response is easily computed as

$$h[n] = \frac{2\sin^2(\pi n/2)}{\pi n} = \begin{cases} 0 & \text{for } n \text{ even} \\ \dfrac{2}{n\pi} & \text{for } n \text{ odd} \end{cases} \qquad (5.43)$$

Clearly $|H(e^{j\omega})| = 1$, so this filter is allpass. It introduces a phase shift of $\pi/2$ in the input signal so that, for instance,

$$h[n] * \cos(\omega_0 n) = -\sin(\omega_0 n) \qquad (5.44)$$

as one can verify from (4.39) and (4.40). More generally, the Hilbert filter is used in communication systems to build efficient demodulation schemes, as we will see later. The fundamental concept is the following: consider a *real* signal $x[n]$ and its DTFT $X(e^{j\omega})$; consider also the signal processed by the Hilbert filter $y[n] = h[n] * x[n]$. This can be defined as

$$A(e^{j\omega}) = \begin{cases} X(e^{j\omega}) & \text{for } 0 \leq \omega < \pi \\ 0 & \text{for } -\pi \leq \omega < 0 \end{cases}$$

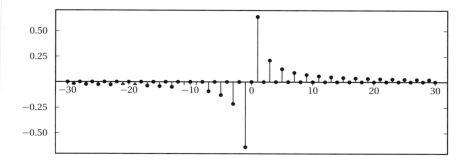

Figure 5.13 Impulse response (portion) of the Hilbert filter.

i.e. $A(e^{j\omega})$ is the positive-frequency part of the spectrum of $x[n]$. Since $x[n]$ is real, its DTFT has symmetry $X(e^{j\omega}) = X^*(e^{-j\omega})$ and therefore we can write

$$X(e^{j\omega}) = A^*(e^{-j\omega}) + A(e^{j\omega})$$

By separating the real and imaginary parts we can always write $A(e^{j\omega}) = A_R(e^{j\omega}) + jA_I(e^{j\omega})$ and so

$$X(e^{j\omega}) = A_R(e^{-j\omega}) - jA_I(e^{-j\omega}) + A_R(e^{j\omega}) + jA_I(e^{j\omega})$$

For the filtered signal, we know that $Y(e^{j\omega}) = H(e^{j\omega})X(e^{j\omega})$ and therefore

$$Y(e^{j\omega}) = jA_R(e^{-j\omega}) + A_I(e^{-j\omega}) - jA_R(e^{j\omega}) + A_I(e^{j\omega})$$

It is, thus, easy to see that

$$x[n] + jy[n] \xleftrightarrow{\text{DTFT}} 2A(e^{j\omega}) \qquad (5.45)$$

i.e. the spectrum of the signal $a[n] = x[n] + jy[n]$ contains only the positive-frequency components of the original signal $x[n]$. The signal $a[n]$ is called the *analytic signal* associated to $x[n]$.

5.7 Realizable Filters

Contrary to ideal filters, realizable filters are LTI systems which can be implemented in practice; this means that there exists an algorithm which computes every output sample with a finite number of operations and using a finite amount of memory storage. Note that the impulse response of a realizable filter need not be finite-support; while FIR filters are clearly realizable we have seen at least one example of realizable IIR filter (i.e. the leaky integrator).

5.7.1 Constant-Coefficient Difference Equations

Let us consider (informally) the possible mathematical description of an LTI system, seen as a "machine" which takes one input sample at a time and produces a corresponding output sample. Linearity in the input-output relationship implies that the description can involve only linear operations, i.e. sums and multiplications by scalars. Time invariance implies that the scalars be constants. Finally, realizability implies that, inside the above mentioned "machine", there can be only a finite number of adders and multipliers (and, correspondingly, a finite number of memory cells). Such a mathematical relationship goes under the name of constant-coefficient difference equation (CCDE).

In its most general form, a constant-coefficient difference equation defines a relationship between an input signal $x[n]$ and an output signal $y[n]$ as

$$\sum_{k=0}^{N-1} a_k y[n-k] = \sum_{k=0}^{M-1} b_k x[n-k] \tag{5.46}$$

In the rest of this book we restrict ourselves to the case in which all the coefficients a_k and b_k are real. Usually, $a_0 = 1$, so that the above equation can easily be rearranged as

$$y[n] = \sum_{k=0}^{M-1} b_k x[n-k] - \sum_{k=1}^{N-1} a_k y[n-k] \tag{5.47}$$

Clearly, the above relation defines each output sample $y[n]$ as a linear combination of past and present input values and past output values. However, it is easy to see that if $a_{N-1} \neq 0$ we can for instance rearrange (5.46) as

$$y[n-N+1] = \sum_{k=0}^{M-1} b'_k x[n-k] - \sum_{k=0}^{N-2} a'_k y[n-k]$$

where $a'_k = a_k/a_{N-1}$ and $b'_k = b_k/a_{N-1}$. With the change of variable $m = n - N + 1$, this becomes

$$y[m] = \sum_{k=N-M}^{N-1} b'_k x[m+k] - \sum_{k=1}^{N-1} a'_k y[m+k] \tag{5.48}$$

which shows that the difference equation can also be computed in another way, namely by expressing $y[m]$ as a linear combination of *future* values of input and output. It is rather intuitive that the first approach defines a causal behavior, while the second approach is anticausal.

5.7.2 The Algorithmic Nature of CCDEs

Contrary to the differential equations used in the characterization of continuous-time systems, difference equations can be used directly to translate the transformation operated by the system into an *explicit algorithmic form*. To see this, and to gain a lot of insight into the properties of difference equations, it may be useful to consider a possible implementation of the system in (5.47), shown as a C code sample in Figure 5.14.

```c
    extern double a[N];      // The a's coefficients
    extern double b[M];      // The b's coefficients

    static double x[M];      // Delay line for x
    static double y[N];      // Delay line for y

    double GetOutput(double input)
    {
      int k;

      // Shift delay line for x:
      for (k = N-1; k > 0; k--)
        x[k] = x[k-1];

      // new input value x[n]:
      x[0] = input;

      // Shift delay line for y:
      for (k = M-1; k > 0; k--)
        y[k] = y[k-1];

      double y = 0;
      for (k = 0; k < M; k++)
        y += b[k] * x[k];
      for (k = 1; k < M; k++)
        y -= a[k] * y[k];

      // New value for y[n]; store in delay line
      return (y[0] = y);
    }
```

Figure 5.14 C code implementation of a generic CCDE.

It is easy to verify that

- the routine effectively implements the difference equation in (5.47);
- the storage required is $(N+M)$;

- each output sample is obtained via $(N+M-1)$ multiplications and additions;

- the transformation is causal.

If we try to compile and execute the code, however, we immediately run into an *initialization* problem: the first time (actually, the first $\max(N, M-1)$ times) we call the function, the delay lines which hold past values of $x[n]$ and $y[n]$ will contain undefined values. Most likely, the compiler will notice this condition and will print a warning message signaling that the static arrays have not been properly initialized. We are back to the problem of setting the initial conditions of the system. *The choice which guarantees linearity and time invariance is called the* zero initial conditions *and corresponds to setting the delay lines to zero before starting the algorithm.* This choice implies that the system response to the zero sequence is the zero sequence and, in this way, linearity and time invariance can be proven as in Section 5.3.2.

5.7.3 Filter Analysis and Design

CCDEs provide a powerful operational view of filtering; in very simple case, such as in Section 5.3.2 or in the case of FIR filters, the impulse response (and therefore its frequency response) can be obtained directly from the filter's equation. This is not the general case however, and to analyze a generic realizable filter from its CCDE, we need to be able to easily derive the transfer function from the CCDE. Similarly, in order to design a realizable filter which meets a set of requirement, we need to devise a procedure which "tunes" the coefficients in the CCDE until the frequency response is satisfactory while preserving stability; in order to do this, again, we need a convenient tool to link the CCDE to the magnitude and phase response. This tool will be introduced in the next Chapter, and goes under the name of z-transform.

Examples

Example 5.1: Radio transmission

AM radio was one of the first forms of telecommunication and remains to this day a ubiquitous broadcast method due to the ease with which a robust receiver can be assembled. From the hardware point of view, an AM transmitter uses a transducer (i.e. a microphone) to convert sound to an electric signal, and then *modulates* this signal into a frequency band which correspond to a region of the electromagnetic spectrum in which propagation is well-behaved (see also Section 12.1.1). An AM receiver simply performs the reverse steps. Here we can neglect the physics of transducers and of antennas and concentrate on an idealized digital AM transmitter.

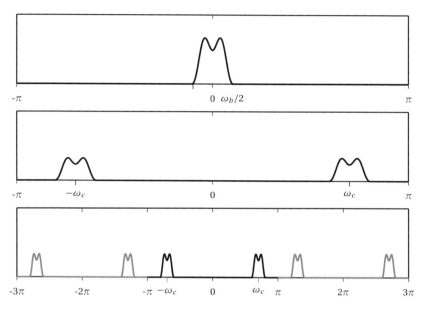

Figure 5.15 AM modulation; original baseband signal (top panel); modulated bandpass signal (middle panel); bandpass signal with explicit spectral repetitions (bottom panel).

Modulation. Suppose $x[n]$ is a real, discrete-time signal representing voice or music. Acoustic signals are a type of lowpass (or *baseband*) signal; while good for our ears (which are baseband receivers) baseband signals are not suitable for direct electromagnetic transmission since propagation in the baseband is poor and since occupancy of the same band would preclude the existence of multiple radio channels. We need to use modulation in order to shift a baseband signal in frequency and transform it into a *bandpass* signal prior to transmission. Modulation is accomplished by multiplying the baseband signal by an oscillatory *carrier* at a given center frequency; note that modulation is *not* a time-invariant operation. Consider the signal

$$y[n] = \text{Re}\{x[n]\, e^{j\omega_c n}\}$$

where ω_c is the carrier frequency. This corresponds to a cosine modulation since

$$y[n] = x[n]\cos(\omega_c n)$$

and (see (4.56)):

$$Y(e^{j\omega}) = \frac{1}{2}\left[X(e^{j(\omega-\omega_c)}) + X(e^{j(\omega+\omega_c)})\right]$$

The complex signal $c[n] = x[n]e^{j\omega_c n}$ is called the *complex bandpass signal* and, while not transmissible in practice, it is a very useful intermediate representation of the modulated signal especially in the case of Hilbert demodulation.

Assume that the baseband signal has spectral support $[-\omega_b/2, \omega_b/2]$ (i.e. assume that its energy is zero for $|\omega| > \omega_b/2$); a common way to express this concept is to say that the *bandwidth* of the baseband signal is ω_b. What is the maximum carrier frequency ω_c that we can use to create a bandpass signal? If we look at the effect of modulation on the spectrum and we take into account its 2π-periodicity as in Figure 5.15 we can see that if we choose too large a modulation frequency the positive passband overlaps with the negative passband of the first repetition. Intuitively, we are trying to modulate too fast and we are falling back into the folding of frequencies larger than 2π which we have seen in Example 2.1. In our case the maximum frequency of the modulated signal is $\omega_c + \omega_b/2$. To avoid overlap with the first repetition of the spectrum, we must guarantee that:

$$\omega_c + \frac{\omega_b}{2} < \pi \qquad (5.49)$$

which limit the maximum carrier frequency to $\omega_c < \pi - \omega_b/2$.

Demodulation. An AM receiver must undo the modulation process; again, assume we're entirely in the discrete-time domain. The first step is to isolate the channel of interest by using a sharp bandpass filter centered on the modulation frequency ω_c (see also Exercise 5.7). Neglecting the impairments introduced by the transmission (noise and distortion) we can initially assume that after filtering the receiver possesses an exact copy of the modulated signal $y[n]$. The original signal $x[n]$ can be retrieved in several ways; an elegant scheme, for instance, is Hilbert demodulation. The idea behind Hilbert demodulation is to reconstruct the complex bandpass signal $c[n]$ from $y[n]$ as $c[n] = y[n] + j(h[n] * y[n])$ where $h[n]$ is the impulse response of the Hilbert filter as given in (5.43). Once this is done, we multiply $c[n]$ by the complex exponential $e^{-j\omega_c n}$ and take the real part. This demodulation scheme will be studied in more detail in Section 12.3.1.

A more classic scheme involves multiplying $y[n]$ by a sinusoidal carrier at the same frequency as the carrier and filtering the result with a lowpass filter with cutoff at $\omega_b/2$. After the multiplication, the signal is

$$\begin{aligned} u[n] &= y[n]\cos\omega_c n \\ &= x[n]\cos^2\omega_c n \\ &= \frac{1}{2}x[n] + \frac{1}{2}x[n]\cos 2\omega_c n \end{aligned}$$

and the corresponding spectrum is therefore

$$U(e^{j\omega}) = \frac{1}{2}X(e^{j\omega}) + \frac{1}{4}\left[X(e^{j(\omega+2\omega_c)}) + X(e^{j(\omega-2\omega_c)})\right]$$

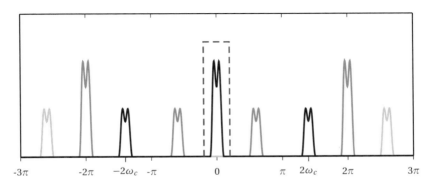

Figure 5.16 Spectrum of the AM demodulated signal $u[n]$, with explicit repetitions (the light gray spectral components are the sidebands from the repetitions at $\pm 4\pi$); the dashed lowpass filter response selects the baseband signal.

This spectrum is shown in Figure 5.16, with explicit repetitions; note that if the maximum frequency condition in (5.49) is satisfied, the components at twice the carrier frequency that may leak into the $[-\pi, \pi]$ interval from the neighboring spectral repetitions do not overlap with the baseband. From the figure, if we choose

$$H(e^{j\omega}) = \begin{cases} 2 & |\omega| < \omega_b/2 \\ 0 & \text{otherwise} \end{cases}$$

then $\hat{X}(e^{j\omega}) = H(e^{j\omega})U(e^{j\omega}) = X(e^{j\omega})$. The component at $\omega = 2\omega_c$ is filtered out and thus the spectrum of the demodulated signal is equal to the spectrum of the original signal. Of course the ideal low-pass is in practice replaced by a realizable IIR or an FIR filter with adequate properties.

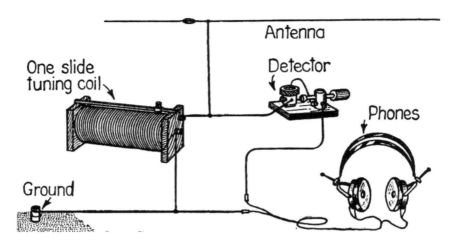

Figure 5.17 Schematics for a galena radio receiver *(from Gernsback's book "Radio For All", 1922)*.

Finally, for the fun of it, we can look at a "digital galena" demodulator. Galena receivers (whose general structure is shown in Figure 5.17) are the simplest (and oldest) type of radio receiver: the antenna and the tuning coil form a variable LC bandpass filter to select the band while a galena crystal, touched by a thin metal wire called the "cat's whisker", acts as a rectifying nonlinearity. A pair of high-impedance headphones is connected between the cat's whisker and ground; the mechanical inertia of the headphones acts as a simple lowpass filter which completes the radio receiver. In a digital simulation of a galena receiver, the antenna and coil are replaced by our sharp digital bandpass filter, at the output of which we find $y[n]$. The rectified signal at the cat's whisker can be modeled as $y_r[n] = |y[n]|$; since $y_r[n]$ is positive, the integration realized by the crude lowpass in the headphone can reveal the baseband envelope and eliminate most of the high frequency content. The process is best understood in the time domain and is illustrated in Figure 5.18. Note that, spectrally, the qualitative effect of the nonlinearity is indeed to bring the bandpass component back to baseband; as for most nonlinear processing, however, no simple analytical form for the baseband spectrum is available.

Example 5.2: Can IIRs see the future?

If we look at the bottom panel of Figure 5.9 we can notice that the group delay is negative for frequencies above approximately $\pi/7$. Does that mean that we can look into the future?
(*Hint:* no.)

Discrete-Time Filters 141

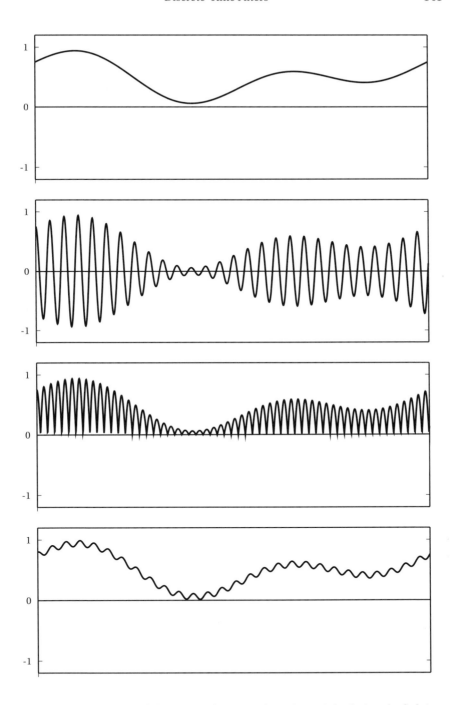

Figure 5.18 Galena demodulation in the time domain; original signal $x[n]$ (top panel); modulated bandpass signal $y[n]$ (second panel); rectified signal $y_r[n]$ (third panel); lowpass-filtered envelope at the headphones (bottom panel).

To see what we mean, consider the effect of group delay on a *narrowband* signal $x[n]$ centered at ω_0; a narrowband signal can be easily constructed by modulating a baseband signal $s[n]$ (i.e. a signal so that $S(e^{j\omega}) = 0$ for $|\omega| > \omega_b$ and ω_b *very* small). Set

$$x[n] = s[n]\cos(\omega_0 n)$$

and consider a real-valued filter $H(e^{j\omega})$ such that for τ small it is $|H(e^{j(\omega_0+\tau)})| \approx 1$ and the antisymmetric phase response is

$$\measuredangle H(e^{j(\omega_0+\tau)}) = \theta - g_d \tau$$
$$\measuredangle H(e^{j(-\omega_0+\tau)}) = -\theta + g_d \tau$$

clearly the group delay of the filter around ω_0 is g_d. If we filter the narrowband signal $x[n]$ with $H(e^{j\omega})$, we can write the DTFT of the output for $0 \leq \omega < \pi$ as

$$Y(e^{j\omega}) = X(e^{j\omega}) e^{j(\theta - g_d(\omega - \omega_0))}$$

since, even though the approximation for $H(e^{j\omega})$ holds only in a small neighborhood of ω_0, $X(e^{j\omega})$ is zero everywhere else so "we don't care". If we write out the expression for the full spectrum we have

$$Y(e^{j\omega}) = \frac{1}{2}\left[S(e^{j(\omega-\omega_0)})e^{j(\theta-g_d(\omega-\omega_0))} + S(e^{j(\omega+\omega_0)})e^{j(-\theta+g_d(\omega-\omega_0))}\right]$$
$$= \frac{1}{2}\left[S(e^{j(\omega-\omega_0)})e^{j\phi} + S(e^{j(\omega+\omega_0)})e^{-j\phi}\right]e^{-jg_d\omega}$$

where we have put $\phi = \theta + g_d \omega_0$. We can recognize by inspection that the first term is simply $s[n]$ modulated by a cosine with a phase offset of ϕ; the trailing linear phase term is just a global delay. If we assume g_d is an integer we can therefore write

$$y[n] = s[n - g_d]\cos(n\omega_0 + \theta) \tag{5.50}$$

so that the effect of the group delay is to delay the narrowband envelope by exactly g_d samples. The analysis still holds even if g_d is not an integer, as we will see in Chapter 9 when we deal with fractional delays.

Now, if g_d is negative, (5.50) seems to imply that the envelope $s[n]$ is *advanced* in time so that a filter with negative group delay is able to produce a copy of the input *before* the input is even applied; we would have a time machine which can look into the future! Clearly there must something wrong but the problem cannot be with the filter since the leaky integrator is an example of a perfectly realizable filter with negative group delay in the stopband. In fact, the inconsistency lies with the hypothesis of having a perfectly narrowband signal: just like the impulse response of an ideal filter is necessarily an infinite two-sided sequence, so any perfectly narrowband signal cannot have an identifiable "beginning". When we think of "applying" the input to the filter, we are implicitly assuming a one-sided (or, more likely, a finite-support) signal and this signal has nonzero spectral components at *all* frequencies. The net effect of these is that the overall delay for the signal will always be nonnegative.

Further Reading

Discrete-time filters are covered in all signal processing books, e.g. a good review is given in *Discrete-Time Signal Processing*, by A. V. Oppenheim and R. W. Schafer (Prentice-Hall, last edition in 1999).

Exercises

Exercise 5.1: Linearity and time-invariance – I. Consider the transformation $\mathcal{H}\{x[n]\} = nx[n]$. Does $\mathcal{H}$ define an LTI system?

Exercise 5.2: Linearity and time-invariance – II. Consider a discrete-time system $\mathcal{H}\{\cdot\}$. When the input is the signal $x[n] = \cos((2\pi/5)n)$, the output is $\mathcal{H}\{x[n]\} = \sin((\pi/2)n)$. Can the system be linear and time-invariant? Explain.

Exercise 5.3: Finite-support convolution. Consider the finite-support signal $h[n]$ defined as

$$h[n] = \begin{cases} 1 & \text{for } |n| \leq M \\ 0 & \text{otherwise} \end{cases}$$

(a) Compute the signal $x[n] = h[n] * h[n]$ for $M = 2$ and sketch the result.

(b) Compute the DTFT of $x[n]$, $X(e^{j\omega})$, and sketch its value in the interval $[0, \pi]$.

(c) Give a qualitative description of how $|X(e^{j\omega})|$ changes as M grows.

(d) Compute the signal $y[n] = x[n] * h[n]$ for $M = 2$ and sketch the result. For a general M, is the behavior of the sequence $y[n]$? (E.g. is it linear in n? Is it quadratic?)

(e) Compute $Y(e^{j\omega})$ and sketch its value.

Exercise 5.4: Convolution – I. Let $x[n]$ be a discrete-time sequence defined as
$$x[n] = \begin{cases} M-n & 0 \leq n \leq M \\ M+n & -M \leq n \leq 0 \\ 0 & \text{otherwise} \end{cases}$$
for some odd integer M.

(a) Show that $x[n]$ can be expressed as the convolution of two discrete-time sequences $x_1[n]$ and $x_2[n]$.

(b) Using the previous results, compute the DTFT of $x[n]$.

Exercise 5.5: Convolution – II. Consider the following discrete-time signals:
$$x[n] = \cos(1.5\,n)$$
$$y[n] = \frac{1}{5}\operatorname{sinc}\left(\frac{n}{5}\right)$$
Compute the convolution: $(x[n])^2 * y[n]$

Exercise 5.6: System properties. Consider the following input-output relations and, for each of the underlying systems, determine whether the system is linear, time invariant, BIBO stable, causal or anti-causal. Characterize the eventual LTI systems by their impulse response.

(a) $y[n] = x[-n]$.

(b) $y[n] = e^{-j\omega n} x[n]$.

(c) $y[n] = \sum_{k=n-n_0}^{n+n_0} x[k]$.

(d) $y[n] = n y[n-1] + x[n]$, such that if $x[n] = 0$ for $n < n_0$, then $y[n] = 0$ for $n < n_0$.

Exercise 5.7: Ideal filters. Derive the impulse response of a bandpass filter with center frequency ω_0 and passband ω_b:
$$H_{bp}(e^{j\omega}) = \begin{cases} 1 & \omega_0 - \omega_b/2 \leq \omega \leq \omega_0 + \omega_b/2 \\ 1 & -\omega_0 - \omega_b/2 \geq \omega \geq -\omega_0 + \omega_b/2 \\ 0 & \text{elsewhere} \end{cases}$$

(*Hint*: consider the following ingredients: a cosine of frequency ω_0, a low-pass filter of bandwidth ω_b and the modulation theorem.)

Exercise 5.8: Zero-phase filtering. Consider an operator $\mathscr{R}$ which turns a sequence into its time-reversed version:

$$\mathscr{R}\{x[n]\} = x[-n]$$

(a) The operator is clearly linear. Show that it is *not* time-invariant.

Suppose you have an LTI filter $\mathscr{H}$ with impulse response $h[n]$ and you perform the following sequence of operations in the followin order:

$$s[n] = \mathscr{H}\{x[n]\}$$
$$r[n] = \mathscr{R}\{s[n]\}$$
$$w[n] = \mathscr{H}\{r[n]\}$$
$$y[n] = \mathscr{R}\{w[n]\}$$

(b) Show that the input-output relation between $x[n]$ and $y[n]$ is an LTI transformation.

(c) Give the frequency response of the equivalent filter realized by the series of transformations and show that it has zero phase.

Exercise 5.9: Nonlinear signal processing. Consider the system $\mathscr{H}$ implementing the input-output relation $y[n] = \mathscr{H}\{x[n]\} = x^2[n]$.

(a) Prove by example that the system is nonlinear.

(b) Prove that the system is time-invariant.

Now consider the following cascade system:

$$x[n] \longrightarrow \boxed{\mathscr{H}} \xrightarrow{y[n]} \boxed{\mathscr{G}} \longrightarrow v[n]$$

where $\mathscr{G}$ is the following ideal highpass filter:

$$G(e^{j\omega}) = \begin{cases} 0 & \text{for } |\omega| < \pi/2 \\ 2 & \text{otherwise} \end{cases}$$

(as per usual, $G(e^{j\omega})$ is 2π-periodic – i.e. prolonged by periodicity outside of $[-\pi, \pi]$). The output of the cascade is therefore $v[n] = \mathscr{G}\{\mathscr{H}\{x[n]\}\}$.

(c) Compute $v[n]$ when $x[n] = \cos(\omega_0 n)$ for $\omega_0 = 3\pi/8$. How would you describe the transformation operated by the cascade on the input?

(d) Compute $v[n]$ as before, with now $\omega_0 = \dfrac{7\pi}{8}$

Exercise 5.10: Analytic signals and modulation. In this exercise we explore a modulation-demodulation scheme commonly used in data transmission systems. Consider two real sequences $x[n]$ and $y[n]$, which represent *two* data streams that we want to transmit. Assume that their spectrum is of lowpass type, i.e. $X(e^{j\omega}) = Y(e^{j\omega}) = 0$ for $|\omega| > \omega_c$. Consider further the following derived signal:

$$c[n] = x[n] + jy[n]$$

and the modulated signal:

$$r[n] = c[n]\, e^{j\omega_0 n}, \qquad \omega_c < \omega_0 < \pi - \omega_c$$

(a) Set $\omega_c = \pi/6$, $\omega_0 = \pi/2$ and sketch $|R(e^{j\omega})|$ for whatever shapes you choose for $X(e^{j\omega})$, $Y(e^{j\omega})$. Verify from your plot that $r[n]$ is an analytic signal.

The signal $r[n]$ is called a *complex bandpass signal*. Of course it cannot be transmitted as such, since it is complex. The transmitted signal is, instead,

$$s[n] = \text{Re}\{r[n]\}$$

This modulated signal is an example of Quadrature Amplitude Modulation (QAM).

(b) Write out the expression for $s[n]$ in terms of $x[n], y[n]$. Now you can see the reason behind the term QAM, since we are modulating with two carriers in quadrature (i.e. out of phase by 90 degrees).

Now we want to recover $x[n]$ and $y[n]$ from $s[n]$. To do so, follow these steps:

(c) Show that $s[n] + j(h[n] * s[n]) = r[n]$, where $h[n]$ is the Hilbert filter. In other words, we have recovered the analytic signal $r[n]$ from its real part only.

(d) Once you have $r[n]$, show how to extract $x[n]$ and $y[n]$.

Chapter 6

The Z-Transform

Mathematically, the z-transform is a mapping between complex sequences and analytical functions on the complex plane. Given a discrete-time signal $x[n]$, the z-transform of $x[n]$ is formally defined as the complex function of a complex variable $z \in \mathbb{C}$

$$X(z) = \mathscr{Z}\{x[n]\} = \sum_{n=-\infty}^{\infty} x[n] z^{-n} \tag{6.1}$$

Contrary to the Fourier transform (as well as to other well-known transforms such as the Laplace transform or the wavelet transform), the z-transform is not an analysis tool *per se*, in that it does not offer a new physical insight on the nature of signals and systems. The z-transform, however, derives its status as a fundamental tool in digital signal processing from two key features:

- Its mathematical formalism, which allows us to easily solve constant-coefficient difference equations as algebraic equations (and this was precisely the context in which the z-transform was originally invented).

- Its close association to the DTFT, which provides us with easy stability criteria for the design and the use of digital filters. (It is evident that the z-transform computed on the unit circle, i.e. for $z = e^{j\omega}$, is nothing but the DTFT of the sequence).

Probably the best approach to the z-transform is to consider it as a clever mathematical *transformation* which facilitates the manipulation of complex sequences; for discrete-time filters, the z-transform bridges the algorithmic side (i.e. the CCDE) to the analytical side (i.e. the spectral properties) in an extremely elegant, convenient and ultimately beautiful way.

6.1 Filter Analysis

To see the usefulness of the z-transform in the context of the analysis and the design of realizable filters, it is sufficient to consider the following two formal properties of the z-transform operator:

- **Linearity:** given two sequences $x[n]$ and $y[n]$ and their respective z-transforms $X(z)$ and $Y(z)$, we have

$$\mathscr{Z}\{\alpha x[n] + \beta y[n]\} = \alpha X(z) + \beta Y(z)$$

- **Time-shift:** given a sequence $x[n]$ and its z-transform $X(z)$, we have

$$\mathscr{Z}\{x[n-N]\} = z^{-N} X(z)$$

In the above, we have conveniently ignored all convergence issues for the z-transform; these will be addressed shortly but, for the time being, let us just make use of the formalism as it stands.

6.1.1 Solving CCDEs

Consider the generic filter CCDE (Constant-Coefficient Difference Equation) in (5.46):

$$y[n] = \sum_{k=0}^{M-1} b_k x[n-k] - \sum_{k=1}^{N-1} a_k y[n-k]$$

If we apply the z-transform operator to both sides and exploit the linearity and time-shifting properties, we have

$$Y(z) = \sum_{k=0}^{M-1} b_k z^{-k} X(z) - \sum_{k=1}^{N-1} a_k z^{-k} Y(z) \tag{6.2}$$

$$= \frac{\sum_{k=0}^{M-1} b_k z^{-k}}{1 + \sum_{k=1}^{N-1} a_k z^{-k}} X(z) \tag{6.3}$$

$$= H(z) X(z) \tag{6.4}$$

$H(z)$ is called the *transfer function* of the LTI filter described by the CCDE. The following properties hold:

- The transfer function of a realizable filter is a *rational transfer function* (i.e. a ratio of finite-degree polynomials in z).

- The transfer function evaluated on the unit circle is the frequency response of the filter. In other words, the z-transform gives us the possibility of obtaining the frequency response of a filter *directly from the underlying CCDE*; in a way, we will no longer need to occupy ourselves with the actual impulse response.

- The transfer function is the z-transform of the filter's impulse response (which follows immediately from the fact that the impulse response is the filter's output when the input is $x[n] = \delta[n]$ and that $\mathscr{Z}\{\delta[n]\} = 1$).

- The result in (6.4) can be extended to general sequences to yield a z-transform version of the convolution theorem. In particular, given the square-summable sequences $x[n]$ and $h[n]$ and their convolution $y[n] = x[n] * h[n]$, we can state that

$$\mathscr{Z}\{y[n]\} = Y(z) = X(z)H(z) \tag{6.5}$$

which can easily be verified using an approach similar to the one used in Section 5.4.2.

6.1.2 Causality

As we saw in Section 5.7.1, a CCDE can be rearranged to express either a causal or a noncausal realization of a filter. This ambiguity is reflected in the z-transform and can be made explicit by the following example. Consider the sequences

$$x_1[n] = u[n] \tag{6.6}$$
$$x_2[n] = \delta[n] - u[-n] \tag{6.7}$$

where $u[n]$ is the unit step. For the first sequence we have

$$X_1(x) = \sum_{n=0}^{\infty} z^{-n} = \frac{1}{1-z^{-1}} \tag{6.8}$$

(again, let us neglect convergence issues for the moment). For the second sequence we have

$$X_2(x) = -\sum_{n=1}^{\infty} z^n = 1 - \frac{1}{1-z} = \frac{1}{1-z^{-1}} \tag{6.9}$$

so that, at least formally, $X_1(z) = X_2(z)$. In other words, the z-transform is not an invertible operator or, more precisely, it is invertible up to a causality specification. If we look more in detail, the sum in (6.8) converges only for $|z| > 1$ while the sum in (6.9) converges only for $|z| < 1$. This is actually a general fact: the values for which a z-transform exists define the causality or anticausality of the underlying sequence.

6.1.3 Region of Convergence

We are now ready to address the convergence issues that we have put aside so far. For any given sequence $x[n]$, the set of points on the complex plane for which $\sum x[n]z^{-n}$ exists and is finite, is called the *region of convergence* (ROC) for the z-transform. In order to study the properties of this region, it is useful to split the sum in (6.1) as

$$X(z) = \sum_{n=-N}^{-1} x[n]z^{-n} + \sum_{n=0}^{M} x[n]z^{-n} \qquad (6.10)$$

$$= \sum_{n=1}^{N} x[n]z^n + \sum_{n=0}^{M} \frac{x[n]}{z^n} \qquad (6.11)$$

$$= X_a(z) + X_c(z) \qquad (6.12)$$

where $N, M \geq 0$ and where both N and M can be infinity. Now, for $X(z_0)$ to exist and be finite, both power series $X_a(z)$ and $X_c(z)$ must converge in z_0; since they are power series, when they do converge, they converge *absolutely*. As a consequence, for all practical purposes, we define the ROC in terms of absolute convergence:[1]

$$z \in \text{ROC}\{X(z)\} \iff \sum_{n=-\infty}^{\infty} \left|x[n]z^{-n}\right| < \infty \qquad (6.13)$$

Then the following properties are easily derived:

- *The ROC has circular symmetry.* Indeed, the sum in (6.13) depends only on the magnitude of z; in other words, if $z_0 \in \text{ROC}$, then the set of complex points $\{z \mid |z| = |z_0|\}$ is also in the ROC, and such a set defines a circle.

- *The ROC for a finite-support sequence is the entire complex plane (with the possible exception of zero and infinity).* For a finite-support se-

---

[1] This definition excludes the points on the boundary of the ROC from the ROC itself, but this has no consequence on the results we will derive and use in what follows.

quence, both N and M in (6.10) are finite. The z-transform is therefore a simple polynomial which exists and is finite for all values of z (except for $z=0$ if $N>0$ and/or $z=\infty$ if $M>0$).

- *The ROC for a causal sequence is a circularly symmetric region in the complex plane extending to infinity* (Fig. 6.1a). For a causal sequence, $M = \infty$ while N is finite (and equal to zero for a strictly causal sequence). In this case, $X_a(z)$ is a finite-degree polynomial and poses no convergence issues (i.e. ROC$\{X_a(z)\} = \mathbb{C} \setminus \{\infty\}$). As for $X_c(z)$, assume $X_c(z_0)$ exists and is finite and take any z_1 so that $|z_1| > |z_0|$; we have that for all n:

$$\left|\frac{x[n]}{z_1^n}\right| \leq \left|\frac{x[n]}{z_0^n}\right|$$

so that $X_c(z)$ is absolutely convergent in z_1 as well.

- *The ROC for an anticausal sequence is a disk in the complex plane, centered in the origin* (Fig. 6.1b). For an anticausal sequence, $N = \infty$ while M is finite so that $X_c(z)$ poses no convergence issues (i.e. ROC$\{X_c(z)\} = \mathbb{C} \setminus \{0\}$). As for $X_a(z)$, assume $X_a(z_0)$ exists and is finite and take any z_1 so that $|z_1| < |z_0|$; we have that for all n:

$$\left|x[n] z_0^n\right| \leq \left|x[n] z_1^n\right|$$

so that $X_a(z)$ is absolutely convergent in z_1 as well.

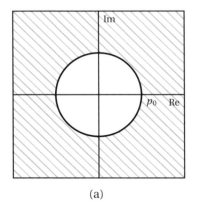
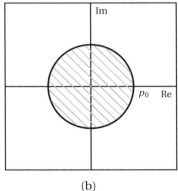

(a)   (b)

Figure 6.1 ROC shapes (hatched area): (a) causal sequence; (b) anticausal (b) sequence.

6.1.4 ROC and System Stability

The z-transform provides us with a quick and easy way to test the stability of a linear system. Recall from Section 5.2.2 that a necessary and sufficient condition for an LTI system to be BIBO stable is the absolute summability of its impulse response. This is equivalent to saying that a system is BIBO stable if and only if the z-transform of its impulse response is absolutely convergent in $|z|=1$. In other words, *a system is BIBO stable if the ROC of its transfer function includes the unit circle.*

6.1.5 ROC of Rational Transfer Functions and Filter Stability

For rational transfer functions, the analysis of the ROC is quite simple; indeed, the only "trouble spots" for convergence are the values for which the denominator of (6.3) is zero. These values are called the *poles* of the transfer functions and clearly they must lie outside of the ROC. As a consequence, we have an extremely quick and practical rule to determine the stability of a realizable filter.

Consider a causal filter:

- Find the roots of the transfer function's denominator (considered as a polynomial in z). These are the system's poles. Call p_0 the pole with the largest magnitude.

- The ROC has circular symmetry, it extends outwards to infinity and it cannot include any pole; therefore the ROC will simply be the area on the complex plane outside of a circle of radius $|p_0|$.

- For the ROC to include the unit circle we must have $|p_0| < 1$. Therefore, in order to have stability, *all the poles must be inside the unit circle.*

For an anticausal system the procedure is symmetrical; once the largest-magnitude pole is known, the ROC will be a disk of radius $|p_0|$ and therefore in order to have stability, all the poles will have to be outside of the unit circle.

6.2 The Pole-Zero Plot

The rational transfer function derived in (6.3) can be written out explicitly in terms of the CCDEs coefficients, as follows:

$$H(z) = \frac{b_0 + b_1 z^{-1} + \cdots + b_{M-1} z^{-(M-1)}}{1 + a_1 z^{-1} + \cdots + a_{N-1} z^{-(N-1)}} \tag{6.14}$$

The transfer function is the ratio of two polynomials in z^{-1} where the degree of the numerator polynomial is $M-1$ and that of the denominator polynomial is $N-1$. As a consequence, the transfer function can be rewritten in factored form as

$$H(z) = b_0 \frac{\prod_{n=1}^{M-1}(1-z_n z^{-1})}{\prod_{n=1}^{N-1}(1-p_n z^{-1})} \qquad (6.15)$$

where the z_n are the $M-1$ complex roots of the numerator polynomial and are called the *zeros* of the system; the p_n are the $N-1$ complex roots of the denominator polynomial and, as we have seen, they are called the poles of the system. Both poles and zeros can have arbitrary multiplicity. Clearly, if $z_i = p_k$ for some i and k (i.e. if a pole and a zero coincide) the corresponding first-order factors cancel each other out and the degrees of numerator and denominator are both decreased by one. In general, it is assumed that such factors have already been removed and that the numerator and denominator polynomials of a given rational transfer function are coprime.

The poles and the zeros of a filter are usually represented graphically on the complex plane as crosses and dots, respectively. This allows for a quick visual assessment of stability which, for a causal system, consists of checking whether all the crosses are inside the unit circle (or, for anticausal systems, outside).

6.2.1 Pole-Zero Patterns

The pole-zero plot can exhibit distinctive patterns according to the properties of the filter.

Real-Valued Filters. If the filter coefficients are real-valued (and this is the only case that we consider in this text book) both the numerator and denominator polynomials are going to have real-valued coefficients. We can now recall a fundamental result from complex algebra: the roots of a polynomial with real-valued coefficients are either real or they occur in complex-conjugate pairs. So, if z_0 is a complex zero of the system, z_0^* is a zero as well. Similarly, if p_0 is a complex pole, so is p_0^*. The pole-zero plot will therefore shows a symmetry around the real axis (Fig. 6.2a).

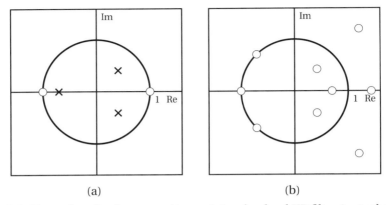

Figure 6.2 Examples of pole-zero patterns: (a) real-valued IIR filter (note the symmetry around the x-axis); (b) linear phase FIR (each zero appears with its reciprocal).

Linear-Phase FIR Filters. First of all, note that the pole-zero plot for an FIR filter is actually just a zero plot, since FIR's have no poles.[2] A particularly important case is that of linear phase FIR filters; as we will see in detail in Section 7.2.2, linear phase imposes some symmetry constraints on the CCDE coefficients (which, of course, coincide with the filter taps). These constraints have a remarkable consequence: if z_0 is a (complex) zero of the system, $1/z_0$ is a zero as well. Since we consider real-valued FIR filters exclusively, the presence of a complex zero in z_0 implies the existence of three other zeros, namely in $1/z_0$, z_0^* and $1/z_0^*$ (Fig. 6.2b). See also the discussion in Section 7.2.2

6.2.2 Pole-Zero Cancellation

We have seen in Section 5.2.1 that the effect of a cascade of two or more filters is that of a single filter whose impulse response is the convolution of all of the filters' impulse responses. By the convolution theorem, this means that the overall transfer function of a cascade of K filters $\mathcal{H}_i$, $i = 1, \ldots, K$ is simply the product of the single transfer functions $H_i(z)$:

$$H(z) = \prod_{i=1}^{K} H_i(z)$$

If all filters are realizable, we can consider the factored form of each $H_i(z)$ as in (6.15). In the product of transfer functions, it may happen that some of

---

[2] Technically, since we use the notation z^{-1} to express a delay, causal FIR filters have a multiple pole in the origin ($z = 0$). This is of no consequence for stability, however, so we will not consider it further.

the poles of a given $H_i(z)$ coincide with the zeros of another transfer function, which leads to a pole-zero cancellation in the overall transfer function. This is a method that can be used (at least theoretically) to stabilize an otherwise unstable filter. If one of the poles of the system (assuming causality) lies outside of the unit circle, this pole can be compensated by cascading an appropriate first- or second-order FIR section to the original filter. In practical realizations, care must be taken to make sure that the cancellation is not jeopardized by numerical precision problems.

6.2.3 Sketching the Transfer Function from the Pole-Zero Plot

The pole-zero plot represents a convenient starting point in order to estimate the shape of the magnitude for a filter's transfer function. The basic idea is to consider the absolute value of $H(z)$, which is a three-dimensional plot ($|H(z)|$ being a real function of a complex variable). To see what happens to $|H(z)|$ it is useful to imagine a "rubber sheet" laid over the complex plane; then,

- every zero corresponds to a point where the rubber sheet is "glued" to the plane,

- every pole corresponds to a "pole" which is "pushing" the rubber sheet up (to infinity),

so that the shape of $|H(z)|$ is that of a very lopsided "circus tent". The magnitude of the transfer function is just the height of this circus tent measured around the unit circle.

In practice, to sketch a transfer function (in magnitude) given the pole-zero plot, we proceed as follows. Let us start by considering the upper half of the unit circle, which maps to the $[0, \pi]$ interval for the ω axis in the DTFT plot; for real-valued filters, the magnitude response is an even function and, therefore, the $[-\pi, 0]$ interval need not be considered explicitly. Then:

1. Check for zeros on the unit circle; these correspond to points on the frequency axis in which the magnitude response is exactly zero.

2. Draw a line from the origin of the complex plane to each pole and each zero. The point of intersection of each line with the unit circle gives the location of a local extremum for the magnitude response.

3. The effect of each pole and each zero is made stronger by their proximity to the unit circle.

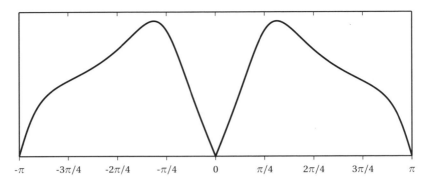

Figure 6.3 Sketch of the magnitude response for the pole-zero plot of Figure 6.2(a).

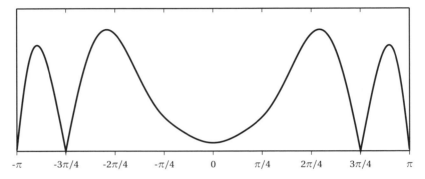

Figure 6.4 Sketch of the magnitude response for the pole-zero plot of Figure 6.2(b).

As an example, the magnitude responses of the pole-zero plots in Figure 6.2 are displayed in Figures 6.3 and 6.4.

6.3 Filtering by Example – Z-Transform

We will quickly revisit the examples of the previous chapter to show the versatility of the z-transform.

Moving Average. From the impulse response in (5.14), the transfer function of the moving average filter is

$$H(z) = \frac{1}{N} \sum_{k=0}^{N-1} z^{-k} = \frac{1}{N} \frac{1-z^{-N}}{1-z^{-1}} \qquad (6.16)$$

from which the frequency response (5.26) is easily derived by setting $z = e^{j\omega}$. It is easy to see that the poles of the filter are on all the roots of unity

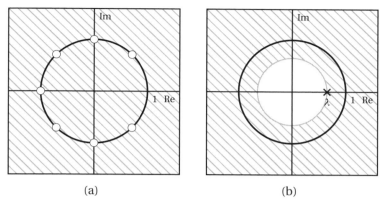

Figure 6.5 Pole-zero plots and ROC: (a) moving average filter with $N=8$; (b) leaky integrator with $\lambda = 0.65$.

except for $z = 1$, where the numerator and denominator in (6.17) cancel each other out. A factored representation of the transfer function for the moving average is therefore

$$H(z) = \frac{1}{N} \prod_{k=1}^{N-1}(1 - W_N^k z^{-k}) \qquad (6.17)$$

and the pole-zero plot (for $N=8$) is shown in Figure 6.5(a). There being no poles, the filter is unconditionally stable.

Leaky Integrator. From the CCDE for the leaky integrator (5.16) we immediately have

$$Y(z) = \lambda z^{-1} Y(z) + (1-\lambda) X(z) \qquad (6.18)$$

from which

$$H(z) = \frac{1-\lambda}{1-\lambda z^{-1}} \qquad (6.19)$$

The transfer function has therefore a single real pole in $z = \lambda$; for a causal realization, this implies that the ROC is the region of the complex plane extending outward from the circle of radius λ. The causal filter is stable if λ lies inside the unit circle, i.e. if $\lambda < 1$. An example of pole-zero plot together with the associated ROC is shown in Figure 6.5(b) for the (stable) case of $\lambda = 0.65$.

Examples

Example 6.1: Transform of periodic functions

The z-transform converges without fuss for infinite-energy sequences which the Fourier transform has some difficulties dealing with. For instance, the

z-transform manages to "bring down" the unit step because of the vanishing power of z^{-n} for $|z| > 1$ and n large and this is the case for all one-sided sequences which grow no more than exponentially. However, if $|z^{-n}| \to 0$ for $n \to \infty$ then necessarily $|z^{-n}| \to \infty$ for $n \to -\infty$ and this may pose a problem for the convergence of the z-transform in the case of two-sided sequences. In particular, the z-transform does not converge in the case of periodic signals since only one side of the repeating pattern is "brought down" while the other is amplified limitlessly. We can circumvent this impasse by "killing" half of the periodic signal with a unit step. Take for instance the one-sided cosine:

$$x[n] = \cos(\omega_0 n) u[n]$$

its z-transform can be derived as

$$X(z) = \sum_{n=-\infty}^{\infty} z^{-n} \cos(\omega_0 n) u[n]$$

$$= \sum_{n=0}^{\infty} z^{-n} \cos(\omega_0 n)$$

$$= \frac{1}{2} \sum_{n=0}^{\infty} e^{j\omega_0 n} z^{-n} + \frac{1}{2} \sum_{n=0}^{\infty} e^{-j\omega_0 n} z^{-n}$$

$$= \frac{1}{2} \left(\frac{1}{1 - e^{j\omega_0} z^{-1}} + \frac{1}{1 - e^{-j\omega_0} z^{-1}} \right)$$

$$= \frac{1 - \cos(\omega_0) z^{-1}}{1 - 2\cos(\omega_0) z^{-1} + z^{-2}}$$

Similar results can be obtained for signals such as $x[n] = \sin(\omega_0 n) u[n]$ or $x[n] = \alpha^n \cos(\omega_0 n) u[n]$.

Example 6.2: The impossibility of ideal filters

The z-transform of an FIR impulse response can be expressed as a simple polynomial $P(z)$ of degree $L-1$ where L is the number of nonzero taps of the filter (we can neglect leading factors of the form z^{-N}). The fundamental theorem of algebra states that such a polynomial has at most $L-1$ roots; as a consequence, the frequency response of an FIR filter can never be identically zero over a frequency interval since, if it were, its z-transform would have an infinite number of roots. Similarly, by considering the polynomial $P(z) - C$, we can prove that the frequency response can never be constant C over an interval which proves the impossibility of achieving ideal (i.e. "brickwall") responses with an FIR filter. The argument can be easily extended to rational transfer functions, confirming the impossibility of a realizable filter whose characteristic is piecewise perfectly flat.

Further Reading

The z-transform is closely linked to the solution of linear, constant coefficient difference equations. For a more complete treatment, see, for example, R. Vich, *Z Transform Theory and Applications* (Springer, 1987), or A. J. Jerri, *Linear Difference Equations with Discrete Transforms Method* (Kluwer, 1996).

Exercises

Exercise 6.1: Interleaving. Consider two two-sided sequences $h[n]$ and $g[n]$ and consider a third sequence $x[n]$ which is built by interleaving the values of $h[n]$ and $g[n]$:

$$x[n] = \ldots, h[-3], g[-3], h[-2], g[-2], h[-1], g[-1], h[0],$$
$$g[0], h[1], g[1], h[2], g[2], h[3], g[3], \ldots$$

with $x[0] = h[0]$.

(a) Express the z-transform of $x[n]$ in terms of the z-transforms of $h[n]$ and $g[n]$.

(b) Assume that the ROC of $H(z)$ is $0.64 < |z| < 4$ and that the ROC of $G(z)$ is $0.25 < |z| < 9$. What is the ROC of $X(z)$?

Exercise 6.2: Properties of the z-transform. Let $x[n]$ be a discrete-time sequence and $X(z)$ be its corresponding z-transform with appropriate ROC.

(a) Prove that the following relation holds:

$$nx[n] \xleftrightarrow{Z} -z\frac{d}{dz}X(z)$$

(b) Using (a), show that

$$(n+1)\alpha^n u[n] \xleftrightarrow{Z} \frac{1}{(1-\alpha z^{-1})^2}, \qquad |z| > |\alpha|$$

(c) Suppose that the above expression corresponds to the impulse response of an LTI system. What can you say about the causality of such a system? What about its stability?

(d) Let $\alpha = 0.8$: what is the spectral behavior of the corresponding filter? What if $\alpha = -0.8$?

Exercise 6.3: Stability. Consider a causal discrete system represented by the following difference equation:

$$y[n] - 3.25\,y[n-1] + 0.75\,y[n-2] = x[n-1] + 3x[n-2]$$

(a) Compute the transfer function and check the stability of this system both analytically and graphically.

(b) If the input signal is $x[n] = \delta[n] - 3\delta[n-1]$, compute the z-transform of the output signal and discuss the stability.

(c) Take an arbitrary input signal that does not cancel the unstable pole of the transfer function and repeat (b).

Exercise 6.4: Pole-zero plot and stability - I. Consider a causal LTI system with the following transfer function:

$$H(z) = \frac{3 + 4.5\,z^{-1}}{1 + 1.5\,z^{-1}} - \frac{2}{1 - 0.5\,z^{-1}}$$

Sketch the pole-zero plot of the transfer function and specify its region of convergence. Is the system stable?

Exercise 6.5: Pole-zero plot and stability - II. Consider the transfer function of an anticausal LTI system

$$H(z) = (1 - z^{-1}) - \frac{1}{1 - 0.5\,z^{-1}}$$

Sketch the pole-zero plot of the transfer function and specify its region of convergence. Is the system stable?

Exercise 6.6: Pole-zero plot and magnitude response. In the following pole-zero plots, multiple zeros at the same location are indicated by the multiplicity number shown to the upper right of the zero. Sketch the magnitude of each frequency response and determine the type of filter.

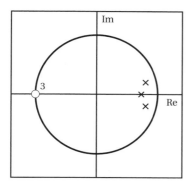
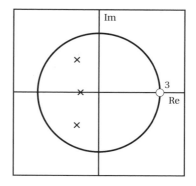
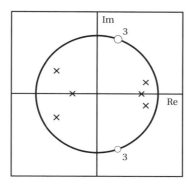

Exercise 6.7: z-transform and magnitude response.
Consider a causal LTI system described by the following transfer function:

$$H(z) = \frac{\frac{1}{6} + \frac{1}{2}z^{-1} + \frac{1}{2}z^{-2} + \frac{1}{6}z^{-3}}{1 + \frac{1}{3}z^{-2}}$$

(a) Sketch the magnitude response $H(e^{j\omega})$ from the z-transform. You can use a numerical package to find the poles and the zeros of the transfer function. What type of filter is $H(z)$?

(b) Sketch the pole-zero plot. Is the system stable?

Now fire up your favorite numerical package (or write some C code) and consider the following length-128 input signal:

$$x[n] = \begin{cases} 0 & n \in [1 \ \ldots \ 50] \\ 1 & n \in [51 \ \ldots \ 128] \end{cases}$$

(c) Plot the magnitude of its DTFT $|X[k]|$.

(d) We want to filter $x[n]$ with $H(z)$ to obtain $y[n]$: Compute and plot $y[n]$ using the Matlab function `filter`. Plot also the magnitude of its DTFT $|Y[k]|$.

(e) Explain qualitatively the form of $y[n]$.

Exercise 6.8: DFT and z-transform. It is immediately obvious that the DTFT of a sequence $x[n]$ is simply its z-transform evaluated on the unit circle, i.e. for $z = e^{j\omega}$. Equivalently, for a finite-length signal **x**, the DFT is simply the z-transform of the finite support extension of the signal evaluated at $z = W_N^{-k}$ for $k = 0, \ldots, N-1$:

$$X[k] = \sum_{n=0}^{N-1} x[n] z^{-n} \bigg|_{z=W_N^{-k}} = \sum_{n=0}^{N-1} x[n] W_N^{nk}$$

By taking advantage of this fact, derive a simple expression for the DFT of the time-reversed signal

$$\mathbf{x}_r = \begin{bmatrix} x[N-1] & x[N-2] & \ldots & x[1] & x[0] \end{bmatrix}^T$$

Exercise 6.9: A CCDE. Consider an LTI system described by the following constant-coefficient difference equation:

$$y[n-1] + 0.25\, y[n-2] = x[n]$$

(a) Write the transfer function of the system.

(b) Plot its poles and zeros, and show the ROC.

(c) Compute the impulse response of the system.

Exercise 6.10: Inverse transform. Write out the discrete-time signal $x[n]$ whose z-transform is

$$X(z) = (1 + 2z)(1 + 3z^{-1})$$

Exercise 6.11: Signal transforms. Consider a discrete-time signal $x[n]$ whose z-transform is

$$X(z) = 1 + z^{-1} + z^{-3} + z^{-4}$$

(a) Compute the DTFT of $x[n]$, $X(e^{j\omega})$. Your final result should be in the form of a real function of ω times a pure phase term.

(b) Sketch the magnitude of $X(e^{j\omega})$ as accurately as you can.

Consider now the length-4 signal $y[n]$:

$$y[n] = x[n], \qquad n = 0, 1, 2, 3$$

(c) Compute, for $y[n]$, the four DFT coefficients $Y[k]$, $k = 0, 1, 2, 3$.

Chapter 7

Filter Design

In discrete-time signal processing, filter design is the art of turning a set of requirements into a well-defined numerical algorithm. The requirements, or *specifications*, are usually formulated in terms of the filter's frequency response; the design problem is solved by finding the appropriate coefficients for a suitable difference equation which implements the filter and by specifying the filter's architecture. Since realizable filters are described by rational transfer functions, filter design can usually be cast in terms of a polynomial optimization procedure for a given error measure. Additional design choices include the *computational cost* of the designed filters, i.e. the number of mathematical operations and storage necessary to compute each output sample. Finally, the structure of the difference equation defines an explicit operational procedure for computing the filter's output values; by arranging the terms of the equation in different ways, we can arrive at different algorithmic structures for the implementation of digital filters.

7.1 Design Fundamentals

As we have seen, a realizable filter is described by a rational transfer function; designing a filter corresponds to determining the coefficients of the polynomials in transfer function with respect to the desired filter characteristics. For an FIR filter of length M, there are M coefficients that need to be determined, and they correspond directly to the filter's impulse response. Similarly, for an IIR filter with a numerator of degree $M-1$ and a denominator of degree $N-1$, there are $M+N-1$ coefficients to determine (since we always assume $a_0 = 1$). The main questions are the following:

- *How do we specify the characteristics of the desired filter?* This question effectively selects the domain in which we will measure the difference (i.e. the *error*) between the desired filter and the achieved implementation. This can be the time domain (where we would be comparing impulse responses) or the frequency domain (where we would be comparing frequency responses). Usually the domain of choice is the frequency domain.

- *What are the criteria to measure the quality of the obtained filter?* This question defines the way in which the above-mentionned error is measured; again, different criteria are possible (such as minimum square error or minimax) and they do depend on the intended application.

- *How do we choose the filter's coefficients in order to obtain the desired filtering characteristics?* This question defines an optimization problem in a parameter space of dimension $M + N - 1$ with the optimality criterion chosen above; it is usually answered by the existence of a numerical recipe which performs the task.

- *What is the best algorithmic structure (software or hardware) to implement a given digital filter?* This last question concerns the algorithmic design of the filter itself; the design is subject to various application-dependent constraints which include computational speed, storage requirement and arithmetic precision. Some of these design choices will be addressed at the end of the Chapter.

As is apparent, real-world filters are designed with a variety of practical requirements in mind, most of which are conflicting. One such requirement, for instance, is to obtain a low "computational price" for the filtering operation; this cost is obviously proportional to the number of coefficients in the filter, but it also depends heavily on the underlying hardware architecture. The tradeoffs between disparate requirements such as cost, precision or numerical stability are very subtle and not altogether obvious; the art of the digital filter designer, although probably less dazzling than the art of the *analog* filter designer, is to determine the best design strategy for a given practical problem.

7.1.1 FIR versus IIR

Filter design has a long and noble history in the analog domain: a linear electronic network can be described in terms of a differential equation linking, for instance, the voltage as a function of time at the input of the network to the voltage at the output. The arrangement of the capacitors, inductances and resistors in the network determine the form of the differential equation, while their values determine its coefficients. A fundamental difference between an analog filter and a digital filter is that the transformation from input to output is almost always considered *instantaneous* (i.e. the propagation effects along the circuit are neglected). In digital filters, on the other hand, the delay is always explicit and is actually the fundamental building block in a processing system. Because of the physical properties of capacitors, which are ubiquitous in analog filters, the transfer function of an analog filter (expressed in terms of its Laplace transform) is "similar" to the transfer function of an IIR filter, in the sense that it contains both poles and zeros. In a sense, IIR filters can be considered as the discrete-time counterpart of classic analog filters. FIR filters, on the other hand, are the flagship of digital signal processing; while one could conceive of an analog equivalent to an FIR, its realization would require the use of analog delay lines, which are costly and impractical components to manufacture. In a digital signal processing scenario, on the other hand, the designer can freely choose between two lines of attack with respect to a filtering problem, IIR or FIR, and therefore it is important to highlight advantages and disadvantages of each.

FIR Filters. The main *advantages* of FIR filters can be summarized as follows:

- unconditional stability;

- precise control of the phase response and, in particular, exact linear phase;

- optimal algorithmic design procedures;

- robustness with respect to finite numerical precision hardware.

While their *disadvantages* are mainly:

- longer input-output delay;

- higher computational cost with respect to IIR solutions.

IIR Filters. IIR filters are often an afterthought in the context of digital signal processing in the sense that they are designed by mimicking established design procedures in the analog domain; their appeal lies mostly in their compact formulation: for a given computational cost, i.e for a given number of operations per input sample, they can offer a much better magnitude response than an equivalent FIR filter. Furthermore, there are a few fundamental processing tasks (such as DC removal, as we will see later) which are the natural domain of IIR filters. The drawbacks of IIR filters, however, mirror in the negative the advantages of FIR's. The main advantages of IIR filters can be summarized as follows:

- lower computational cost with respect to an FIR with similar behavior;
- shorter input-output delay;
- compact representation.

While their disadvantages are mainly:

- stability is not guaranteed;
- phase response is difficult to control;
- design is complex in the general case;
- sensitive to numerical precision.

For these reasons, in this book, we will concentrate mostly on the FIR design problem and we will consider of IIR filters only in conjunction with some specific processing tasks which are often encountered in practice.

7.1.2 Filter Specifications and Tradeoffs

A set of filter specifications represents a set of guidelines for the design of a realizable filter. Generally, the specifications are formulated in the frequency domain and are cast in the form of boundaries for the magnitude of the frequency response; less frequently, the specifications will take the phase response into account as well.

A set of filter specifications is best illustrated by example: suppose our goal is to design a half-band lowpass filter, i.e. a lowpass filter with cutoff frequency $\pi/2$. The filter will possess a *passband*, i.e. a frequency range over which the input signal is unaffected, and a *stopband*, i.e. a frequency range where the input signal is annihilated. In order to turn these requirements into specifications the following practical issues must be taken into account:

- **Transition band.** The range of frequencies between passband and stopband is called the *transition band*. We should know by now (and we shall see again shortly) that we cannot obtain an instantaneous transition in a realizable filter[1]. Therefore, we must be willing to allow for a gap between passband and stopband where we renounce control over the frequency response; suppose we estimate that we can tolerate a transition band width up to 20% of the total bandwidth: since the cutoff is supposed to be at 0.5π, the transition band will thus extend from 0.4π to 0.6π.

- **Tolerances.** Similarly, we cannot impose a strict value of 1 for the passband and a value of 0 for the stopband (again, this has to do with the fact that the rational transfer function, being analytical, cannot be a constant over an interval). So we must allow for *tolerances*, i.e. minimum and maximum values for the frequency response over passband and stopband (while, in the transition band, we don't care). In our example, suppose that after examining the filter usage scenario we decide we can afford a 10% error in the passband and a 30% error in the stopband. (Note that these are *huge* tolerances, but they make the plots easier to read).

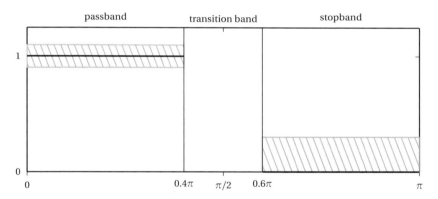

Figure 7.1 Typical lowpass filter specifications.

These specifications can be represented graphically as in Figure 7.1; note that, since we are dealing with real-valued filter coefficients, it is sufficient to specify the desired frequency response only over the $[0, \pi]$ interval, the magnitude response being symmetric. The filter design problem consists now in finding the minimum size FIR or IIR filter which fulfills the required

---

[1] To get an initial intuition as to why this is, consider that an instantaneous (vertical) transition constitutes a jump discontinuity in the frequency response. But the frequency response is just the transfer function computed on the unit circle and, for a realizable filter, the transfer function is a rational function which must be continuous.

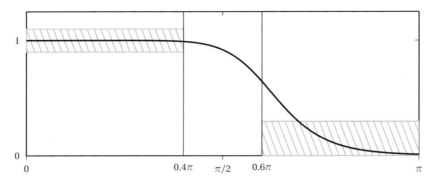

Figure 7.2 Example of monotonic filter which does not satisfies the specifications.

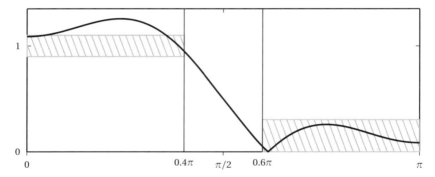

Figure 7.3 Example of FIR filter which does not satisfies the specifications.

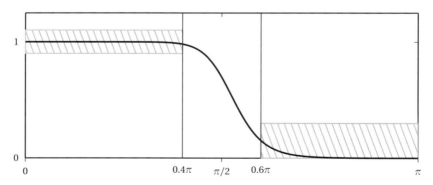

Figure 7.4 Example of monotonic filter which satisfies (and exceeds) the specifications.

specifications. As an example, Figure 7.2 shows an IIR filter which does not fulfill the specifications since the stopband error is above the maximum error at the beginning of the stopband. Similarly, Figure 7.3 shows an FIR filter which breaks the specifications in the passband. Finally, Figure 7.4 shows a monotonic IIR filter which matches and exceeds the specifications (i.e. the error is always smaller than the maximum error).

7.2 FIR Filter Design

In this section we will explore two fundamental strategies for FIR filter design, the window method and the minimax (or Parks-McClellan) method. Both methods seek to minimize the error between a desired (and often ideal) filter transfer function and the transfer function of the designed filter; they differ in the error measure which is used in the minimization. The window method is completely straightforward and it is often used for quick designs. The minimax method, on the other hand, is the procedure of choice for accurate, optimal filters. Both methods will be illustrated with respect to the design of a lowpass filter.

7.2.1 FIR Filter Design by Windowing

We have already seen in Section 5.6 that if there are no constraints (not even realizability) the best lowpass filter with cutoff frequency ω_c is the ideal lowpass. The impulse response is therefore the inverse Fourier transform of the desired transfer function:

$$\begin{aligned} h[n] &= \frac{1}{2\pi} \int_{-\pi}^{\pi} H(e^{j\omega}) e^{j\omega n} \, d\omega \\ &= \frac{1}{2\pi} \int_{-w_c}^{w_c} e^{j\omega n} \, d\omega \\ &= \frac{1}{2\pi j n} \left[e^{j\omega_c n} - e^{-j\omega_c n} \right] \\ &= \frac{\sin(\omega_c n)}{\pi n} \\ &= \frac{\omega_c}{\pi} \operatorname{sinc}\left(\frac{\omega_c}{\pi} n \right) \end{aligned}$$

The resulting filter, as we saw, is an ideal filter and it cannot be represented by a rational transfer function with a finite number of coefficients.

Impulse Response Truncation. A simple idea to obtain a realizable filter is to take a finite number of samples from the ideal impulse response and use them as coefficients of a (possibly rather long) FIR filter:[2]

$$\hat{h}[n] = \begin{cases} h[n] & -N \leq n \leq N \\ 0 & \text{otherwise} \end{cases} \quad (7.1)$$

This is a $(2N+1)$-tap FIR obtained by truncating an ideal impulse response (Figs 5.10 and 5.11). Note that the filter is noncausal, but that it can be made causal by using an N-tap delay; it is usually easier to design FIR's by considering a noncausal version first, especially if the resulting impulse response is symmetric (or antisymmetric) around $n=0$. Although this approximation was derived in a sort of "intuitive" way, it actually satisfies a very precise approximation criterion, namely the minimization of the mean square error (MSE) between the original and approximated filters. Denote by E_2 this error, that is

$$E_2 = \int_{-\pi}^{\pi} \left| H(e^{j\omega}) - \hat{H}(e^{j\omega}) \right|^2 d\omega$$

We can apply Parseval's theorem (see (4.59)) to obtain the equivalent expression in the discrete-time domain:

$$E_2 = 2\pi \sum_{n \in \mathscr{Z}} \left| h[n] - \hat{h}[n] \right|^2$$

If we now recall that $\hat{h}[n] = 0$ for $|n| > N$, we have

$$E_2 = 2\pi \left[\sum_{n=-N}^{N} \left| h[n] - \hat{h}[n] \right|^2 + \sum_{n=N+1}^{\infty} \left| h[n] \right|^2 + \sum_{n=-\infty}^{-N-1} \left| h[n] \right|^2 \right]$$

Obviously the last two terms are nonnegative and independent of $\hat{h}[n]$. Therefore, the minimization of E_2 with respect to $\hat{h}[n]$ is equivalent to the minimization of the first term only, and this is easily obtained by letting

$$\hat{h}[n] = h[n], \qquad \text{for } n = -N, \ldots, N$$

In spite of the attractiveness of such a simple and intuitive solution, there is a major drawback. If we consider the frequency response of the approximated filter, we have

$$\hat{H}(e^{j\omega}) = \sum_{n=-N}^{N} h[n] e^{-j\omega}$$

---

[2] Here and in the following the "hat" notation will denote an approximated or otherwise *derived* filter.

which means that $\hat{H}(e^{j\omega})$ is an approximation of $H(e^{j\omega})$ obtained by using only $2N+1$ Fourier coefficients. Since $H(e^{j\omega})$ has a jump discontinuity in ω_c, $\hat{H}(e^{j\omega})$ incurs the well-known Gibbs phenomenon around ω_c. The Gibbs phenomenon states that, when approximating a discontinuous function with a finite number of Fourier coefficients, the maximum error in an interval around the jump discontinuity is actually independent of the number of terms in the approximation and is always equal to roughly 9% of the jump. In other words, we have no control over the maximum error in the magnitude response. This is apparent in Figure 7.5 where $|\hat{H}(e^{j\omega})|$ is plotted for increasing values of N; the maximum error does not decrease with increasing N and, therefore, there are no means to meet a set of specifications which require less than 9% error in either stopband or passband.

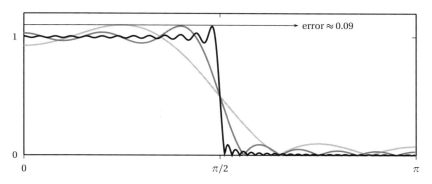

Figure 7.5 Gibbs phenomenon in lowpass approximation; magnitude of the approximated lowpass filter for $N = 4$ (light gray), $N = 10$ (dark gray) and $N = 50$ (black).

The Rectangular Window. Another way to look at the resulting approximation is to express $\hat{h}[n]$ as

$$\hat{h}[n] = h[n]w[n] \tag{7.2}$$

with

$$w[n] = \text{rect}\left(\frac{n}{N}\right) = \begin{cases} 1 & -N \leq n \leq N \\ 0 & \text{otherwise} \end{cases} \tag{7.3}$$

$w[n]$ is called a rectangular *window* of length $(2N+1)$ taps, which in this case is centered at $n = 0$.

We know from the modulation theorem in (5.22) that the Fourier transform of (7.2) is the convolution (in the space of 2π-periodic functions) of the Fourier transforms of $h[n]$ and $w[n]$:

$$\hat{H}(e^{j\omega}) = \frac{1}{2\pi} \int_{-\pi}^{\pi} H(e^{j\omega}) W(e^{j(\omega-\theta)}) d\theta$$

It is easy to compute $W(e^{j\omega})$ as

$$W(e^{j\omega}) = \sum_{n=-N}^{N} e^{-j\omega n} = \frac{\sin\left(\omega\left(N+\frac{1}{2}\right)\right)}{\sin\left(\frac{\omega}{2}\right)} \tag{7.4}$$

An example of $W(e^{j\omega})$ for $N=6$ is shown in Figure 7.6. By analyzing the form of $W(e^{j\omega})$ for arbitrary N, we can determine that:

- the first zero crossing of $W(e^{j\omega})$ occurs at $\omega = 2\pi/(2N+1)$;
- the width of the main lobe of the magnitude response is $\Delta = 4\pi(2N+1)$;
- there are multiple *sidelobes*, an oscillatory effect around the main lobe and there are up to $2N$ sidelobes for a $2N+1$-tap window.

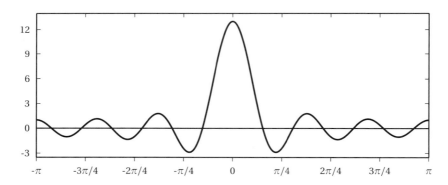

Figure 7.6 Fourier transform of the rectangular window for $N=6$.

In order to understand the shape of the approximated filter, let us go back to the lowpass filter example and try to visualize the effect of the convolution in the Fourier transform domain. First of all, since all functions are 2π-periodic, everything happens circularly, i.e. what "goes out" on the right of the $[-\pi,\pi]$ interval "pops" immediately up on the left. The value at ω_0 of $\hat{H}(e^{j\omega})$ is the integral of the product between $H(e^{j\omega})$ and a version of

$W(e^{j\omega})$ circularly shifted by ω_0. Since $H(e^{j\omega})$ is zero except over $[-\omega_c, \omega_c]$, where it is one, this value is actually:

$$\hat{H}(e^{j\omega_0}) = \frac{1}{2\pi} \int_{-\omega_c}^{\omega_c} W(e^{j(\omega-\omega_0)}) d\theta$$

When ω_0 is such that the first right sidelobe of $W(e^{j\omega})$ is *outside* of the $[-\omega_c, \omega_c]$ interval, then the integral reaches its maximum value, since the sidelobe is negative and it's the largest. The maximum value is dependent on the shape of the window (a rectangle in this case) but not on its length. Hence the Gibbs phenomenon.

To recap, the windowing operation on the ideal impulse response, i.e. the circular convolution of the ideal frequency response with $W(e^{j\omega})$, produces two main effects:

- The sharp transition from passband to stopband is smoothed by the convolution with the main lobe of width Δ.

- Ripples appear both in the stopband and the passband due to the convolution with the sidelobes (the largest ripple being the Gibbs phenomenon).

The sharpness of the transition band and the size of the ripples are dependent on the shape of the window's Fourier transform; indeed, by carefully designing the shape of the windowing sequence we can trade mainlobe width for sidelobe amplitude and obtain a more controlled behavior in the frequency response of the approximation filter (although the maximum error can *never* be arbitrarily reduced).

Other Windows. In general, the recipe for filter design by windowing involves two steps: the analytical derivation of an ideal impulse response followed by a suitable windowing to obtain an FIR filter. The ideal impulse response $h[n]$ is obtained from the desired frequency response $H(e^{j\omega})$ by the usual DTFT inversion formula

$$h[n] = \frac{1}{2\pi} \int_{-\pi}^{\pi} H(e^{j\omega}) e^{j\omega n} d\omega$$

While the analytical evaluation of the above integral may be difficult or impossible in the general case, for frequency responses $H(e^{j\omega})$ which are *piecewise linear*, the computation of $h[n]$ can be carried out in an exact (if nontrivial) way; the result will be a linear combination of modulated sinc and sinc-squared sequences.[3] The FIR approximation is then obtained by applying a finite-length window $w[n]$ to the ideal impulse response as in (7.2).

---

[3] For more details one can look at the Matlab `fir1` function.

The shape of the window can of course be more sophisticated than the simple rectangular window we have just encountered and, in fact, a hefty body of literature is devoted to the design of the "best" possible window. In general, a window should be designed with the following goals in mind:

- the window should be short, as to minimize the length of the FIR and therefore its computational cost;

- the spectrum of the window should be concentrated in frequency around zero as to minimize the "smearing" of the original frequency response; in other words, the window's main lobe should be as narrow as possible (it is clear that for $W(e^{j\omega}) = \delta(\omega)$ the resulting frequency response is identical to the original);

- the unavoidable sidelobes of the window's spectrum should be small, so as to minimize the rippling effect in the resulting frequency response (Gibbs phenomenon).

It is clear that the first two requirements are openly in conflict; indeed, the width of the main lobe Δ is inversely proportional to the length of the window (we have seen, for instance, that for the rectangular window $\Delta = 4\pi/M$, with M, the length of the filter). The second and third requirements are also in conflict, although the relationship between mainlobe width and sidelobe amplitude is not straightforward and can be considered a design parameter. In the frequency response, reduction of the sidelobe amplitude implies that the Gibbs phenomenon is decreased, but at the "price" of an enlargement of the filter's transition band. While a rigorous proof of this fact is beyond the scope of this book, consider the simple example of a triangular window (with N odd):

$$w_t[n] = w[n] = \begin{cases} \dfrac{N-n}{N} & |n| < N \\ 0 & \text{otherwise} \end{cases} \qquad (7.5)$$

It is easy to verify that $w_t[n] = w[n] * w[n]$, with $w[n] = \text{rect}(2n/(N-1))$ (i.e. the triangle can be obtained as the convolution of a half-support rectangle with itself) so that, as a consequence of the convolution theorem, we have

$$W_t(e^{j\omega}) = W(e^{j\omega})W(e^{j\omega}) = \left[\frac{\sin(\omega N/2)}{\sin(\omega/2)}\right]^2 \qquad (7.6)$$

The net result is that the amplitude of the sidelobes is quadratically reduced but the amplitude of the mainlobe Δ is roughly doubled with respect to an equivalent-length rectangular window; this is displayed in Figure 7.7 for a 17-point window (values are normalized so that both frequency responses are equal in $\omega = 0$). Filters designed with a triangular window therefore exhibit a much wider transition band.

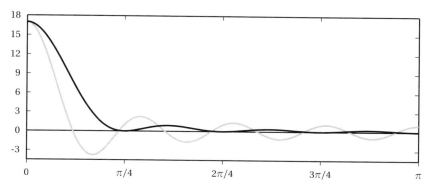

Figure 7.7 Fourier transform of the 17-point rectangular window (gray) vs. an equal-length triangular window (black).

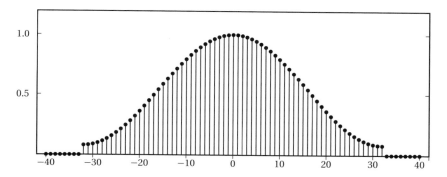

Figure 7.8 Hamming window ($N = 32$).

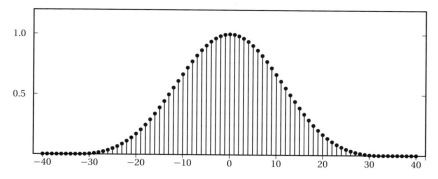

Figure 7.9 Blackman window ($N = 32$).

Other commonly used windows admit a simple parametric closed form representation; the most important are the Hamming window (Fig. 7.8):

$$w(n) = 0.54 - 0.46\cos\left(2\pi\frac{n+N}{2N}\right), \qquad |n| \leq N-1$$

and the Blackman window (Fig. 7.9):

$$w(n) = 0.42 - 0.5\cos\left(2\pi\frac{n+N}{2N}\right) + 0.08\cos\left(4\pi\frac{n+N}{2N}\right), \quad |n| \leq N-1$$

The magnitude response of both windows is plotted in Figure 7.10 (on a log scale so as to enhance the difference in sidelobe amplitude); again, we can remark the tradeoff between mainlobe width (translating to a wider transition band in the designed filter) and sidelobe amplitude (influencing the maximum error in passband and stopband).

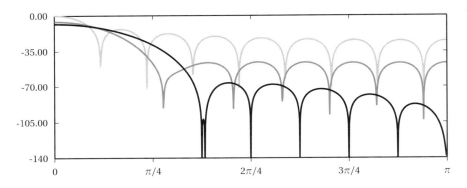

Figure 7.10 Magnitude response (dB scale) of the 17-point rectangular (light gray), Hamming (dark gray) and Blackman (black) windows.

Limitations of the Window Method. Lack of total control on passband and stopband error is the main limitation inherent to the window method; this said, the method remains a fundamental staple of practical signal processing as it yields perfectly usable filters via a quick, flexible and simple procedure. The error characteristic of a window-designed filter can be particularly aggravating in sensitive applications such as audio processing, where the peak in the stopband error can introduce unacceptable artifacts. In order to improve on the filter performance, we need to completely revise our design approach. A more suitable optimization criterion may, for instance, be the *minimax* criterion, where we aim to explicitly minimize the *maximum* approximation error over the entire frequency support; this is thoroughly analyzed in the next section. We can already say, however, that while the minimum square error is an integral criterion, the minimax is a pointwise criterion; or, mathematically, that the MSE and the minimax are respectively $L_2([-\pi,\pi])$- and $L_\infty([-\pi,\pi])$-norm minimizations for the error function $E(\omega) = \hat{H}(e^{j\omega}) - H(e^{j\omega})$. Figure 7.11 illustrates the typical result of applying both criteria to the ideal lowpass problem. As can be seen, the minimum square and minimax solutions are very different.

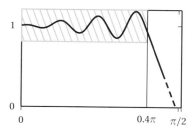 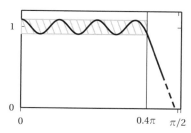

Figure 7.11 Error shapes in passband for MSE and minimax optimization methods.

7.2.2 Minimax FIR Filter Design

As we saw in the opening example, FIR filter design by windowing minimizes the overall mean square error between the desired frequency response and the actual response of the filter. Since this might lead to a very large error at frequencies near the transition band, we now consider a different approach, namely the design by minimax optimization. This technique minimizes the maximum allowable error in the filter's magnitude response, both in the passband and in the stopband. Optimality in the minimax sense requires therefore the explicit stating of a set of *tolerances* in the prototypical frequency response, in the form of design specifications as seen in Section 7.1.2. Before tackling the design procedure itself, we will need a series of intermediate results.

Generalized Linear Phase. In Section 5.4.3, we introduced the concept of linear phase; a filter with linear phase response is particularly desirable since the phase response translates to just a time delay (possibly fractional) and we can concentrate on the magnitude response only. We also introduced the notion of group delay and showed that linear phase corresponds to constant group delay. Clearly, the converse is not true: a frequency response of the type

$$H(e^{j\omega}) = |H(e^{j\omega})| e^{-j\omega d + j\alpha}$$

has constant group delay but differs from a linear phase system by a constant phase factor $e^{j\alpha}$. We will call this type of phase response *generalized linear phase*. Important cases are those for which $\alpha = 0$ (strictly linear phase) and $\alpha = \pi/2$ (generalized linear phase used in differentiators).

FIR Filter Types. Consider a causal, M-tap FIR filter with impulse response $h[n]$, $n = 0, 1, \ldots, M - 1$; in the following, we are interested in filters whose impulse response is *symmetric or antisymmetric around the "midpoint"*. If the number of taps is odd, the midpoint of the impulse response coincides with the center tap $h[(M-1)/2]$; if the number of taps is even, on the other hand, the midpoint is still at $(M-1)/2$ but this value does not coincide with a tap since it is located "right in between" taps $h[M/2-1]$ and $h[M/2]$. Symmetric and antisymmetric FIR filters are important since their frequency response has generalized linear phase. The delay introduced by these filters is equal to $(M-1)/2$ samples; clearly, this is an integer delay if M is odd, and it is fractional (half a sample more) if M is even. There are four different possibilities for linear phase FIR impulse responses, which are listed here with their corresponding generalized linear phase parameters:

Type	Nb. of Taps	Symmetry	Delay	Phase
Type I	odd	symmetric	integer	$\alpha = 0$
Type II	even	symmetric	fractional	$\alpha = 0$
Type III	odd	antisymmetric	integer	$\alpha = \pi/2$
Type IV	even	antisymmetric	fractional	$\alpha = \pi/2$

The generalized linear phase of (anti)symmetric FIRs is easily shown. Consider for instance a Type I filter, and define $C = (M-1)/2$, the location of the center tap; we can compute the transfer function of the shifted impulse response $h_d[n] = h[n + C]$, which is now symmetric around zero. i.e. $h_d[-n] = h_d[n]$:

$$H_d(z) = \sum_{n=-C}^{C} h_d[n] z^{-n}$$

$$= h_d[0] + \sum_{n=-C}^{-1} h_d[n] z^{-n} + \sum_{n=1}^{C} h_d[n] z^{-n}$$

$$= h_d[0] + \sum_{n=1}^{C} h_d[n](z^n + z^{-n}) \tag{7.7}$$

By undoing the time shift we obtain the original Type I transfer function:

$$H(z) = z^{-\frac{M-1}{2}} H_d(z) \tag{7.8}$$

Filter Design

On the unit circle (7.7) becomes

$$H_d(e^{j\omega}) = h_d[0] + \sum_{n=1}^{C} h_d[n](e^{j\omega n} + e^{-j\omega n})$$

$$= h_d[0] + 2\sum_{n=1}^{C} h_d[n]\cos n\omega \tag{7.9}$$

which is a *real* function; the original Type I frequency response is obtained from (7.8):

$$H(e^{j\omega}) = \left[h\left[\frac{M-1}{2}\right] + 2\sum_{n=(M+1)/2}^{M-1} h[n]\cos n\omega \right] e^{-j\omega \frac{M-1}{2}}$$

which is clearly linear phase with delay $d = (M-1)/2$ and $\alpha = 0$. The generalized linear phase of the other three FIR types can be shown in exactly the same way.

Zero Locations. The symmetric structures of the four types of FIR filters impose some constraints on the locations of the zeros of the transfer function. Consider again a Type I filter; from (7.7) it is easy to see that $H_d(z^{-1}) = H_d(z)$; by using (7.8) we therefore have

$$\begin{cases} H(z) = z^{-\frac{M-1}{2}} H_d(z) \\ H(z^{-1}) = z^{\frac{M-1}{2}} H_d(z) \end{cases}$$

which leads to

$$H(z^{-1}) = z^{M-1} H(z) \tag{7.10}$$

It is easy to show that the above relation is also valid for Type II filters, while for Type III and Type IV (antisymmetric filters) we have

$$H(z^{-1}) = -z^{M-1} H(z) \tag{7.11}$$

These relations mean that if z_0 is a zero of a linear phase FIR, then so is z_0^{-1}. This result, coupled with the usual fact that all complex zeros come in conjugate pairs, implies that if z_0 is a zero of $H(z)$, then:

- If $z_0 = \rho \in \mathbb{R}$ then ρ and $1/\rho$ are zeros.
- If $z_0 = \rho e^{j\theta}$ then $\rho e^{j\theta}, (1/\rho)e^{j\theta}, \rho e^{-j\theta}$ and $(1/\rho)e^{-j\theta}$ are zeros.

Consider now equation (7.10) again; if we set $z = -1$,

$$H(-1) = (-1)^{M-1} H(-1) \tag{7.12}$$

for Type II filters, $M-1$ is an odd number, which leads to the conclusion that $H(-1) = 0$; in other words, Type II filters *must* have a zero at $\omega = \pi$. Similar results can be demonstrated for the other filter types, and they are summarized below:

Filter Type	Relation	Constraint on Zeros
Type I	$H(z^{-1}) = z^{M-1}H(z)$	No constraints
Type II	$H(z^{-1}) = z^{M-1}H(z)$	Zero at $z = -1$ (i.e. $\omega = \pi$)
Type III	$H(z^{-1}) = -z^{M-1}H(z)$	Zeros at $z = \pm 1$ (i.e. at $\omega = 0$, $\omega = \pi$)
Type IV	$H(z^{-1}) = -z^{M-1}H(z)$	Zero at $z = 1$ (i.e. $\omega = 0$)

These constraints are important in the choice of the filter type for a given set of specifications. Type II and Type III filters are not suitable in the design of highpass filters, for instance; similarly, Type III and Type IV filters are not suitable in in the design of lowpass filters.

Chebyshev Polynomials. Chebyshev polynomials are a family of orthogonal polynomials $\{T_k(x)\}_{k \in \mathbb{N}}$ which have, amongst others, the following interesting property:

$$\cos n\omega = T_n(\cos \omega) \tag{7.13}$$

in other words, the cosine of an integer multiple of an angle ω can be expressed as a polynomial in the variable $\cos \omega$. The first few Chebyshev polynomials are

$$T_0(x) = 1$$
$$T_1(x) = x$$
$$T_2(x) = 2x^2 - 1$$
$$T_3(x) = 4x^3 - 3x$$
$$T_4(x) = 8x^4 - 8x^2 + 1$$

and, in general, they can be derived from the recursion formula:

$$T_{k+1}(x) = 2x\,T_k(x) - T_{k-1}(x) \tag{7.14}$$

From the above table it is easy to see that we can write, for instance,

$$\cos(3\omega) = 4\cos^3 \omega - 3\cos \omega$$

The interest in Chebyshev polynomials comes from the fact that the zero-centered frequency response of a linear phase FIR can be expressed as a linear combination of cosine functions, as we have seen in detail for Type I filters in (7.9). By using Chebyshev polynomials we can rewrite such a response as just one big polynomial in the variable $\cos\omega$. Let us consider an explicit example for a length-7 Type I filter with nonzero coefficients $h[n] = [d\ c\ b\ a\ b\ c\ d]$; we can state that

$$H_d(e^{j\omega}) = a + 2b\cos\omega + 2c\cos 2\omega + 2d\cos 3\omega$$

and by using the first four Chebyshev polynomials we can write

$$H_d(e^{j\omega}) = a + 2b\cos\omega + 2c(2\cos^2\omega - 1) + 2d(4\cos^3\omega - 3\cos\omega)$$
$$= (a - 2c) + (2b - 6d)\cos\omega + 4c\cos^2\omega + 8d\cos^3\omega \quad (7.15)$$

In this case, $H_d(e^{j\omega})$ turns out to be a third degree polynomial in the variable $\cos\omega$. This is the case for any Type I filter, for which we can always write

$$H_d(e^{j\omega}) = \sum_{k=0}^{(M-1)/2} c_k \cos^k\omega \quad (7.16)$$

$$= P(x)\Big|_{x=\cos\omega} \quad (7.17)$$

where $P(x)$ is a polynomial of degree $(M - 1)/2$ whose coefficients c_k are derived as linear combinations of the original filter coefficients a_k as illustrated in (7.15). For the other types of linear phase FIR, a similar representation can be obtained after a few trigonometric manipulations. The general expression is

$$H_d(e^{j\omega}) = f(\omega) \sum_{k=0}^{L} c_k \cos^k\omega$$

$$= f(\omega) P(x)\Big|_{x=\cos\omega} \quad (7.18)$$

where the c_k are still linear combinations of the original filter coefficients and where $f(\omega)$ is a weighting trigonometric function. Both $f(\omega)$ and the polynomial degree K vary as a function of the filter type.[4] In the following Sections, however, we will concentrate only on the design of Type I filters, so these details will be overlooked; in practice, since the design is always

---

[4] For the sake of completeness, here is a summary of the details:

Filter Type	L	$f(\omega)$
Type I	$(M-1)/2$	1
Type II	$(M-2)/2$	$\cos(\omega/2)$
Type III	$(M-3)/2$	$\sin(\omega)$
Type IV	$(M-2)/2$	$\sin(\omega/2)$

carried out using numerical packages, the appropriate formulation for the filter expression is taken care of automatically.

Polynomial Optimization. Going back to the filter design problem, we stipulate that the FIR filters are (generalized) linear phase, so we can concentrate on the real frequency response of the zero-centered filter, which is represented by the trigonometric polynomial (7.18). Moreover, since the impulse response is real and symmetric, the aforementioned real frequency response is also symmetric around $\omega = 0$. The filter design procedure can thus be carried out only for values of ω over the interval $[0, \pi]$, with the other half of the spectrum obtained by symmetry. For these values of ω, the variable $x = \cos \omega$ is mapped onto the interval $[1, -1]$ and the mapping is invertible. Therefore, *the filter design problem becomes a problem of polynomial approximation over intervals.*

To illustrate the procedure by example, consider once more the set of filter specifications in Figure 7.1 and suppose we decide to use a Type I filter. Recall that we required the prototype filter to be lowpass, with a transition band from $\omega_p = 0.4\pi$ to $\omega_s = 0.6\pi$; we further stated that the tolerances for the realized filter's magnitude must not exceed 10% in the passband and 1% in the stopband. This implies that the maximum magnitude error between the prototype filter and the FIR filter response $H(e^{j\omega})$ must not exceed $\delta_p = 0.1$ in the interval $[0, \omega_p]$ and must not exceed $\delta_s = 0.01$ in the interval $[\omega_s, \pi]$. We can formulate this as follows: the frequency response of the desired filter is

$$H_D(e^{j\omega}) = \begin{cases} 1 & \omega \in [0, \omega_p] \\ 0 & \omega \in [\omega_s, \pi] \end{cases}$$

(note that $H_D(e^{j\omega})$ is not specified in the transition band). Since the tolerances on passband and stopband are different, they can be expressed in terms of a weighting function $H_W(\omega)$ such that the tolerance on the error is constant over the two bands:

$$H_W(\omega) = \begin{cases} 1 & \omega \in [0, \omega_p] \\ \dfrac{\delta_p}{\delta_s} = 10 & \omega \in [\omega_s, \pi] \end{cases} \quad (7.19)$$

With this notation, the filter specifications amount to the following:

$$\max_{\omega \in [0, \omega_p] \cup [\omega_s, \pi]} \left\{ H_W(\omega) \left| H_d(e^{j\omega}) - H_D(e^{j\omega}) \right| \right\} \leq \delta_p = 0.1 \quad (7.20)$$

and the question now is to find the coefficients for $h[n]$ (their number M and their values) which minimize the above error. Note that we leave the transition band unconstrained (i.e. it does not affect the minimization of the error).

The next step is to use (7.18) to reformulate the above expression as a polynomial optimization problem. To do so, we replace the frequency response $H_d(e^{j\omega})$ with its polynomial equivalent and set $x = \cos \omega$; the passband interval $[0, \omega_p]$ and the stopband interval $[\omega_s, \pi]$ are mapped into the intervals for x:

$$I_p = [\cos \omega_p, 1]$$
$$I_s = [-1, \cos \omega_s]$$

respectively; similarly, the desired response becomes:

$$D(x) = \begin{cases} 1 & \omega \in I_p \\ 0 & \omega \in I_s \end{cases} \qquad (7.21)$$

and the weighting function becomes:

$$W(x) = \begin{cases} 1 & \omega \in I_p \\ \delta_p/\delta_s & \omega \in I_s \end{cases} \qquad (7.22)$$

The new set of specifications are shown in Figure 7.12. Within this polynomial formulation, the optimization problem becomes:

$$\max_{x \in I_p \cup I_s} \{W(x)|P(x) - D(x)|\} = \max\{|E(x)|\} \leq \delta_p \qquad (7.23)$$

where $P(x)$ is the polynomial representation of the FIR frequency response as in (7.18).

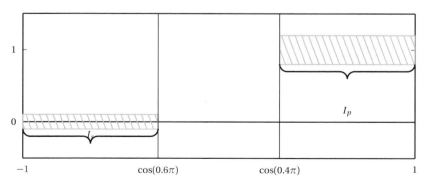

Figure 7.12 Filter specifications as in Figure 7.1 formulated here in terms of polynomial approximation, i.e. for $x = \cos \omega$, $\omega \in [0, \pi]$.

Alternation Theorem. The optimization problem stated by (7.23) can be solved by using the following theorem:

Theorem 7.1 *Consider a set $\{I_k\}$ of closed, disjoint intervals on the real axis and their union $I = \bigcup_k I_k$. Consider further:*

- *a polynomial $P(x)$ of degree L, $P(x) = \sum_{n=0}^{L} a_n x^n$;*
- *a desired function $D(x)$, continuous over I;*
- *a positive weighting function $W(x)$.*

Consider now the approximation error function

$$E(x) = W(x)[D(x) - P(x)]$$

and the associated maximum approximation error over the set of closed intervals

$$E_{\max} = \max_{x \in I}\{|E(x)|\}$$

Then $P(x)$ is the unique *order-L polynomial which minimizes $E_{\max}$ if and only if there exist at least $L+2$ successive values x_i in I such that $|E(x_i)| = E_{\max}$ and*

$$E(x_i) = -E(x_{i+1})$$

In other words, the error function must have at least $L+2$ alternations between its maximum and minimum values. Such a function is called equiripple.

Going back to our lowpass filter example, assume we are trying to design a 9-tap optimal filter. This theorem tells us that if we found a polynomial $P(x)$ of degree 4 such that the error function (7.23) over I_p and I_s as is shown in Figure 7.13 (6 alternations), then the polynomial would be the *optimal*

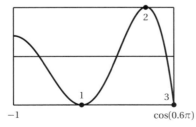
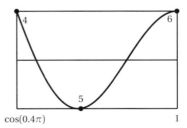

Figure 7.13 Approximation error function $E(x)$ for a 9-tap lowpass prototype; alternations are marked by a dot.

and *unique* solution. Note that the extremal points (i.e. the values of the error function at the edges of the optimization intervals) *do* count in the number of alternations since the intervals I_k are closed.

The above theorem may seem a bit far-fetched since it does not tell us how to find the coefficients but it only gives us a test to verify their optimality. This test, however, is at the core of an *iterative* algorithm which refines the polynomial from an initial guess to the point when the optimality condition is met. Before considering the optimization procedure more in detail, we will state without formal proof, three consequences of the alternation theorem as it applies to the design of Type I lowpass filters:

- The minimum number of alternations for an optimal M-tap lowpass filter is $L+2$, with $L=(M-1)/2$; this is the result of the alternation theorem. The *maximum* number of alternation, however, is $L+3$; filters with $L+3$ alternation are called extraripple filters.

- Alternations always take place at $x = \cos \omega_p$ and $x = \cos \omega_s$ (i.e. at $\omega = \omega_p$ and $\omega = \omega_s$.

- If the error function has a local maximum or minimum, its absolute value at the extremum must be equal to $E_{\max}$ except possibly in $x = 0$ or $x = 1$. In other words, all local maxima and minima of the frequency response must be alternation, except in $\omega = 0$ or $\omega = \pi$.

- If the filter is extraripple, the extra alternation occurs at either $\omega = 0$ or $\omega = \pi$.

Optimization Procedure. Finally, by putting all the elements together, we are ready to state an algorithmic optimization procedure for the design of optimal minimax FIR filters; this procedure is usually called the Parks-McClellan algorithm. Remember, we are trying to determine a polynomial $P(x)$ such that the approximation error in (7.23) is equiripple; for this, we need to determine both the degree of the polynomial and its coefficients. For a given degree L, for which the resulting filter has $2L+1$ taps, the L coefficients are found by an iterative procedure which successively refines an initial guess for the $L+2$ alternation points x_i until the error is equiripple.[5] After the iteration has converged, we need to check that the corresponding

---

[5] Details about this crucial optimization step can be found in the bibliographic references. While a thorough discussion of the algorithm is beyond the scope of the book, we can mention that at each iteration the new set of candidate extremal points is obtained by exchanging the old set with the ordinates of the current local maxima. This trick is known as the Remez exchange algorithm and that is why, in Matlab, the Parks-McClellan algorithm is named `remez`.

$E_{\max}$ satisfies the upper bound imposed by the specifications; when this is not the case, the degree of the polynomial (and therefore the length of the filter) must be increased and the procedure must be restarted. Once the conditions on the error are satisfied, the filter coefficients can be obtained by inverting the Chebyshev expansion.

As a final note, an initial guess for the number of taps can be obtained using the empirical formula by Kaiser; for an M-tap FIR $h[n]$, $n = 0,\ldots, M-1$:

$$M \simeq \frac{-10 \log_{10}(\delta_p \delta_s) - 13}{2.324\,\Omega} + 1$$

where δ_p is the passband tolerance, δ_s is the stopband tolerance and $\Omega = \omega_s - \omega_p$ is the width of the transition band.

The Final Design. We now summarize the design steps for the specifications in Figure 7.1. We use a Type I FIR. We start by using Kaiser's formula to obtain an estimate of the number of taps: since $\delta_p \delta_s = 10^{-3}$ and $\Omega = 0.2\pi$, we obtain $M = 12.6$ which we round up to 13 taps. At this point we can use any numerical package for filter design to run the Parks-McClellan algorithm. In Matlab this would be

```
[h, err] = remez(12, [0 0.4 0.6 1], [1 1 0 0], [1 10]);
```

The resulting frequency response is plotted in Figure 7.14; please note that we are plotting the frequency responses of the zero-centered filter $h_d[n]$, which is a real function of ω. We can verify that the filter has indeed $(M-1)/2 = 6$ alternation by looking at enlarged picture of the passband and the stopband, as in Figure 7.15. The maximum error as returned by Matlab is however 0.102 which is larger than what our specifications called for,

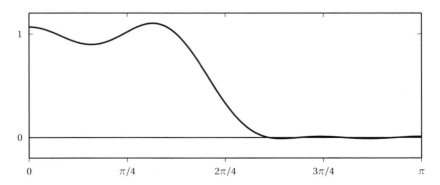

Figure 7.14 An optimal 13-tap Type I filter which does not meet the error specifications.

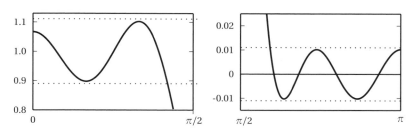

Figure 7.15 Details of passband and stopband of the frequency response in Figure 7.14.

i.e. 0.1. We are thus forced to increase the number of taps; since we are using a Type I filter, the next choice is $M = 15$. Again, the error turns out to be larger than 0.1, since in this case we have $E_{max} = 0.1006$. The next choice, $M = 17$, finally yields an error $E_{max} = 0.05$, which exceeds the specifications by a factor of 2. It is the designer's choice to decide whether the computational gains of a shorter filter ($M = 15$) outweigh the small excess error. The impulse response and the frequency response of the 17-tap filter are plotted in Figure 7.16 and Figure 7.17. Figure 7.18 shows the zero locations for the filter; note the typical linear-phase zero pattern as well as the zeros on the unit circle in the stopband.

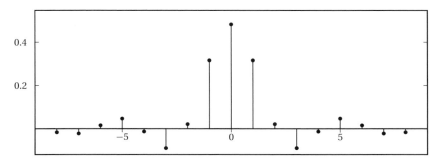

Figure 7.16 Impulse response of the 17-tap filter meeting the specifications.

Other Types of Filters. The Parks-McClellan optimal FIR design procedure can be made to work for arbitrary filter types as well, such as highpass and bandpass, but also for more sophisticated frequency responses. The constraints imposed by the zero locations as we saw on page 181 determine the type of filter to use; once the desired response $H_D(e^{j\omega})$ is expressed as a trigonometric function, the optimization algorithm can take its course. For arbitrary frequency responses, however, the fact that the transition bands are left unconstrained may lead to unacceptable peaks which render the filter useless. In these cases, visual inspection of the obtained response is mandatory and experimentation with different filter lengths and tolerance may improve the final result.

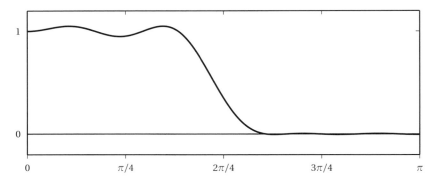

Figure 7.17 Frequency response of the 17-tap filter meeting the specifications.

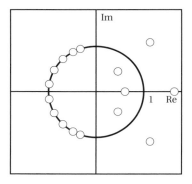

Figure 7.18 Pole-zero plot for the equiripple FIR in Figure 7.17.

7.3 IIR Filter Design

As we mentioned earlier, no optimal procedure exists for the design of IIR filters. The fundamental reason is that the optimization of the coefficients of a rational transfer function is a highly nonlinear problem and no satisfactory algorithm has yet been developed for the task. This, coupled with the impossibility of obtaining an IIR with linear phase response[6] makes the design of the IIR filter a much less formalized art. Many IIR designed

---

[6] It can be proved rigorously that an infinite impulse response with linear phase is necessarily not realizable – think of a sinc filter, for instance.

techniques are described in the literature and their origin is usually in tried-and-true analog filter design methods. In the early days of digital signal processing, engineers would own voluminous books with exhaustive lists of capacitance and inductance values to be used for a given set of (analog) filter specifications. The idea behind most digital IIR filter design techniques was to be able to make use of that body of knowledge and to devise formulas which would translate the analog design into a digital one. The most common such method is known as *bilinear transformation*. Today, the formal step through an analog prototype has become unnecessary and numerical tools such as Matlab can provide a variety of routines to design an IIR.

Here we concentrate only on some basic IIR filters which are very simple and which are commonly used in practice.

7.3.1 All-Time Classics

There are a few applications in which simple IIR structures are the design of choice. These filters are so simple and so well behaved that they are a fundamental tool in the arsenal of any signal processing engineer.

DC Removal and Mean Estimation. The DC component of a signal is its mean value; a signal with zero mean is also called an AC signal. This nomenclature comes from electrical circuit parlance: DC is shorthand for *direct current*, while AC stands for *alternating current*;[7] you might be familiar with these terms in relation to the current provided by a battery (constant and hence DC) and the current available from a mains socket (alternating at 50 or 60 Hz and therefore AC).

For a given sequence $x[n]$, one can always write

$$x[n] = x_{\text{AC}}[n] + x_{\text{DC}}$$

where x_{DC} is the mean of the sequence values. Please note the followings:

- The DC value of a finite-support signal is the value of its Fourier transform at $\omega = 0$ times the length of the signal's support.

- The DC value of an infinite-support signal must be zero for the signal to be absolutely summable or square summable.

In most signal processing applications, where the input signal comes from an acquisition device (such as a sampler, a soundcard and so on), it is important to remove the DC component; this is because the DC offset is often

---
[7] And AC/DC for Heavy Metal...

a random offset caused by ground mismatches between the acquisition device and the associated hardware. In order to eliminate the DC component we need to first estimate it, i.e. we need to estimate the mean of the signal.

For finite-length signals, computation of the mean is straightforward since it involves a finite number of operations. In most cases, however, we do not want to wait until the end of the signal before we try to remove its mean; what we need is a way to perform DC removal *on line*. The approach is therefore to obtain, at each instant, an *estimate* of the DC component from the past signal values, with the assumption that the estimate converges to the real mean of the signal. In order to obtain such an estimate, i.e. in order to obtain the average value of the past input samples, both approaches detailed in Section 5.3 are of course valid (i.e. the Moving Average and the Leaky Integrator filters) . We have seen, however, that the leaky integrator provides a superior cost/benefit tradeoff and therefore the output of a leaky integrator with λ very close to one (usually 10^{-3}) is the estimate of choice for the DC component of a signal. The closer λ is to one, the more accurate the estimation; the speed of convergence of the estimate however becomes slower and slower as $\lambda \to 1$. This can easily be seen from the group delay at $\omega = 0$, which is

$$\mathrm{grd}\{H(1)\} = \frac{\lambda}{1-\lambda}$$

Resonator Filter. Let us look again at how the leaky integrator works. Consider its z-transform:

$$H(z) = \frac{1-\lambda}{1-\lambda z^{-1}}$$

and notice that what we really want the filter to do is to extract the zero-frequency component (i.e. the frequency component that does not oscillate, that is, the DC component). To do so, we placed a pole near $z = 1$, which of course corresponds to $z = e^{j\omega}$ for $\omega = 0$. Since the magnitude response of the filter exhibits a peak near a pole, and since the peak will be higher, the closer the pole is to the unit circle, we are in fact amplifying the zero-frequency component; this is apparent from the plot of the filter's frequency response in Figure 5.9. The numerator, $1-\lambda$, is chosen such that the magnitude of the filter at $\omega = 0$ is one; the net result is that the zero-frequency component will pass unmodified while all the other frequencies will be attenuated. The value of a filter's magnitude at a given frequency is often called the *gain*.

The very same approach can now be used to extract a signal component at *any* frequency. We will use a pole whose magnitude is still close to one (i.e. a pole near the unit circle) but whose phase is that of the frequency we

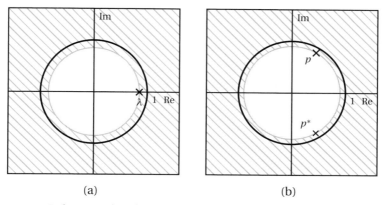

Figure 7.19 Pole-zero plots for the leaky integrator and the simple resonator.

want to extract. We will then choose a numerator so that the magnitude is unity at the frequency of interest. The one extra detail is that, since we want a real-valued filter, we must place a complex conjugate pole as well. The resulting filter is called a resonator and a typical pole-zero plot is shown in Figure 7.19. The z-transform of a resonator at frequency ω_0 is therefore determined by the pole $p = \lambda e^{j\omega_0}$ and by its conjugate:

$$H(z) = \frac{G_0}{(1-pz^{-1})(1-p^*z^{-1})} = \frac{G_0}{1-(2\lambda\cos\omega_0)z^{-1}+\lambda^2 z^{-2}} \quad (7.24)$$

The numerator value G_0 is computed so that the filter's gain at $\pm\omega_0$ is one; since in this case $|H(e^{j\omega_0})| = |H(e^{-j\omega_0})|$, we have

$$G_0 = (1-\lambda)\sqrt{1+\lambda^2-2\lambda\cos 2\omega_0}$$

The magnitude and phase of a resonator with $\lambda = 0.9$ and $\omega_0 = \pi/3$ are shown in Figure 7.20.

A simple variant on the basic resonator can be obtained by considering the fact that the resonator is just a bandpass filter with a very narrow passband. As for all bandpass filters, we can therefore place a zero at $z = \pm 1$ and sharpen its midband frequency response. The corresponding z-transform is now

$$H(z) = G_1 \frac{1-z^{-2}}{1-(2\lambda\cos\omega_0)z^{-1}+\lambda^2 z^{-2}}$$

with

$$G_1 = \frac{G_0}{\sqrt{2(1-\cos 2\omega_0)}}$$

The corresponding magnitude response is shown in Figure 7.21.

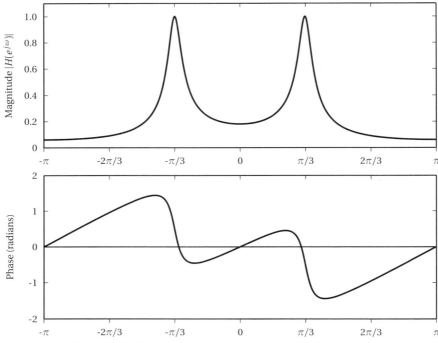

Figure 7.20 Frequency response of the simple resonator.

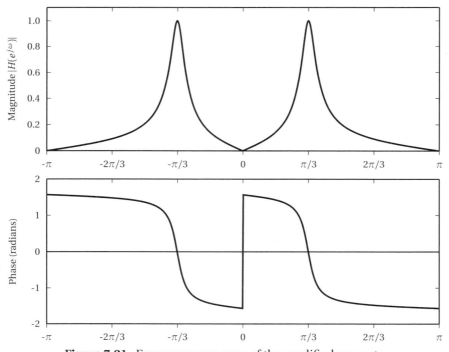

Figure 7.21 Frequency response of the modified resonator.

7.4 Filter Structures

We have seen in Section 5.7.2 a practical implementation of a constant-coefficient difference equation (written in C). That was just one particular way of translating Equation (5.46) into a numerical procedure; in this Section we explore other alternatives for both FIR and IIR and introduce the concept of computational efficiency for filters.

The cost of a numerical filter is dependent on the number of operations per output sample and on the storage (memory) required in the implementation. If we consider a generic CCDE, it is easy to see that the basic building blocks which make up the recipe for a realizable filter are:

- an addition operator for sequence values, implementing $y[n] = x_1[n] + x_2[n]$;

- a scalar multiplication operator, implementing $y[n] = \alpha x[n]$;

- a unit delay operator, implementing $y[n] = x[n-1]$. Note that the unit delay operator is nothing but a memory cell, holding the previous value of a time-varying quantity.

By properly combining these elements and by exploiting the different possible decomposition of a filter's rational transfer function, we can arrive at a variety of different working implementations of a filter. To study the possibilities at hand, instead of relying on a specific programming language, we will use self explanatory block diagrams.

Cascade Forms. Recall that a rational transfer function $H(z)$ can always be written out as follows:

$$H(z) = b_0 \frac{\prod_{n=1}^{M-1}(1 - z_n z^{-1})}{\prod_{n=1}^{N-1}(1 - p_n z^{-1})} \tag{7.25}$$

where the z_n are the $M-1$ (complex) roots of the numerator polynomial and the p_n are the $N-1$ (complex) roots of the denominator polynomial. Since the coefficients of the CCDE are assumed to be real, complex roots for both polynomials always appear in complex-conjugate pairs. A pair of first-order terms with complex-conjugate roots can be combined into a second-order term with real coefficients:

$$(1 - az^{-1})(1 - a^*z^{-1}) = 1 - 2\operatorname{Re}\{a\}z^{-1} + |a|^2 z^{-2} \tag{7.26}$$

As a consequence, the transfer function can be factored into the product of first- and second-order terms in which the coefficients are all strictly real; namely:

$$H(z) = b_0 \frac{\prod_{n=1}^{M_r}(1 - z_n z^{-1}) \prod_{n=1}^{M_c}(1 - 2\operatorname{Re}\{z_n\} z^{-1} + |z_n|^2 z^{-2})}{\prod_{n=1}^{N_r}(1 - p_n z^{-1}) \prod_{n=1}^{N_c}(1 - 2\operatorname{Re}\{p_n\} z^{-1} + |p_n|^2 z^{-2})} \qquad (7.27)$$

where M_r is the number of real zeros, M_c is the number of complex-conjugate zeros and $M_r + 2M_c = M$ (and, equivalently, for the poles, $N_r + 2N_c = N$). From this representation of the transfer function we can obtain an alternative structure for a filter; recall that if we apply a series of filters in sequence, the overall transfer function is the product of the single transfer functions. Working backwards, we can interpret (7.27) as the cascade of smaller sections. The resulting structure is called a *cascade* and it is particularly important for IIR filters, as we will see later.

Parallel Forms. Another interesting rewrite of the transfer function is based on a partial fraction expansion of the type:

$$H(z) = \sum_n D_n z^{-n} + \sum_n \frac{A_n}{1 - p_n z^{-1}} \\ + \sum_n \frac{B_n + C_n z^{-1}}{(1 - p_n z^{-1})(1 - p_n^* z^{-1})} \qquad (7.28)$$

where the multiplicity of the three types of terms as well as the relative coefficients are dependent (in a non-trivial way) on the original filter coefficients. This generates a parallel structure of filters, whose outputs are summed together. The first branch corresponds to the first sum and it is an FIR filter; a further set of branches are associated to each term in the second sum, each one of them a first order IIR; the last set of branches is a collection of second order sections, one for each term of the third sum.

7.4.1 FIR Filter Structures

In an FIR transfer function all the denominator coefficients a_n other than a_0 are zero; we have therefore:

$$H(z) = b_0 + b_1 z^{-1} + \cdots + b_{M-1} z^{-(M-1)}$$

where, of course, the coefficients correspond to the nonzero values of the impulse response $h[n]$, i.e. $b_n = h[n]$. Using the constitutive elements outlined above, we can immediately draw a block diagram of an FIR filter as in

Figure 7.22. In practice, however, additions are distributed as shown in Figure 7.23; this kind of implementation is called a *transversal filter*. Further, ad-hoc optimizations for FIR structures can be obtained in the the case of symmetric and antisymmetric linear phase filters; these are considered in the exercises.

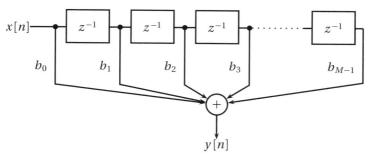

Figure 7.22 Direct FIR implementation.

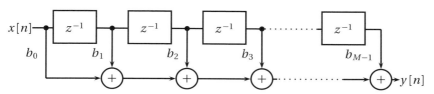

Figure 7.23 Transversal FIR implementation.

7.4.2 IIR Filter Structures

For an IIR filter, all the a_n and b_n in (5.46) are nonzero. One possible implementation based on the direct form of the transfer function is given in Figure 7.24. This implementation is called *Direct Form I* and it can immediately be seen that the C-code implementation in Section 5.7.2 realizes a Direct Form I algorithm. Here, for simplicity, we have assumed $N = M$ but obviously we can set some a_n or b_n to zero if this is not the case.

By the commutative properties of the z-transform, we can invert the order of computation to turn the Direct Form I structure into the structure shown in Figure 7.25 (shown for a second order section); we can then combine the parallel delays together to obtain the structure in Figure 7.26. This implementation is called *Direct Form II*; its obvious advantage is the reduced number of the required delay elements (hence of memory storage).

198   Filter Structures

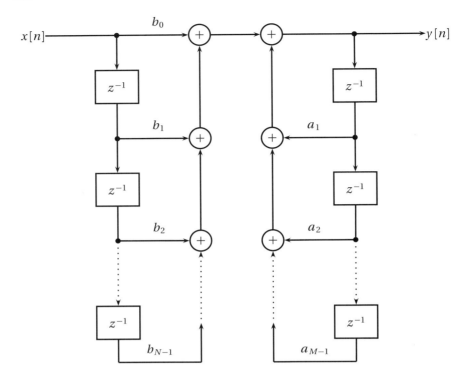

Figure 7.24 Direct Form implementation of an IIR filter.

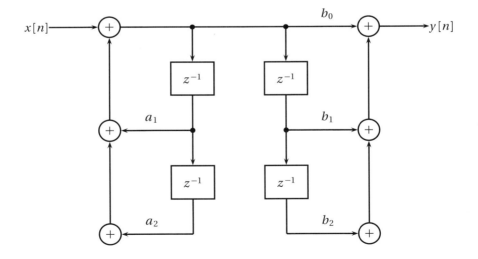

Figure 7.25 Direct form I with inverted order.

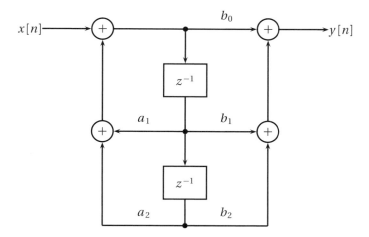

Figure 7.26 Direct Form II implementation of a second-order section.

The second order filter

$$H(z) = \frac{1 + b_1 z^{-1} + b_2 z^{-2}}{1 - a_1 z^{-1} - a_2 z^{-2}}$$

which gives rise to the second order section displayed in Figure 7.26, is particularly important in the case of cascade realizations. Consider the factored form of $H(z)$ as in (7.27): if we combine the complex conjugate poles and zeros, and group the real poles and zeros in pairs, we can create a modular structure composed of second order sections. For instance, Figure 7.27 represents a 4th order system. Odd order systems can be obtained by setting some of the a_n or b_n to zero.

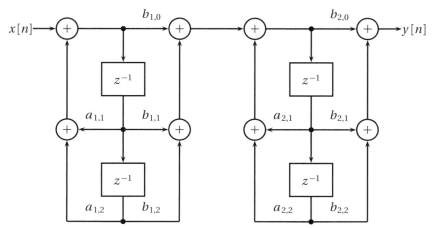

Figure 7.27 4th order IIR: cascade implementation.

7.4.3 Some Remarks on Numerical Stability

A very important issue with digital filters is their numerical behavior for a given implementation. Two key questions are:

- Assume the computations are made with (basically) infinite precision but that the filter coefficients are represented internally with finite precision. How good is the resulting filter? Is it still stable?

- If computations are also made with finite precision arithmetic (which implies rounding and truncation error), what is the resulting numerical behavior of the system?

One important difference is that, in the first case, the system is at least guaranteed to be linear; in the second case, however, we can have non-linear effects such as overflows and limit cycles.

Precision and computational issues are very hard to analyze. Here, we will just note that the direct form implementation is more sensible to precision errors than the cascade form, which is why the cascade form is usually preferred in practice. Moreover, alternative filter structures such as the *lattice* are designed to provide robustness for systems with low numerical precision, albeit at a higher computational cost.

7.5 Filtering and Signal Classes

The filtering structures that we have shown up to now are general-purpose architectures which apply to the most general class of discrete-time signals, (infinite) sequences. We now consider the other two main classes of discrete-time signals, namely finite-length signals and periodic sequences, and show that specialized filtering algorithms can be advantageously put to use.

7.5.1 Filtering of Finite-Length Signals

The convolution sum in (5.3) is defined for infinite sequences. For a finite-length signal of length N we may choose to write simply:

$$y[n] = \mathscr{H}\{x[n]\} = \sum_{k=0}^{N-1} x[k]h[n-k] \tag{7.29}$$

i.e. we let the summation index span only the indices for which the signal is defined. It can immediately be seen, however, that in so doing we are actually computing $y[n] = \tilde{x}[n] * h[n]$, where $\tilde{x}[n]$ is the finite support extension of $x[n]$ as in (2.24)); that is, by using (7.29), we are *implicitly* assuming a finite support extension for the input signal.

Even when the input is finite-length, the output of an LTI system is not necessarily a finite-support sequence. When the impulse response is FIR, however, the output has finite support; specifically, if the input sequence has support N and the impulse response has support L, the support of the output is $N + L - 1$.

7.5.2 Filtering of Periodic Sequences

For periodic sequences, the convolution sum in (5.3) is well defined so there is no special care to be taken. It is easy to see that, for any LTI system, an N-periodic input produces an N-periodic output. A case of particular interest is the following: consider a length-N signal $x[n]$ and its N-periodic extension $\tilde{x}[n]$. Consider then a filter whose impulse response is FIR with a length-N support; if we call $h[n]$ the length-N signal obtained by considering only the values of the impulse response over its finite support, the impulse response of the filter is $\bar{h}[n]$ (see (2.24)). In this case we can write

$$\tilde{y}[n] = \sum_{k=-\infty}^{\infty} \tilde{x}[k]\bar{h}[n-k] = \sum_{k=0}^{N-1} h[k]\tilde{x}[(n-k) \bmod N] \quad (7.30)$$

Note that in the last sum, only the first period of $\tilde{x}[n]$ is used; we can therefore define the sum just in terms of the two N-point signals $x[n]$ and $h[n]$:

$$\tilde{y}[n] = \sum_{k=0}^{N-1} h[k]x[(n-k) \bmod N] \quad (7.31)$$

The above summation is called the *circular convolution* of $x[n]$ and $h[n]$ and is sometimes indicated as

$$\tilde{y}[n] = x[n] \circledast h[n]$$

Note that, for periodic sequences, the convolution as defined in (5.8) and the circular convolution coincide. The circular convolution, just like the standard convolution operator, is associative and commutative:

$$x[n] \circledast h[n] = h[n] \circledast x[n]$$
$$(h[n] + f[n]) \circledast x[n] = h[n] \circledast x[n] + f[n] \circledast x[n]$$

as is easily proven.

Consider now the output of the filter, expressed using the commutative property of the circular convolution:

$$\tilde{y}[n] = \sum_{k=0}^{N-1} x[k]h[(n-k) \bmod N]$$

Since the output sequence $\tilde{y}[n]$ is itself N-periodic we can consider the finite-length signal $y[n] = \tilde{y}[n]$, $n = 0,\ldots,N-1$, i.e. the first period of the output sequence. The circular convolution can now be expressed in matrix form as

$$\mathbf{y} = \mathbf{Hx} \qquad (7.32)$$

where $\mathbf{y}, \mathbf{x}$ are the usual vector notation for the finite-length signals $y[n], x[n]$ and where

$$\mathbf{H} = \begin{bmatrix} h[0] & h[N-1] & h[N-2] & \ldots & h[2] & h[1] \\ h[1] & h[0] & h[N-1] & \ldots & h[3] & h[2] \\ \vdots & \vdots & \vdots & \ddots & \vdots & \vdots \\ h[N-1] & h[N-2] & h[N-3] & \ldots & h[1] & h[0] \end{bmatrix} \qquad (7.33)$$

The above matrix is called a *circulant matrix*, since each row is obtained by a right circular shift of the previous row. A fundamental result, whose proof is left as an exercise, is that the length-N DFT basis vectors $\mathbf{w}^{(k)}$ defined in (4.3) *are left eigenvectors of $N \times N$ circulant matrices*:

$$\left(\mathbf{w}^{(k)}\right)^T \mathbf{H} = H[k]\mathbf{w}^{(k)}$$

where $H[k]$ is the k-th DFT coefficient of the length-N signal $h[n]$, $n = 0,\ldots,N-1$. If we now take the DFT of (7.32) then

$$\mathbf{Y} = \mathbf{WHx} = \mathbf{\Gamma Wx} = \mathbf{\Gamma X}$$

with

$$\mathbf{\Gamma} = \mathrm{diag}(H[0], H[1], \ldots, H[N-1])$$

or, in other words

$$Y[k] = H[k]X[k] \qquad (7.34)$$

We have just proven a finite-length version of the convolution theorem; to repeat the main points:

- The convolution of an N-periodic sequence with a N-tap FIR impulse response is equal to the periodic convolution of two finite-length signals of length N, where the first signal is one period of the input and the second signal is the values of the impulse response over the support.

- The periodic convolution can be expressed as a matrix-vector product in which the matrix is circulant.
- The DFT of the circular convolution is simply the product of the DFTs of the two finite-length signals; in particular, (7.34) can be used to easily prove the commutativity and distributivity of the circular convolution.

The importance of this particular case of filtering stems from the following fact: the matrix-vector product in (7.32) requires $O(N^2)$ operations. The same product can however be written as

$$\mathbf{y} = \frac{1}{N}\mathbf{W}^H\mathbf{\Gamma}\mathbf{W}\mathbf{x} = \text{DFT}^{-1}\{\mathbf{\Gamma}\,\text{DFT}\{\mathbf{x}\}\}$$

which, by using the FFT algorithm, requires approximately $N + 2N\log_2 N$ operations and is therefore much more efficient even for moderate values of N. Practical applications of this idea are the *overlap-save* and *overlap-add* filtering methods, for a thorough description of which see [?]. The basic idea is that, in order to filter a long input sequence with an N-tap FIR filter, the input is broken into consecutive length-N pieces and each piece, considered as the main period of a periodic sequence, is filtered using the FFT strategy above. The difference between the two methods is in the subtle technicalities which allow the output pieces to bind together in order to give the correct final result.

Finally, we want to show that we could have quickly arrived at the same results just by considering the formal DTFTs of the sequences involved; this is an instance of the power of the DTFT formalism. From (4.43) and (4.44) we obtain:

$$\begin{aligned}Y(e^{j\omega}) &= \tilde{H}(e^{j\omega})\tilde{X}(e^{j\omega}) \\ &= \left(\sum_{k=0}^{N-1} H[k]\Lambda\left(\omega - \frac{2\pi}{N}k\right)\right)\left(\frac{1}{N}\sum_{k=0}^{N-1} X[k]\tilde{\delta}\left(\omega - \frac{2\pi}{N}k\right)\right) \\ &= \frac{1}{N}\sum_{k=0}^{N-1} H[k]X[k]\tilde{\delta}\left(\omega - \frac{2\pi}{N}k\right) \end{aligned} \quad (7.35)$$

where the last equality results from the sifting property of the Dirac delta (see (4.31)) and the fact that $\Lambda(0) = 1$. In the last expression, the DTFT of a periodic sequence whose DFS coefficients are given by $H[k]X[k]$, is easily recognezed.

Examples

Example 7.1: The Goertzel filter

Consider the IIR structure shown in Figure 7.28; the filter is called a Goertzel filter, and its single coefficient (which is also the only pole of the system) is the k-th power of the N-th root of unity $W_N = e^{-j2\pi/N}$. Note that, contrary to what we have seen so far, this is a *complex-valued* filter; the analysis of this type of structure however is identical to that of a normal real-valued scheme.

As we said, the only pole of the filter is *on* the unit circle, so the system is not stable. We can nevertheless compute its impulse response, a task which is trivial in the case of a one-pole IIR; we assume zero initial conditions and we use the difference equation directly: by setting $x[n] = \delta[n]$ in

$$y[n] = x[n] + W_N^{-k} y[n-1]$$

and by working out the first few iterations, we obtain

$$h[n] = W_N^{-kn} u[n]$$

Note that the impulse response is N-periodic (a common trait of sequences whose poles are on the unit circle).

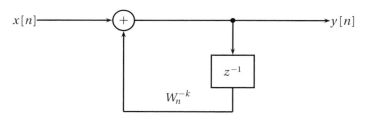

Figure 7.28 The Goertzel filter.

Assume now we have a length-N signal $x[n]$ and we build a finite-support extension $\tilde{x}[n]$ so that $\tilde{x}[n] = 0$ for $n < 0$, $n \geq N$ and $\tilde{x}[n] = x[n]$ otherwise. If we process such a signal with the Goertzel filter we have

$$y[0] = x[0]$$
$$y[1] = x[1] + W_N^{-k} x[0]$$
$$y[2] = x[2] + W_N^{-k} x[1] + W_N^{-2k} x[0]$$
$$y[3] = x[3] + W_N^{-k} x[2] + W_N^{-2k} x[1] + W_N^{-3k} x[0]$$
$$\vdots$$

so that finally:

$$y[N] = \sum_{n=0}^{N-1} W_N^{-k(N-n)} x[n] = \sum_{n=0}^{N-1} x[n] W_N^{nk} = X[k]$$

that is, the output at time $n = N$ is the k-th DFT coefficient of $x[n]$. The Goertzel filter is therefore a little machine which allows us to obtain one specific Fourier coefficient without needing to compute the whole DFT. As a filter, its usage is nonstandard, since its delay element must be manually reset to zero initial conditions after each group of N iterations. Goertzel algorithm is used in digital detectors of DTMF tones.

Example 7.2: Filtering and numerical precision

Digital filters are implemented on general-purpose microprocessors; the precision of the arithmetics involved in computing the output values depends on the intrinsic word length in the digital architecture, i.e. in the number of bits used to represent both the data and the filter coefficients. To illustrate some of the issues related to numeric precision consider the case of an *allpass* filter. The magnitude response of an allpass filter is *constant* over the entire $[-\pi, \pi]$ interval, hence the name. Such filters are often used in cascade with other filters to gain control on the overall phase response of the system.

Consider the filter described by the following difference equation:

$$y[n] = \alpha y[n-1] - \alpha x[n] + x[n-1]$$

with $0 < \alpha < 1$. The transfer function $H(z)$ is

$$H(z) = \frac{-\alpha + z^{-1}}{1 - \alpha z^{-1}} = -\alpha \frac{1 - (1/\alpha)z^{-1}}{1 - \alpha z^{-1}}$$

and the filter is indeed allpass since:

$$\begin{aligned}|H(z)|^2 &= H(z)H^*(z) \\ &= \frac{-\alpha + z^{-1}}{1 - \alpha z^{-1}} \cdot \frac{-\alpha + (z^{-1})^*}{1 - \alpha(z^{-1})^*} \\ &= \frac{\alpha^2 - \alpha \operatorname{Re}\{z^{-1}\} + |z^{-1}|^2}{\alpha^2 |z^{-1}|^2 - \alpha \operatorname{Re}\{z^{-1}\} + 1}\end{aligned}$$

for $z = e^{j\omega}$ (and therefore $|z^{-1}| = 1$):

$$|H(e^{j\omega})|^2 = |H(e^{j\omega})| = 1$$

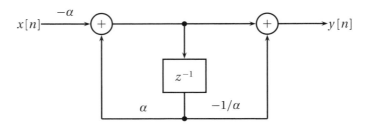

Figure 7.29 Allpass filter Direct Form II implementation.

The filter can be implemented in Direct Form II as in Figure 7.29. Note that the two coefficients of the filter are α and $1/\alpha$ so that, if α is small then $1/\alpha$ will be big, and vice versa. This creates a problem in a digital architecture in which the internal representation has only a small number of bits. Call $\mathcal{Q}\{\cdot\}$ the operator which associates a real number to the closest value in the architecture's internal representation; the process is called *quantization* and we will study it in more detail in Chapter 10. The transfer function with quantized coefficients becomes

$$H_Q(z) = \mathcal{Q}\{-\alpha\}\frac{1-\mathcal{Q}\{1/\alpha\}z^{-1}}{1-\mathcal{Q}\{\alpha\}z^{-1}} = \frac{-\hat{\alpha}+\beta z^{-1}}{1-\hat{\alpha}z^{-1}}$$

where $\beta = \mathcal{Q}\{\alpha\}\mathcal{Q}\{1/\alpha\}$. If the quantization is done with too few bits, $\beta \neq 1$ and the filter characteristic is no longer allpass. Suppose for instance that the filter uses four bits to store its coefficients using an unsigned fixed point 2.2 format; the 16 possible values are listed in Table 7.1.

Table 7.1 Binary-decimal conversion table for fixed-point 2.2 notation.

binary	decimal	binary	decimal
0000	0.00	1000	2.00
0001	0.25	1001	2.25
0010	0.50	1010	2.50
0011	0.75	1011	2.75
0100	1.00	1100	3.00
0101	1.25	1101	3.25
0110	1.50	1110	3.50
0111	1.75	1111	3.75

If $\alpha = 0.4$ we have that $\mathcal{Q}\{0.4\} = 0010 = 0.5$, $\mathcal{Q}\{1/0.4\} = \mathcal{Q}\{2.5\} = 1010 = 2.5$ and therefore $\beta = 0101 = 1.25 \neq 1$.

It is important to point out that the numerical stability of a filter is dependent on the chosen realization. If we rewrite the allpass difference equation as

$$y[n] = \alpha(y[n-1] - x[n]) + x[n-1]$$

we can a block diagram as in Figure 7.30 which, although a non-canonical form, implements the filter with no quantization issues *independently of α*. Note that the price we pay for robustness is the fact that we have to use two delays instead of one.

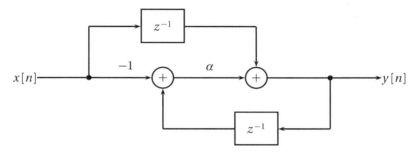

Figure 7.30 Allpass filter Direct Form II implementation.

Example 7.3: A guitar synthesizer

We have encountered the Karplus-Strong algorithm in Example 2.2. A practical implementation of the algorithm is shown in Figure 7.31; it is a quite peculiar filter structure since it has no input! Indeed assume there are N delays in cascade and neglect for a moment the filter $H(z)$; the structure forms a feedback loop in which the N values *contained in the delay units at power-up* are endlessly cycled at the output. By loading the N delay units with all sorts of finite-length sequences we can obtain a variety of different sounds; by changing N we can change the fundamental frequency of the note.

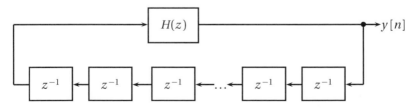

Figure 7.31 Karplus-Strong implementation.

The detailed analysis of a waveform generated by the device in Figure 7.31 is complicated by the fact that the filter does not have zero initial conditions.

Intuitively, however, we can easily appreciate that we can use the filter to simulate a natural decay in the waveform; imagine $H(z) = \alpha$ with $0 < \alpha < 1$: at each passage through the feedback loop the values in the delay line are scaled down exponentially. More complicated filters can be used to simulate different types of acoustic decay as long as $|H(e^{j\omega})| < 1$ over the entire $[-\pi, \pi]$ interval.

Further Reading

Filter design has been a very popular topic in signal processing, with a large literature, a variety of software designs, and several books devoted to the topic. As examples, we can mention R. Hamming's *Digital Filters* (Dover, 1997), and T. W. Parks and C. S. Burrus, *Digital Filter Design* (Wiley-Interscience, 1987), the latter being specifically oriented towards implementations on a digital signal processor (DSP). All classic signal-processing books cover the topic, for example Oppenheim and Schafer's book *Discrete-Time Signal Processing* (Prentis Hall, 1999) gives both structures and design methods for various digital filters.

Exercises

Exercise 7.1: Discrete-time systems and stability. Consider the system in the picture below. Assume a causal input ($x[n] = 0$ for $n < 0$) and zero initial conditions.

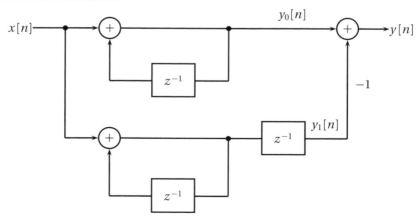

(a) Find the constant-coefficients difference equations linking $y_0[n]$, $y_1[n]$ and $y[n]$ to the input $x[n]$.

(b) Find $H_0(z)$, $H_1(z)$ and $H(z)$, the transfer functions relating the input $x[n]$ to the signals $y_0[n]$, $y_1[n]$ and $y[n]$, respectively.

(c) Consider the relationship between the input and the output; is the system BIBO stable?

(d) Is the system stable internally? (i.e. are the subsystems described by $H_0(z)$ and $H_1(z)$ stable?)

(e) Consider the input $x[n] = u[n]$, where, as usual, $u[n] = 1$ for $n \geq 0$ and $u[n] = 0$ for $n < 0$. How do $y_0[n]$, $y_1[n]$ and $y[n]$ evolve over time? Sketch their values.

(f) Suppose the above system is implemented in practice with finite-precision arithmetic (say 16 bits). Would it work as planned?

Exercise 7.2: Filter properties – I. Assume $\mathcal{G}$ is a stable, causal IIR filter with impulse response $g[n]$ and transfer function $G(z)$. Which of the following statements is/are true for *any* choice of $G(z)$?

(a) The inverse filter, $1/G(z)$, is stable.

(b) The inverse filter is FIR.

(c) The DTFT of $g[n]$ exists.

(d) The cascade $G(z)G(z)$ is stable.

Exercise 7.3: Filter properties – II. Consider $G(z)$, the transfer function of a causal stable LTI system. Which of the following statements is/are true for *any* such $G(z)$?

(a) The zeros of $G(z)$ are inside the unit circle.

(b) The ROC of $G(z)$ includes the curve $|z| = 0.5$.

(c) The system $H(z) = (1 - 3z^{-1})G(z)$ is stable.

(d) The system is an IIR filter.

Exercise 7.4: Fourier transforms and filtering. Consider the following signal:

$$x[n] = \begin{cases} (-1)^{(n/2)+1} & \text{for } n \text{ even} \\ 0 & \text{for } n \text{ odd} \end{cases}$$

(a) Sketch $x[n]$ in time.

(b) Which is the most appropriate Fourier representation for $x[n]$? (DFT, DFS, DTFT?) Explain your choice and compute the transform.

(c) Compute the DTFT of $x[n]$, $X(e^{j\omega})$, and plot its magnitude and phase.

(d) Consider a filter with the impulse response

$$h[n] = \frac{\sin n}{\pi n}$$

and compute $y[n] = x[n] * h[n]$.

Exercise 7.5: FIR filters. Consider the following set of complex numbers:

$$z_k = e^{j\pi(1-2^{-k})}, \qquad k = 1, 2, \ldots, M$$

For $M = 4$,

(a) Plot z_k, $k = 1, 2, 3, 4$, on the complex plane.

(b) Consider an FIR whose transfer function $H(z)$ has the following zeros:

$$\{z_1, z_2, z_1^*, z_2^*, -1\}$$

and write out explicitly the expression for $H(z)$.

(c) How many nonzero taps does the impulse response $h[n]$ have at most?

(d) Sketch the magnitude of $H(e^{j\omega})$.

(e) What can you say about this filter: What FIR type is it? (I, II, etc.)

Is it lowpass, bandpass, highpass?

Is it equiripple?

Is this a "good" filter? (By "good" we mean a filter which is close to 1 in the passband, close to zero in the stopband and which has a narrow transition band.)

Exercise 7.6: Linear-phase FIR filter structure. Assume $H(z)$ is a Type III FIR filter. One of the zeros of the filter is at $z_0 = 0.8 + 0.1j$. You should be able to specify another five zero locations for the filter. Which are they?

Exercise 7.7: FIR filters analysis – I. Consider a causal FIR lowpass filter with the following transfer function:

$$H(z) = 0.005 + 0.03 z^{-1} + 0.11 z^{-2} + 0.22 z^{-3} + 0.27 z^{-4} + \\ + 0.22 z^{-5} + 0.11 z^{-6} + 0.03 z^{-7} + 0.005 z^{-8}$$

whose magnitude response is plotted in the following figure between 0 and π:

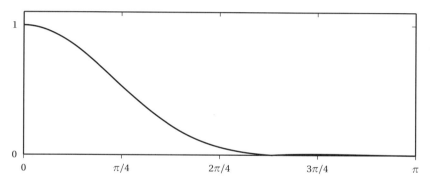

while the following figure displays an enlarged view of the passband and stopband:

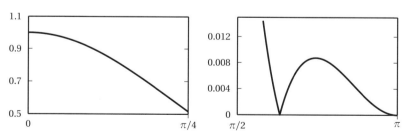

(a) Is the filter linear phase? If so, what type is it (I, II, III, IV)?

(b) What is the group delay of the filter?

(c) Is the filter optimal (in the sense of Parks-McClellan)? Justify your answer.

We will now explore some filters which can be obtained from $H(z)$:

(d) Sketch the magnitude response of a filter $g[n]$ whose taps are as follows:

$$g[n] = \{0.005, 0, -0.11, 0, 0.27, 0, -0.11, 0, 0.005\}$$

(i.e. $g[0] = 0.005$, $g[1] = 0$, $g[2] = -0.11$, etc.)

We now want to obtain a linear phase highpass filter $f[n]$ from $h[n]$ and the following design is proposed:

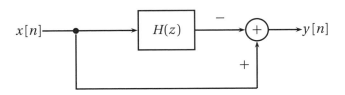

However the design is faulty:

(e) From the impulse response of the above system, show that the resulting filter is *not* linear phase.

(f) Clearly, the designer's idea was to obtain a system with magnitude $|F(e^{j\omega})| = |1 - |H(e^{j\omega})||$ (note the magnitude signs around $H(e^{j\omega})$); this, however, is not the magnitude response of the above system. Write out the actual magnitude response.
(*Hint:* it is easier to consider the squared magnitude response and, since $H(e^{j\omega})$ is linear phase, to express $H(e^{j\omega})$ as a real term $A(e^{j\omega}) \in \mathbb{R}$, multiplied by a pure phase term.)

Now it is your turn to design a highpass filter:

(g) How would you modify the above design to obtain a linear phase highpass filter?

(h) Sketch the magnitude response of the resulting filter.

Exercise 7.8: FIR filters analysis – II. Consider a generic N-tap Type I FIR filter. Since the filter is linear phase, its frequency response can be expressed as

$$H(e^{j\omega}) = A(e^{j\omega}) H_r(e^{j\omega})$$

where $H_r(e^{j\omega})$ is a real function of ω and $A(e^{j\omega})$ is a pure phase term.

(a) Specify $A(e^{j\omega})$ so that the filter is causal; i.e. find the phase term for which the impulse response $h[n]$ is nonzero only for $0 \leq n \leq N-1$.

Now consider a specific N-tap Type-I FIR filter designed with the Parks-McClellan algorithm. The real part $H_r(e^{j\omega})$ for the causal filter is plotted in the following figure.

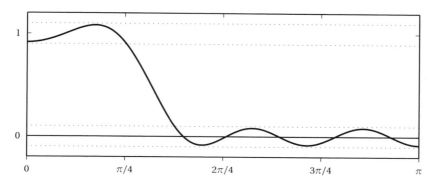

(b) What is the number of coefficients N for this specific filter? (The filter is *not* extraripple.)

(c) One zero of the filter's transfer function is at $z_0 = -1.8 + 0.4j$. Sketch the *complete* pole-zero plot for the filter.

We now modify the *causal* filter $H(e^{j\omega})$ to obtain a new causal filter $H_1(e^{j\omega})$; the real part of the new frequency response is plotted as follows:

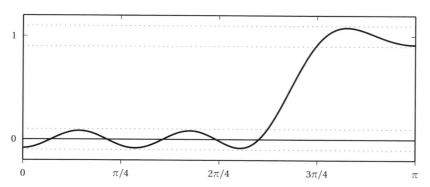

(d) Assume you know the original impulse response $h[n]$; show how you can modify $h[n]$ to obtain $h_1[n]$.

(e) What type is the new filter?

Exercise 7.9: IIR filtering. Consider a causal IIR filter with the following transfer function:

$$H(z) = \frac{1+z^{-1}}{1-1.6\cos(2\pi/7)z^{-1}+0.64z^{-2}}$$

(a) Sketch the pole-zero plot of the filter and the ROC of its transfer function.

(b) Sketch the magnitude of its frequency response.

(c) Draw at least two different block diagrams which implement the filter (e.g. direct forms I and II).

(d) Compute the first five values (for $n = 0, 1, \ldots, 5$) of the signal $y[n] = h[n] * x[n]$, where $x[n] = \delta[n] + 2\delta[n-1]$. Assume zero initial conditions.

Exercise 7.10: Generalized linear phase filters. Consider the filter given by $H(z) = 1 - z^{-1}$.

(a) Show that $H(z)$ is a generalized linear phase filter, i.e. that it can be written as

$$H(e^{j\omega}) = |H(e^{j\omega})| e^{-j(\omega d - \alpha)}$$

Give the corresponding group delay d and the phase factor α.

(b) What type of filter is it (I, II, III or IV)? Explain.

(c) Give the expression of $h[n]$ and show that it satisfies

$$\sum_n h[n]\sin(\omega(n-d)+\alpha) = 0$$

for all ω.

(d) More generally, show that any generalized linear phase filter $h[n]$ must satisfy

$$\sum_n h[n]\sin(\omega(n-d)+\alpha) = 0$$

for all ω. The above expression is, thus, a necessary condition for a filter to be generalized linear phase.

Exercise 7.11: Echo cancellation. In data communication systems over phone lines (such as voiceband modems), one of the major problems is represented by *echos*. Impedance mismatches along the analog line create delayed and attenuated replicas of the transmitted signal. These replicas are added to the original signal and represent one type of distortion.

Assume a simple situation where a single echo is created; the transmitted signal is $x[n]$ and, because of the echo, the received signal is

$$y[n] = x[n] - \alpha x[n-D]$$

where α is the attenuation factor (with $0 < \alpha < 1$) and D is the echo delay (assume D is an integer).

(a) Write the transfer function $H(z)$ of the "echo system", i.e. the system which produces $y[n]$ from $x[n]$.

(b) Sketch the pole-zero plot for $H(z)$ for $\alpha = 0.1$ and $D = 12$ (for our purposes, assume $(0.826)^{12} = 0.1$).

(c) Sketch the squared magnitude response $|H(e^{j\omega})|^2$.

Now assume we have a good estimate of α and D; we want to design a causal echo cancellation filter, i.e. a filter with causal impulse response $g[n]$ such that $y[n] * g[n] = x[n]$.

(d) Write the expression for $G(z)$.

(e) Sketch its pole-zero plot and its ROC for the same values of α and D as before.

(f) What is the practical difficulty in implementing this echo cancellation system?

Exercise 7.12: FIR filter design – I. Is it a good idea to use a Type III FIR to design a lowpass filter? Briefly explain.

Exercise 7.13: FIR filter design – II. Suppose you want to design a linear phase FIR approximation of a Hilbert filter. Which FIR type would you use? Why? Discuss advantages and disadvantages.

Chapter 8

Stochastic Signal Processing

In the previous Chapters, the signals we considered were all *deterministic* signals in the sense that they could either be expressed in analytic form (such as $x[n] = (1-\lambda)\lambda^n$) or they could be explicitly described in terms of their samples, such as in the case of finite-length signals. When designing a signal processing system, however, it is very rare that we know exactly the expression for the set of all the possible input signals (in some sense, if we did, we would not need a processing system at all.) Fortunately, very often this set can be characterized in terms of the *statistical* properties of its member signals; this entails leaving the domain of deterministic quantities and entering the world of stochastic processes. A detailed and rigorous treatment of statistical signal processing is beyond the scope of this book; here, we only consider elementary concepts and restrict ourselves to the discrete-time case. We will be able to derive that, fundamentally, in the case of stationary random signals, the standard signal processing machinery that we have seen so far (and especially the usual filter design techniques) is still applicable with very intuitive results. To establish a coherent notation, we start by briefly reviewing some elementary concepts of probability theory.

8.1 Random Variables

Probability Distribution. Consider a real-valued random variable X taking values over $\mathbb{R}$. The random variable[1] is characterized by its *cumulative distribution function* F_X (cdf) which is defined as

$$F_X(\alpha) = P[X \leq \alpha], \qquad \alpha \in \mathbb{R}$$

---
[1] Note that in this Chapter, random quantities will be indicated by *uppercase* variables.

that is, $F_X(\alpha)$ measures the probability that X takes values less than or equal to α. The *probability density function* (pdf) is related to the cdf (assuming that F_X is differentiable) as

$$f_X(\alpha) = \frac{d F_X(\alpha)}{d\alpha}, \qquad \alpha \in \mathbb{R}$$

and thus

$$F_X(\alpha) = \int_{-\infty}^{\alpha} f_X(x)\,dx, \qquad \alpha \in \mathbb{R}$$

Expectation and Second Order Statistics. For random variables, a fundamental concept is that of *expectation*, defined as follows:

$$\mathrm{E}[X] = \int_{-\infty}^{\infty} x f_X(x)\,dx$$

The expectation operator is linear; given two random variables X and Y, we have

$$\mathrm{E}[aX + bY] = a\mathrm{E}[X] + b\mathrm{E}[Y]$$

Furthermare, given a function $g : \mathbb{R} \mapsto \mathbb{R}$, we have

$$\mathrm{E}[g(X)] = \int_{-\infty}^{\infty} g(x) f_X(x)\,dx$$

The expectation of a random variable is called its *mean*, and we will indicate it by m_X. The expectation of the product of two random variables defines their *correlation*:

$$R_{XY} = \mathrm{E}[XY]$$

The variables are uncorrelated if

$$\mathrm{E}[XY] = \mathrm{E}[X]\mathrm{E}[Y]$$

Sometimes, the "centralized" correlation, or *covariance*, is used, namely

$$K_{XY} = \mathrm{E}[(X - m_X)(Y - m_Y)]$$
$$= \mathrm{E}[XY] - \mathrm{E}[X]\mathrm{E}[Y]$$

Again, the two variables are uncorrelated if and only if their covariance is zero. Note that if two random variables are independent, then they are also uncorrelated. The converse, however, is not true; in other words, statistical independence is a stronger condition than decorrelation.[2]

The *variance* of a random variable X, denoted by σ_X^2, is defined as

$$\sigma_X^2 = E[(X - m_X)^2]$$

The square root of the variance, σ_X, is often called the *standard deviation* of X.

Example: Gaussian Random Variable. A Gaussian random variable is described by the probability density function:

$$f(x) = \frac{1}{\sqrt{2\pi\sigma^2}} e^{-\frac{(x-m)^2}{2\sigma^2}}, \qquad x \in \mathbb{R} \qquad (8.1)$$

which is known as the *normal* distribution. Clearly, the mean of the Gaussian variable is m, while its variance is σ^2. The normalization factor $1/\sqrt{2\pi\sigma^2}$ ensures that, as for all random variables, the integral of the pdf over the entire real line is equal to one.

8.2 Random Vectors

Probability Distribution. A random vector $\mathbf{X}$ is a collection of N random variables $[X_0 \; X_1 \; \ldots \; X_{N-1}]^T$, with a cumulative distribution function $F_\mathbf{X}$ given by

$$F_\mathbf{X}(\boldsymbol{\alpha}) = \mathscr{P}[X_i \leq \alpha_i, \; i = 0, 1, \ldots, N-1]$$

where $\boldsymbol{\alpha} = [\alpha_0 \; \alpha_1 \; \ldots \; \alpha_{N-1}]^T \in \mathbb{R}^N$. The pdf is obtained, assuming differentiability, as

$$f_\mathbf{X}(\boldsymbol{\alpha}) = \frac{\partial^N}{\partial \alpha_0, \partial \alpha_1, \ldots, \partial \alpha_{N-1}} F_\mathbf{X}(\alpha_0, \alpha_1, \ldots, \alpha_{N-1})$$

With respect to vector random variables, two key notions are:

- **independent elements**: a collection of N random variables is independent if and only if the joint pdf has the form:

$$f_{X_0 X_1 \cdots X_{N-1}}(x_0, x_1, \ldots, x_{N-1}) = f_{X_0}(x_0) \cdot f_{X_1}(x_1) \cdots f_{X_{N-1}}(x_{N-1}) \quad (8.2)$$

---

[2] A special case is that of Gaussian random variables, for which independence and decorrelation are equivalent.

- **i.i.d. elements**: a collection of N random variables is *independent and identically distributed* (i.i.d.) if the variables are independent and each random variable has the same distribution:

$$f_{X_i}(x_i) = f(x_i), \qquad i = 0, 1, \ldots, N-1$$

Random vectors represent the generalization of finite-length, discrete-time signals to the space of random signals.

Expectation and Second Order Statistics. For random vectors, the definitions given, in the case of random variables, extend immediately to the multidimensional case. The mean of a N-element random vector $\mathbf{X}$ is simply the N-element vector:

$$\begin{aligned} E[\mathbf{X}] &= \begin{bmatrix} E[X_0] & E[X_1] & \ldots & E[X_{N-1}] \end{bmatrix}^T \\ &= \begin{bmatrix} m_{X_0} & m_{X_1} & \ldots & m_{X_{N-1}} \end{bmatrix}^T \\ &= \mathbf{m_X} \end{aligned}$$

The correlation of two N-element random vectors is the $N \times N$ matrix:

$$\mathbf{R_{XY}} = E[\mathbf{XY}^T]$$

where the expectation operator is applied individually to all the elements of the matrix $\mathbf{XY}^T$. The covariance is again:

$$\mathbf{K_{XY}} = E[(\mathbf{X} - \mathbf{m_X})(\mathbf{Y} - \mathbf{m_Y})^T]$$

and it coincides with the correlation for zero-mean random vectors. Note that the general element $\mathbf{R_{XY}}(k, l)$ indicates the correlation between the k-th element of $\mathbf{X}$ and the l-th element of $\mathbf{Y}$. In particular, $\mathbf{R_{XX}}(k, l)$ indicates the correlation between elements of the random vector $\mathbf{X}$; if the elements are uncorrelated, then the correlation matrix is diagonal.

Example: Jointly Gaussian Random Vector. An important type of vector random variable is the Gaussian random vector of dimension N. To define its pdf, we need a length-N vector $\mathbf{m}$ and a positive definite matrix $\mathbf{\Lambda}$ of size $N \times N$. Then, the N-dimensional Gaussian pdf is given by

$$f(\mathbf{x}) = \frac{1}{\sqrt{(2\pi)^N |\mathbf{\Lambda}|}} e^{-\frac{1}{2}(\mathbf{x}-\mathbf{m})^T \mathbf{\Lambda}^{-1}(\mathbf{x}-\mathbf{m})}, \qquad \mathbf{x} \in \mathbb{R}^N \tag{8.3}$$

where $|\mathbf{\Lambda}|$ is the determinant of $\mathbf{\Lambda}$. Clearly, $\mathbf{m}$ is the vector of the means of the single elements of the Gaussian vector while $\mathbf{\Lambda}$ is the autocorrelation matrix. A diagonal matrix implies the decorrelation of the random vector's elements; in this case, since all the elements are Gaussian variables, this also means that the elements are independent. Note how, for $N = 1$ and $\mathbf{\Lambda} = \sigma^2$, this reduces to the usual Gaussian distribution of (8.1).

8.3 Random Processes

Probability Distribution. Intuitively, a discrete-time *random process* is the infinite-dimensional generalization of a vector random variable, just like a discrete-time sequence is the infinite generalization of a finite-length signal. For a random process (also called a stochastic process) we use the notation $X[n]$ to indicate the n-th random variable which is the n-th value (sample) of the sequence.[3] Note however that the pdf associated to the random process is the *joint* distribution of the entire set of samples in the sequence; in general, therefore, the statistical properties of each sample depend on the global stochastic description of the process and this accounts for local and long-range dependencies in the random data. In fact, consider a random process $\{X[n], n \in \mathbb{Z}\}$; any finite subset of random variables from $X[n]$ is a vector random variable $\mathbf{X} = \begin{bmatrix} X[i_0] & X[i_1] & \ldots & X[i_{k-1}] \end{bmatrix}^T$, $k \in \mathbb{N}$. The statistical description of a random process involves specifying the joint pdf for $\mathbf{X}$ for *all* k-tuples of time indices i_k and *all* $k \in \mathbb{N}$, i.e. all the pdfs of the form

$$f_{X[i_0]X[i_1]\cdots X[i_{k-1}]}(x_0, x_1, \ldots, x_{k-1}) \tag{8.4}$$

Clearly, the most general form of random process possesses a statistical description which is difficult to use. At the other extreme, the simplest form of stochastic process is the i.i.d. process. For an i.i.d. process we have that the elements of $\mathbf{X}$ are i.i.d. for all k-tuples of time indices i_k and all $k \in \mathbb{N}$, that is

$$f_{X[i_0]X[i_1]\cdots X[i_{k-1}]}(x_0, x_1, \ldots, x_{k-1}) = \prod_{i=0}^{k-1} f(x_i) \tag{8.5}$$

where $f(x)$ is called the pdf of the i.i.d. process.

Second Order Description. The mean of a process $X[n]$, $n \in \mathbb{Z}$ is simply $\mathrm{E}[X[n]]$ which, in general, depends on the index n. The *correlation* (also called the autocorrelation) of $X[n]$ is defined as

$$R_X[l, k] = \mathrm{E}[X[l] X[k]], \qquad l, k \in \mathbb{Z}$$

while its *covariance* (also called autocovariance)[4] is

$$K_X[l, k] = \mathrm{E}\left[(X[l] - m_{X[l]})(X[k] - m_{Y[k]})\right]$$
$$= R_X[l, k] - m_{X[l]} m_{X[k]}, \qquad l, k \in \mathbb{Z}$$

---

[3] Again, in this Chapter we use uppercase variables to stress the random nature of a stochastic signal. This will not be strictly enforced in later Chapters.

[4] The term autocovariance is rarely used; the "auto-" forms of correlation and covariance are meant to stress the difference with the *cross*-correlation and *cross*-covariance), which involve two distinct random processes.

Finally, given two random processes $X[n]$ and $Y[n]$, their *cross-correlation* is defined as

$$R_{XY}[l,k] = \mathrm{E}\big[X[l]\,Y[k]\big] \tag{8.6}$$

Mean and variance of a random process represent a *second order description* of the process since their computation requires knowledge of only the second order joint pdf of the process (i.e. of the pdfs in (8.4) involving only two indices i_k). A second order description is physically meaningful since it can be associated to the mean value and mean power of the random process, as we will see.

Stationary Processes. A very important class of random processes are the *stationary processes*, for which the probabilistic behavior is constant over time. Stationarity, in the *strict sense*, implies that the full probabilistic description of the process is time-invariant; for example, any i.i.d. process is also a strict-sense stationary process. Stationarity can be restricted to *n-th order stationarity*, meaning that joint distributions (and therefore expectations) involving up to n samples are invariant with respect to a time shift. The case $n = 2$ is particulary important and it is called *wide-sense stationarity* (WSS). For a WSS process, the mean and the variance are constant over time:

$$\mathrm{E}[X[n]] = m_X, \qquad n \in \mathbb{Z} \tag{8.7}$$

$$\mathrm{E}\big[(X[n] - m_X)^2\big] = \sigma_X^2, \qquad n \in \mathbb{Z} \tag{8.8}$$

and the autocorrelation and covariance depend only on the time lag $(l - k)$:

$$R_X[l,k] = r_X[l-k], \qquad l,k \in \mathbb{Z} \tag{8.9}$$
$$K_X[l,k] = k_X[l-k], \qquad l,k \in \mathbb{Z} \tag{8.10}$$

Finally, note that if $X[n]$ and $Y[n]$ are both stationary processes, then their cross-correlation depends only also on the time lag:

$$R_{XY}[l,k] = r_{XY}[l-k]$$

Ergodicity. In the above paragraphs, it is important to remember that expectations are taken with respect to an *ensemble of realizations* of the process under analysis. To visualize the concept, imagine having a black box which, at the turn of a switch, can generate a realization of a discrete-time random process $X[n]$. In order to estimate the mean of the process at time n_0, i.e. $\mathrm{E}[X[n_0]]$, we need to collect as many realizations as possible and then estimate the mean at time n_0 by averaging the values of the process at

n_0 across realizations. For stationary processes, it may seem intuitive that instead of averaging across realizations, we can average across successive samples of the same realization. This is not true in the general case, however. Consider for instance the process

$$X[n] = \alpha$$

where α is a random variable. Clearly the process is stationary since each realization of this process is a constant discrete-time signal, but the value of the constant changes for each realization. If we try to estimate the mean of the process from a single realization, we obtain no information on the distribution of α; that can be achieved only by looking at several independent realizations.

The class of processes for which it is legitimate to estimate expectations from a single realization is the class of *ergodic processes*. For ergodic processes we can, for instance, take the time average of the samples of a single realization and this average converges to the ensemble average or, in other words, it represents a precise estimate of the true mean of the stochastic process. The same can be said for expectations involving the product of process samples, such as in the computation of the variance or of the correlation.

Ergodicity is an extremely useful concept in the domain of stochastic signal processing since it allows us to extract useful statistical information from a single realization of the process. More often than not, experimental data is difficult or expensive to obtain and it is not practical to repeat an experiment over and over again to compute ensemble averages; ergodicity is the way out this problem, and it is often just assumed (sometimes without rigorous justification).

Example: Gaussian Random Processes. A Gaussian random process is one for which any set of samples is a jointly Gaussian random vector. A fundamental property of a Gaussian random process is that, if it is wide-sense stationary, then it is also stationary in the strict sense. This means that second order statistics are a sufficient representation for Gaussian random processes.

8.4 Spectral Representation of Stationary Random Processes

Given a stationary random process, we are interested in characterizing its "energy distribution" in the frequency domain. Note that we have used quotes around the term energy: since a stationary process does not decay in time (because of stationarity), it is rather intuitive that its energy is infinite (very much like a periodic signal). In other words, the sum:

$$\sum_{n=-M}^{M} X^2[n]$$

diverges in expectation. Signals which are not square-summable are not absolutely summable either, and therefore their Fourier transform does not exist in the standard sense. In order to derive a spectral representation for a random process we thus need to look for an alternative point of view.

8.4.1 Power Spectral Density

In Section 2.1.6 we introduced the notion of a *power signal*, particularly in relation to the class of periodic sequences; while the total energy of a power signal may be infinite, its energy over any *finite* support is always finite and it is proportional to the length of the support. In this case, the limit:

$$\lim_{M \to \infty} \frac{1}{2M+1} \sum_{n=-M}^{M} |x[n]|^2$$

is finite and it represents the signal's average power (in time). Stationary random processes are themselves power signals if their variance is finite; indeed (assuming a zero-mean process), we have

$$\mathrm{E}\left[\frac{1}{2M+1} \sum_{n=-M}^{M} X^2[n]\right] = \frac{1}{2M+1} \sum_{n=-M}^{M} \mathrm{E}[X^2[n]]$$

$$= \frac{1}{2M+1} \sum_{n=-M}^{M} \sigma^2$$

$$= \sigma^2$$

so that the average power (in expectation) for a stationary process is given by its variance.

For signals (stochastic or not) whose power is finite but whose energy is not, a meaningful spectral representation is obtained by considering the so-called *power spectral density* (PSD). We know that, for a square-summable sequence, the square magnitude of the Fourier transform represents the global spectral energy distribution. Since the energy of a power signal is finite over a finite-length observation window, the *truncated* Fourier transform

$$X_M(e^{j\omega}) = \sum_{n=-M}^{M} x[n] e^{-j\omega n} \tag{8.11}$$

exists, is finite, and its magnitude is the energy distribution of the signal over the time interval $[-M, M]$. The power spectral density is defined as

$$P(e^{j\omega}) = \lim_{M \to \infty} \left\{ \frac{1}{2M+1} |X_M(e^{j\omega})|^2 \right\} \tag{8.12}$$

and it represents the distribution of *power* in frequency (and therefore its physical dimensionality is expressed as units of energy over units of time over units of frequency). Obviously, the PSD is a 2π-periodic *real* and *non-negative* function of ω.

It can be shown that the PSD of an N-periodic *stationary* signal $\tilde{s}[n]$ is given by the formula:

$$P(e^{j\omega}) = \sum_{k=0}^{N-1} |\tilde{S}[k]|^2 \, \tilde{\delta}\left(\omega - \frac{2\pi}{N}k\right)$$

where all the $\tilde{S}[k]$ are the N DFS coefficients of $s[n]$; this is rather intuitive since, for a periodic signal, the power is distributed only over the harmonics of the fundamental frequency. Conversely, the PSD of a finite-energy deterministic signal is obviously zero since its power is zero.

8.4.2 PSD of a Stationary Process

For stationary random processes the situation is rather interesting. If we rewrite (8.11) for the WSS random process $X[n]$, the quantity:

$$|X_M(e^{j\omega})|^2 = \left| \sum_{n=-M}^{M} X[n] e^{-j\omega n} \right|^2$$

which we could call a "local energy distribution", is now a random variable itself parameterized by ω. We can therefore consider its mean value and we have

$$\begin{aligned}
\mathrm{E}\left[|X_M(e^{j\omega})|^2\right] &= \mathrm{E}[X_M^*(e^{j\omega}) X_M(e^{j\omega})] \\
&= \mathrm{E}\left[\sum_{n=-M}^{M} X[n] e^{j\omega n} \sum_{m=-M}^{M} X[m] e^{-j\omega m} \right] \\
&= \sum_{n=-M}^{M} \sum_{m=-M}^{M} \mathrm{E}[X[n]X[m]] \, e^{-j\omega(m-n)} \\
&= \sum_{n=-M}^{M} \sum_{m=-M}^{M} r_X[m-n] \, e^{-j\omega(m-n)}
\end{aligned}$$

Now, with the change of variable $k = n - m$ and some simple considerations on the structure of the above sum, we obtain:

$$E\left[\left|X_M(e^{j\omega})\right|^2\right] = \sum_{k=-2M}^{2M} (2M+1-|k|)\, r_X[k]\, e^{-j\omega k}$$

The power spectral density is obtained by plugging the above expression into (8.12), which gives

$$P_X(e^{j\omega}) = \lim_{M\to\infty}\left\{\frac{1}{2M+1} E\left[\left|X_M(e^{j\omega})\right|^2\right]\right\}$$

$$= \lim_{M\to\infty} \sum_{k=-2M}^{2M} \left(1 - \frac{|k|}{2M+1}\right)\left(r_X[k]\, e^{-j\omega k}\right)$$

$$= \lim_{M\to\infty} \sum_{k=-\infty}^{\infty} w_k(M)\, r_X[k]\, e^{-j\omega k} \qquad (8.13)$$

where we have set

$$w_k(M) = \begin{cases} 1 - \dfrac{|k|}{2M+1} & |k| \leq 2M \\ 0 & |k| > 2M \end{cases} \qquad (8.14)$$

Since $\left|w_k(M) r_X[k]\, e^{-j\omega k}\right| \leq \left|r_X[k]\right|$, if the autocorrelation is absolutely summable then the sum (8.13) converges uniformly to a continuous function of M. We can therefore move the limiting operation inside the sum; now the key observation is that the weighting term $w_k(M)$, considered as a function of k parametrized by M, converges in the limit to the constant one (Eq. (8.14)):

$$\lim_{M\to\infty} w_k(M) = 1$$

We finally obtain:

$$P_X(e^{j\omega}) = \sum_{k=-\infty}^{\infty} r_X[k]\, e^{-j\omega k} \qquad (8.15)$$

This fundamental result means that the *power spectral density of a WSS process is the discrete-time Fourier transform of its autocorrelation.* Similarly, we can define the cross-power spectral density between two WSS processes $X[n]$ and $Y[n]$ as

$$P_{XY}(e^{j\omega}) = \sum_{t=-\infty}^{\infty} r_{XY}[k]\, e^{-j\omega k} \qquad (8.16)$$

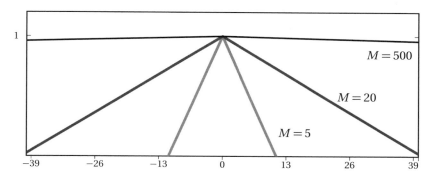

Figure 8.1 Weighting function $w_k(M)$ in (8.14) as a function of k.

8.4.3 White Noise

A WSS random process $W[n]$ whose mean is zero and whose samples are uncorrelated is called *white noise*. The autocorrelation of a white noise process is therefore:

$$r_W[n] = \sigma_W^2 \, \delta[n] \qquad (8.17)$$

where σ_W^2 is the variance (i.e. the expected power) of the process. The power spectral density of a white noise process is simply:

$$P_W(e^{j\omega}) = \sigma_W^2 \qquad (8.18)$$

Please note:

- The probability distribution of a white noise process can be any, provided that it is always zero mean.

- The joint probability distribution of a white noise process need not be i.i.d.; if it is i.i.d., however, then the process is strict-sense stationary and it is also called a strictly white process.

- White noise is an ergodic process, so that its pdf can be estimated from a single realization.

8.5 Stochastic Signal Processing

In stochastic signal processing, we are considering the outcome of a filtering operation which involves a random process; that is, given a linear time-invariant filter with impulse response $h[n]$, we want to describe the output signal as follows:

$$Y[n] = \sum_{k=-\infty}^{\infty} h[k] X[n-k]$$

Note that $Y[n]$ and $X[n]$ denote random variables and are thus capitalized, while $h[n]$ is a deterministic impulse response and is therefore lowercase. In the following, we will assume a stable LTI filter and a wide-sense stationary (WSS) input process.

Time-Domain Analysis. The expected value of the filter's output is

$$m_{Y[n]} = \mathrm{E}[Y[n]] = \mathrm{E}\left[\sum_{k=0}^{\infty} h[k] X[n-k]\right]$$
$$= \sum_{k=-\infty}^{\infty} h[k] \mathrm{E}[X[n-k]]$$
$$= \sum_{k=-\infty}^{\infty} h[k] m_{n-k} \qquad (8.19)$$

where m_n is the mean of $X[n]$. For a WSS input, obviously $\mathrm{E}[X[n]] = m_X$ for all n, and therefore the output has a constant expected value:

$$m_Y = m_X \sum_{k=-\infty}^{\infty} h[k] = m_X H(e^{j0})$$

If the input is WSS, it is fairly easy to show that the output is also WSS; in other words, *LTI filtering preserves wide-sense stationarity*. The autocorrelation of the output process $Y[n]$ depends only on the time difference:

$$R_Y[n, m] = r_Y[n-m]$$

and it can be shown that:

$$r_Y[n] = \sum_{k=-\infty}^{\infty} \sum_{i=-\infty}^{\infty} h[k] h[i] r_Y[n-i+k]$$

or, more concisely,

$$r_Y[n] = h[n] * h[-n] * r_X[n] \qquad (8.20)$$

Similarly, the cross-correlation between input and output is

$$r_{XY}[n] = h[n] * r_X[n] \qquad (8.21)$$

Frequency-Domain Analysis. It is immediately obvious from (8.20) that the power spectral density of the output process $Y[n]$ is

$$P_Y(e^{j\omega}) = |H(e^{j\omega})|^2 P_X(e^{j\omega}) \tag{8.22}$$

where $H(e^{j\omega})$ is, as usual, the frequency response of the filter. Similarly, from (8.21) we obtain

$$P_{XY}(e^{j\omega}) = H(e^{j\omega}) P_X(e^{j\omega}) \tag{8.23}$$

The above result is of particular interest in the practical problem of estimating the characteristics of an unknown filter; this is a particular instance of a *spectral estimation problem.* Indeed, if we inject white noise of known variance σ^2 into an unknown LTI system $\mathcal{H}$, equation (8.23) becomes:

$$P_{XY}(e^{j\omega}) = H(e^{j\omega}) \sigma^2$$

By numerically computing the cross-correlation between input and output, we can therefore derive an estimation of the frequency response of the system.

The total power of a stochastic process $X[n]$ is the variance of the process itself, $\sigma_X^2 = r_X[0]$; from the PSD, this can be obtained by the usual DTFT inversion formula as

$$\sigma_X^2 = \frac{1}{2\pi} \int_{-\pi}^{\pi} P_X(e^{j\omega}) d\omega \tag{8.24}$$

which, for a filtered process, specializes to

$$\sigma_Y^2 = \frac{1}{2\pi} \int_{-\pi}^{\pi} |H(e^{j\omega})|^2 P_X(e^{j\omega}) d\omega \tag{8.25}$$

Examples

Example 8.1: Intuition behind power spectra

The *empirical average* for a random variable X is the simple average:

$$\hat{m}_X = \frac{1}{N} \sum_{i=1}^{N} X_i$$

where $x_1, x_2, \ldots, x_N$ are N independent "trials" (think coin tossing). Similarly, we can obtain an estimation for the variance as

$$\hat{\sigma}_X^2 = \frac{1}{N} \sum_{i=1}^{N} (X_i - \hat{m}_X)^2$$

For an ergodic process, we can easily obtain a similar empirical estimate for the covariance: we replace the N trials with N successive samples of the process so that (assuming zero mean):

$$\hat{r}_X[n] = \frac{1}{N} \sum_{i=1}^{N} X[i]\, X[i-n]$$

The empirical autocorrelation has the form of an inner product between two displaced copies of the same sequence. We saw in Section 5.2 that this represents a measure of self-similarity. In white processes the samples are so independent that for even the smallest displacement ($n=1$) the process is totally "self-dissimilar". Consider now a process with a lowpass power spectrum; this means that the autocorrelation varies slowly with the index and we can deduce that the process possesses a long-range self-similarity, i.e. it is smooth. Similarly, a highpass power spectrum implies a jumping autocorrelation, i.e. a process whose self-similarity varies in time.

Example 8.2: Top secret filters

Stochastic processes are a fundamental tool in *adaptive signal processing*, a more advanced type of signal processing in which the system changes over time to better match the input in the pursuit of a given goal. A typical example is audio coding, for instance, in which the signal is compressed by algorithms which are modified as a function of the type of content. Another prototypical application is *denoising*, in which we try to remove spurious noise from a signal. Since these systems are adaptive, they are best described as a function of a probabilistic model for the input.

In this book stochastic processes will be used mainly as a tool to study the effects of quantization (hence the rather succinct treatment heretofore). We can nonetheless try and "get a taste" of adaptive processing by considering one of the fundamental results in the art of denoising, namely the Wiener filter. This filter was developed by Norbert Wiener during World War II as a tool to smooth out the tracked trajectories of enemy airplanes and aim the antiaircraft guns at the most likely point the target would pass by next. Because of this sensitive application, the theory behind the filter remained classified information until well after the end of the war.

The problem, in its essence, is shown in Figure 8.2 and it begins with a signal corrupted by additive noise:

$$X[n] = S[n] + W[n]$$

both the clean signal $S[n]$ and the noise $W[n]$ are assumed to be zero-mean stationary processes; assume further that they are jointly stationary and in-

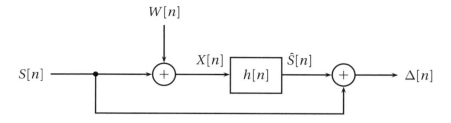

Figure 8.2 Denoising with a Wiener filter.

dependent. We want to design a filter $h[n]$ to clean up the signal as much as possible; if the filter's output is

$$\hat{S}[n] = h[n] * X[n]$$

then the error between the original and the estimation can be expressed as

$$\Delta[n] = S[n] - \hat{S}[n]$$

It can be shown that the minimization of the expected square error corresponds to an *orthogonality condition* between the error and the filter's input:[5]:

$$E[\Delta[n] X[m]] = 0 \qquad (8.26)$$

intuitively this means that anything contained in the error could *not* have been predicted from the input, so the filter is doing the absolute best it can. From the orthogonality condition we can derive the optimal filter; from (8.26), we have

$$E[S[n] X[m]] = E[\hat{S}[n] X[m]]$$

and, by using (8.21),

$$r_{SX}[n] = h[n] * r_X[n] \qquad (8.27)$$

By invoking the independence of signal and noise and their zero mean we have

$$r_{SX}[n] = r_S[n]$$
$$r_X[n] = r_S[n] + r_W[n]$$

so that in the end

$$H(e^{j\omega}) = \frac{P_S(e^{j\omega})}{P_S(e^{j\omega}) + P_W(e^{j\omega})} \qquad (8.28)$$

---

[5] It should be $E[\Delta[n]X^*[n]] = 0$ but we assume $X[n]$ real for simplicity.

the frequency response attenuates the input where the noise is powerful while the signal is not, and it leaves the input almost unchanged otherwise, hence the data-dependence of its expression.

The optimal filter above was derived for an arbitrary infinite two-sided impulse response. Wiener's contribution was mainly concerned with the design of a *causal* response; the derivation is a little complicated and we will not detail it here. A third, interesting design choice is imposing that the impulse response be an N-tap FIR. In this case (8.27) becomes

$$\sum_{k=0}^{N-1} h[k]\, r_X[n-k] = r_S[n]$$

and by picking N successive values for n we can build the system of equations:

$$\begin{bmatrix} r_X[0] & r_X[1] & \ldots & r_X[N-1] \\ r_X[1] & r_X[0] & \ldots & r_X[N-2] \\ r_X[2] & r_X[1] & \ldots & r_X[N-3] \\ \vdots & \vdots & \ddots & \vdots \\ r_X[N-1] & r_X[N-2] & \ldots & r_X[0] \end{bmatrix} \begin{bmatrix} h[0] \\ h[1] \\ h[2] \\ \vdots \\ h[N-1] \end{bmatrix} = \begin{bmatrix} r_S[0] \\ r_S[1] \\ r_S[2] \\ \vdots \\ r_S[N-1] \end{bmatrix}$$

where the Toeplitz nature of the matrix comes from the fact that $r_X[-n] = r_X[n]$. This is a classical Yule-Walker system of equations and it is a fundamental staple of adaptive signal processing.

Further Reading

A good introductory reference on the subject is E. Parzen's classic *Stochastic Processes* (Society for Industrial Mathematics, 1999). For adaptive signal processing, see P. M. Clarkson's *Optimal and Adaptive Signal Processing* (CRC Press, 1993). For an introduction to probability, see the textbook: D. P. Bertsekas and J. N. Tsitsiklis, *Introduction to Probability* (Athena Scientific, 2002). The classic book by A. Papoulis, *Probability, Random Variables, and Stochastic Processes* (McGraw Hill, 2002) still serves as a good, engineering-oriented introduction. A more contemporary treatment of stochastic processes can be found in P. Thiran's excellent class notes for the course "Processus Stochastiques pour les Communications", given at the Swiss Federal Institute of Technology (EPFL) in Lausanne.

Exercises

Exercise 8.1: Filtering a random process – I. Consider a zero-mean white random process $X[n]$ with autocorrelation function $r_x[m] = \sigma^2 \delta[m]$. The process is filtered with a 2-tap FIR filter whose impulse response is $h[0] = h[1] = 1$. Compute the values of the autocorrelation for the output random process $Y[n] = X[n] * h[n]$ from $n = -3$ to $n = 3$.

Exercise 8.2: Filtering a random process – II. Consider a zero-mean white random process $X[n]$ with autocorrelation function $r_x[m] = \sigma^2 \delta[m]$. The process is filtered with leaky integrator $(H(z) = (1-\lambda)/(1-\lambda z^{-1}))$ producing the signal $Y[n] = X[n] * h[n]$. Compute the values of the input-output cross-correlation from $n = -3$ to $n = 3$.

Exercise 8.3: Power spectral density. Consider the stochastic process defined by

$$Y[n] = X[n] + \beta X[n-1]$$

where $\beta \in \mathbb{R}$ and $X[n]$ is a zero-mean wide-sense stationary process with autocorrelation function given by

$$R_X[k] = \sigma^2 \alpha^{|k|}$$

for $|\alpha| < 1$.

(a) Compute the power spectral density $P_Y(e^{j\omega})$ of $Y[n]$.

(b) For which values of β does $Y[n]$ corresponds to a white noise? Explain.

Exercise 8.4: Filtering a sequence of independent random variables. Let $X[n]$ be a real signal modeled as the outcome of a sequence of i.i.d. real random variables that are Gaussian, centered (zero mean), with variance $\sigma_X^2 = 3$. We wish to process this sequence with the following filter:

$$h[1] = \frac{1}{2}, \quad h[2] = \frac{1}{4}, \quad h[3] = \frac{1}{4}, \quad h[n] = 0 \;\; \forall\, n \neq 1, 2, 3$$

Moreover, at the output of the filter, we add a white Gaussian noise $Z[n]$ with unitary variance. The system is shown in the following diagram:

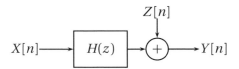

Consider now an input signal of length N.

(a) Write a numerical routine (using Matlab or any other numerical package) to generate $X[1],\ldots,X[N]$.

(b) Write a numerical routine to generate $Z[1],\ldots,Z[N]$.

(c) Write a numerical routine that performs the filtering operation on $X[1],\ldots,X[N]$.

(d) Write a numerical routine to estimate the power spectral density of $Y[1],\ldots,Y[N]$.

(e) Compute the theoretical power spectral density of $Y[1],\ldots,Y[N]$ and compare it with the estimated power spectral density obtained numerically.

Chapter 9

Interpolation and Sampling

Signals (in signal processing) are nothing but mathematical models capturing the essence of a flow of information. Discrete-time signals are the model of choice in two archetypal processing situations: the first, which encompasses the long-established tradition of observing physical phenomena, captures the process of repeatedly measuring the value of a physical quantity at successive instants in time for *analysis* purposes (precipitation levels, stock values, etc.). The second, which is much more recent and dates back to the first digital processors, is the ability to *synthesize* discrete-time signals by means of iterative numerical algorithms (mathematical simulations, computer music, etc.). Discrete-time is the mechanized playground of digital machines.

Continuous-time signals, on the other hand, leverage on a view of the world in which physical phenomena have, potentially, an infinitely small granularity, in the sense that measurements can be arbitrarily dense. In this continuous-time paradigm, real-world phenomena are modeled as *functions of a real variable*; the definition of a signal over the real line allows for infinitely small subdivisions of the function's domain and, therefore, infinitely precise localization of its values. Whether philosophically valid[1] or physically valid,[2] the continuous-time paradigm is an indispensable model in the analysis of analog signal processing systems.

We will now study the mathematical description of the (porous) interface between continuous-time and discrete time. The tools that we will introduce, will allow us to cross this boundary, back and forth, with little or no loss of information for the signals involved.

---

[1] Remember Zeno's paradoxes...

[2] The shortest unit of time at which the usual laws of gravitational physics still hold is called *Planck time* and is estimated at 10^{-43} seconds. Apparently, therefore, the universe works in discrete-time...

9.1 Preliminaries and Notation

Interpolation. Interpolation comes into play when discrete-time signals need to be converted to continuous-time signals. The need arises at the interface between the digital world and the analog world; as an example, consider a discrete-time waveform synthesizer which is used to drive an analog amplifier and loudspeaker. In this case, it is useful to express the input to the amplifier as a function of a real variable, defined over the entire real line; this is because the behavior of analog circuitry is best modeled by continuous-time functions. We will see that at the core of the interpolation process is the association of a physical time duration T_s to the intervals between samples of the discrete-time sequence. The fundamental questions concerning interpolation involve the spectral properties of the interpolated function with respect to those of the original sequence.

Sampling. Sampling is the method by which an underlying continuous-time phenomenon is "reduced" to a discrete-time sequence. The simplest sampling system just records the value of a physical variable at repeated instants in time and associates the value to a point in a discrete-time sequence; in the following, we refer to this scheme as the "naive" sampling operator. Other sampling methods exist (and we will see the most important one) but, in all cases, a correspondence is established between time instants in continuous time and points in the discrete-time sequence. In the following, we only consider *uniform sampling*, in which the time instants are uniformly spaced T_s seconds apart. T_s is called the *sampling period* and its inverse, F_s is called the *sampling frequency* of a sampling system. The fundamental question of sampling is whether any information is lost in the sampling process. If the answer is in the negative (at least for a given class of signals), this means that all the processing tools developed in the discrete-time domain can be applied to continuous-time signals as well, after sampling.

Table 9.1 Notation used in the Chapter.

Name	Description	Units	Relations
T_s	Sampling period	seconds	$T_s = 1/F_s$
F_s	Sampling frequency	hertz	$F_s = 1/T_s$
Ω_s	Sampling frequency (angular)	rad/sec	$\Omega_s = 2\pi F_s = 2\pi/T_s$
Ω_N	Nyquist frequency (angular)	rad/sec	$\Omega_N = \Omega_s/2 = \pi/T_s$

Notation. In the rest of this Chapter we will encounter a series of variables which are all interrelated and whose different forms will be used interchangeably according to convenience. They are summarized as a quick reference in Table 9.1.

9.2 Continuous-Time Signals

Interpolation and sampling constitute the bridges between the discrete- and continuous-time worlds. Before we proceed to the core of the matter, it is useful to take a quick tour of the main properties of continuous-time signals, which we simply state here without formal proofs.

Continuous-time signals are modeled by complex functions of a real variable t which usually represents time (in seconds) but which can represent other physical coordinates of interest. For maximum generality, no special requirement is imposed on functions modeling signals; just as in the discrete-time case, the functions can be periodic or aperiodic, or they can have a finite support (in the sense that they are nonzero over a finite interval only). A common condition, on an aperiodic signal, is that its modeling function be square integrable; this corresponds to the reasonable requirement that the signal have finite energy.

Inner Product and Convolution. We have already encountered some examples of continuous-time signals in conjunction with Hilbert spaces; in Section 3.2.2, for instance, we introduced the space of square integrable functions over an interval and we will shortly introduce the space of bandlimited signals. For inner product spaces, whose elements are functions on the real line, we use the following inner product definition:

$$\langle f(t), g(t) \rangle = \int_{-\infty}^{\infty} f^*(t) g(t) \, dt \qquad (9.1)$$

The *convolution* of two real continuous-time signals is defined as usual from the definition of the inner product; in particular:

$$(f * g)(t) = \langle f(t - \tau), g(\tau) \rangle \qquad (9.2)$$

$$= \int_{-\infty}^{\infty} f(t - \tau) g(\tau) \, d\tau \qquad (9.3)$$

The convolution operator, in continuous time, is linear and time invariant, as can be easily verified. Note that, in discrete-time, convolution represents the operation of filtering a signal with a continuous-time LTI filter, whose impulse response is of course a continuous-time function.

Frequency-Domain Representation of Continuous-Time Signals.

The Fourier transform of a continuous-time signal $x(t)$ and its inversion formula are defined as[3]

$$X(j\Omega) = \int_{-\infty}^{\infty} x(t) e^{-j\Omega t} \, dt \tag{9.4}$$

$$x(t) = \frac{1}{2\pi} \int_{-\infty}^{\infty} X(j\Omega) e^{j\Omega t} \, d\Omega \tag{9.5}$$

The convergence of the above integrals is assured for functions which satisfy the so-called Dirichlet conditions. In particular, the FT is always well defined for square integrable (finite energy), continuous-time signals. The Fourier transform in continuous time is a linear operator; for a list of its properties, which mirror those that we saw for the DTFT, we refer to the bibliography. It suffices here to recall the conservation of energy, also known as Parseval's theorem:

$$\int_{-\infty}^{\infty} |x(t)|^2 \, dt = \frac{1}{2\pi} \int_{-\infty}^{\infty} |X(j\Omega)|^2 \, d\Omega$$

The FT representation can be formally extended to signals which are not square summable by means of the Dirac delta notation as we saw in Section 4.4.2. In particular we have

$$\text{FT}\{e^{j\Omega_0 t}\} = \delta(\Omega - \Omega_0) \tag{9.6}$$

from which the Fourier transforms of sine, cosine, and constant functions can easily be derived. Please note that, in continuous-time, the FT of a complex sinusoid is *not* a train of impulses but just a single impulse.

The Convolution Theorem.

The convolution theorem for continuous-time signal exactly mirrors the theorem in Section 5.4.2; it states that if $h(t) = (f * g)(t)$ then the Fourier transforms of the three signals are related by $H(j\Omega) = F(j\Omega)G(j\Omega)$. In particular we can use the convolution theorem to compute

$$(f * g)(t) = \frac{1}{2\pi} \int_{-\infty}^{\infty} F(j\Omega) G(j\Omega) e^{j\Omega t} \, d\Omega \tag{9.7}$$

---

[3] The notation $X(j\Omega)$ mirrors the specialized notation that we used for the DTFT; in this case, by writing $X(j\Omega)$ we indicate that the Fourier transform is just the (two-sided) Laplace transform $X(s) = \int x(t) e^{-st} \, dt$ computed on the imaginary axis.

9.3 Bandlimited Signals

A signal whose Fourier transform is nonzero only, over a finite frequency interval, is called *bandlimited*. In other words, the signal $x(t)$ is bandlimited if there exists a frequency Ω_N such that:[4]

$$X(j\Omega) = 0 \qquad \text{for } |\Omega| \geq \Omega_N$$

Such a signal will be called Ω_N-bandlimited and Ω_N is often called the *Nyquist frequency*. It may be useful to mention that, symmetrically, a continuous-time signal which is nonzero, over a finite time interval only, is called a *time-limited* signal (or finite-support signal). A fundamental theorem states that a bandlimited signal cannot be time-limited, and vice versa. While this can be proved formally and quite easily, here we simply give the intuition behind the statement. The time-scaling property of the Fourier transform states that:

$$\text{FT}\{f(at)\} = \frac{1}{a} F\left(j\frac{\Omega}{a}\right) \qquad (9.8)$$

so that the more "compact" in time a signal is, the wider its frequency support becomes.

The Sinc Function. Let us now consider a prototypical Ω_N-bandlimited signal $\varphi(t)$ whose Fourier transform is a real *constant* over the interval $[-\Omega_N, \Omega_N]$ and zero everywhere else. If we define the rect function as follows (see also Section 5.6):

$$\text{rect}(x) = \begin{cases} 1 & |x| \leq 1/2 \\ 0 & |x| > 1/2 \end{cases}$$

we can express the Fourier transform of the prototypical Ω_N-bandlimited signal as

$$\Phi(j\Omega) = \frac{\pi}{\Omega_s} \text{rect}\left(\frac{\Omega}{2\Omega_N}\right) \qquad (9.9)$$

where the leading factor is just a normalization term. The time-domain expression for the signal is easily obtained from the inverse Fourier transform as

$$\varphi(t) = \frac{\sin \Omega_N t}{\Omega_N t} = \text{sinc}\left(\frac{t}{T_s}\right) \qquad (9.10)$$

---

[4] The use of $\geq$ instead of $>$ is a technicality which will be useful in conjunction with the sampling theorem.

where we have used $T_s = \pi/\Omega_N$ and defined the sinc function as

$$\text{sinc}(x) = \begin{cases} \dfrac{\sin(\pi x)}{\pi x} & x \neq 0 \\ 1 & x = 0 \end{cases}$$

The sinc function is plotted in Figure 9.6. Note the following:

- The function is symmetric, $\text{sinc}(x) = \text{sinc}(-x)$.

- The sinc function is zero for all integer values of its argument, except in zero. This feature is called the *interpolation property* of the sinc, as we will shortly see more in detail.

- The sinc function is square integrable (it has finite energy) but it is not absolutely integrable (hence the discontinuity of its Fourier transform).

- The decay is slow, asymptotic to $1/x$.

- The scaled sinc function represents the impulse response of an ideal, continuous-time lowpass filter with cutoff frequency Ω_N.

9.4 Interpolation

Interpolation is a procedure whereby we convert a discrete-time sequence $x[n]$ to a continuous-time function $x(t)$. Since this can be done in an arbitrary number of ways, we have to start by formulating some requirements on the resulting signal. At the heart of the interpolating procedure, as we have mentioned, is the association of a physical time duration T_s to the interval between the samples in the discrete-time sequence. An intuitive requirement on the interpolated function is that its values at multiples of T_s be equal to the corresponding points of the discrete-time sequence, i.e.

$$x(t)\Big|_{t=nT_s} = x[n]$$

The interpolation problem now reduces to "filling the gaps" between these instants.

9.4.1 Local Interpolation

The simplest interpolation schemes create a continuous-time function $x(t)$ from a discrete-time sequence $x[n]$, by setting $x(t)$ to be equal to $x[n]$ for $t = nT_s$ and by setting $x(t)$ to be some linear combination of neighboring sequence values when t lies in between interpolation instants. In general, the local interpolation schemes can be expressed by the following formula:

$$x(t) = \sum_{n=-\infty}^{\infty} x[n] I\left(\frac{t - nT_s}{T_s}\right) \tag{9.11}$$

where $I(t)$ is called the interpolation function (for linear functions the notation $I_N(t)$ is used and the subscript N indicates how many discrete-time samples, besides the current one, enter into the computation of the interpolated values for $x(t)$). The interpolation function must satisfy the fundamental *interpolation properties*:

$$\begin{cases} I(0) = 1 \\ I(k) = 0 \quad \text{for } k \in \mathbb{Z} \setminus \{0\} \end{cases} \tag{9.12}$$

where the second requirement implies that, no matter what the support of $I(t)$ is, its values should not affect other interpolation instants. By changing the function $I(t)$, we can change the type of interpolation and the properties of the interpolated signal $x(t)$.

Note that (9.11) can be interpreted either simply as a linear combination of shifted interpolation functions or, more interestingly, as a "mixed domain" convolution product, where we are convolving a discrete-time signal $x[n]$ with a continuous-time "impulse response" $I(t)$ scaled in time by the interpolation period T_s.

Zero-Order Hold. The simplest approach for the interpolating function is the piecewise-constant interpolation; here the continuous-time signal is kept constant between discrete sample values, yielding

$$x(t) = x[n], \quad \text{for } \left(n - \frac{1}{2}\right) T_s \leq t < \left(n + \frac{1}{2}\right) T_s$$

and an example is shown in Figure 9.1; it is apparent that the resulting function is far from smooth since the interpolated function is discontinuous. The interpolation function is simply:

$$I_0(t) = \text{rect}(t)$$

and the values of $x(t)$ depend only on the current discrete-time sample value.

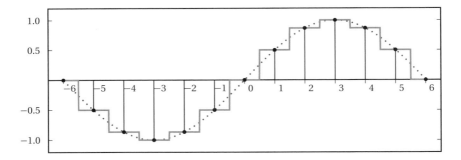

Figure 9.1 Interpolation of a discrete-time sinusoid with the zero-order hold. Note the discontinuities introduced by this simple scheme.

First-Order Hold. A linear interpolator (sometimes called a first-order hold) simply connects the points corresponding to the samples with straight lines. An example is shown in Figure 9.2; note that now $x(t)$ depends on two consecutive discrete-time samples, across which a connecting straight line is drawn. From the point of view of smoothness, this interpolator already represents an improvement over the zero-order hold: indeed the interpolated function is now continuous, although its first derivative is not. The first-order hold can be expressed in the same notation as in (9.11) by defining the following triangular function:

$$I_1(t) = \begin{cases} 1 - |t| & \text{if } |t| < 1 \\ 0 & \text{otherwise} \end{cases}$$

which is shown in Figure 9.3.[5] It is immediately verifiable that $I_1(t)$ satisfies the interpolation properties (9.12).

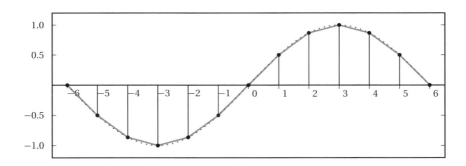

Figure 9.2 Interpolation of a discrete-time sinusoid with the first-order hold. Note that the first derivative is discontinuous.

---

[5] Note that $I_1(t) = (I_0 * I_0)(t)$.

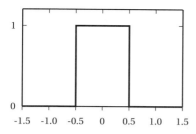 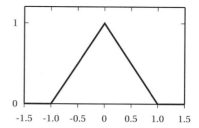

Figure 9.3 Interpolation functions for the zero-order (left) and first-order interpolators (right).

Higher-Order Interpolators. The zero- and first-order interpolators are widely used in practical circuits due to their extreme simplicity. Note that the interpolating functions $I_0(t)$ and $I_1(t)$ are alsol knows as the B-spline functions of order zero and one respectively. These schemes can be extended to higher order interpolation functions and, in general, $I_N(t)$ is a N-th order polynomial in t. The advantage of the local interpolation schemes is that, for small N, they can easily be implemented in practice as *causal* interpolation schemes (locality is akin to FIR filtering); their disadvantage is that, because of the locality, their N-th derivative is discontinuous. This discontinuity represents a lack of smoothness in the interpolated function; from a spectral point of view this corresponds to a high frequency energy content, which is usually undesirable.

9.4.2 Polynomial Interpolation

The lack of smoothness of local interpolations is easily eliminated when we need to interpolate just a *finite* number of discrete-time samples. In fact, in this case the task becomes a classic polynomial interpolation problem for which the optimal solution has been known for a long time under the name of *Lagrange interpolation*. Note that a polynomial interpolating a finite set of samples is a maximally smooth function in the sense that it is continuous, together with all its derivatives.

Consider a length $(2N+1)$ discrete-time signal $x[n]$, with $n = -N, \ldots, N$. Associate to each sample an abscissa $t_n = nT_s$; we know from basic algebra that there is one and only one polynomial $P(t)$ of degree $2N$ which passes through all the $2N+1$ pairs $(t_n, x[n])$ and this polynomial is the Lagrange interpolator. The coefficients of the polynomial could be found by solving the set of $2N+1$ equations:

$$\{P(t_n) = x[n]\}_{n=-N,\ldots,N} \tag{9.13}$$

but a simpler way to determine the expression for $P(t)$ is to use the set of $2N+1$ *Lagrange polynomials* of degree $2N$:

$$L_n^{(N)}(t) = \prod_{\substack{k=-N \\ k \neq n}}^{N} \frac{(t-t_k)}{(t_n-t_k)}$$

$$= \prod_{\substack{k=-N \\ k \neq n}}^{N} \frac{t/T_s - k}{n-k}, \qquad n = -N, \ldots, N \qquad (9.14)$$

The polynomials $L_n^{(N)}(t)$ for $T_s = 1$ and $N = 2$ (i.e. interpolation of 5 points) are plotted in Figure 9.4. By using this notation, the *global* Lagrange interpolator for a given set of abscissa/ordinate pairs can now be written as a simple linear combination of Lagrange polynomials:

$$P(t) = \sum_{n=-N}^{N} x[n] L_n^{(N)}(t) \qquad (9.15)$$

and it is easy to verify that this is the unique interpolating polynomial of degree $2N$ in the sense of (9.13). Note that *each* of the $L_n^{(N)}(t)$ satisfies the interpolation properties (9.12) or, concisely (for $T_s = 1$):

$$L_n^{(N)}(m) = \delta[n-m]$$

The interpolation formula, however, cannot be written in the form of (9.11) since the Lagrange polynomials are not simply shifts of a single prototype function. The continuous time signal $x(t) = P(t)$ is now a *global* interpo-

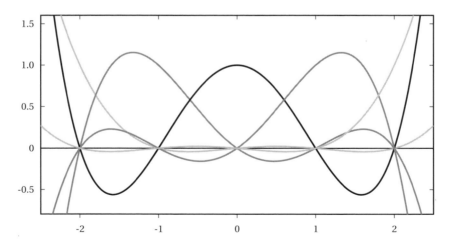

Figure 9.4 Lagrange interpolation polynomials $L_n^{(2)}(t)$ for $T = 1$ and $n = -2, \ldots, 2$. Note that $L_n^{(N)}(t)$ is zero for t integer except for $t = n$, where it is 1.

lating function for the finite-length discrete-time signal $x[n]$, in the sense that it depends on *all* samples in the signal; as a consequence, $x(t)$ is maximally smooth $(x(t) \in C^\infty)$. An example of Lagrange interpolation for $N = 2$ is plotted in Figure 9.5.

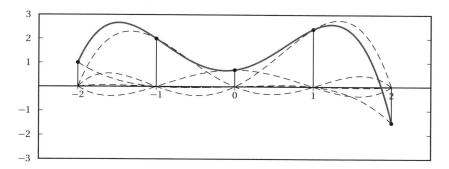

Figure 9.5 Maximally smooth Lagrange interpolation of a length-5 signal.

9.4.3 Sinc Interpolation

The beauty of local interpolation schemes lies in the fact that the interpolated function is simply a linear combination of shifted versions of the *same* prototype interpolation function $I(t)$; unfortunately, this has the disadvantage of creating a continuous-time function which lacks smoothness. Polynomial interpolation, on the other hand, is perfectly smooth but it only works in the finite-length case and it requires different interpolation functions with different signal lengths. Yet, both approaches can come together in a convenient mathematical way and we are now ready to introduce the maximally smooth interpolation scheme for infinite discrete-time signals.

Let us take the expression for the Lagrange polynomial of degree N in (9.14) and consider its limit for N going to infinity. We have

$$\lim_{N \to \infty} L_n^{(N)}(t) = \prod_{\substack{k=-\infty \\ k \neq n}}^{\infty} \frac{t/T_s - k}{n - k} = \prod_{\substack{m=-\infty \\ m \neq 0}}^{\infty} \frac{t/T_s - n + m}{m}$$

$$= \prod_{\substack{m=-\infty \\ m \neq 0}}^{\infty} \left(1 + \frac{t/T_s - n}{m}\right)$$

$$= \prod_{m=1}^{\infty} \left(1 - \left(\frac{t/T_s - n}{m}\right)^2\right) \quad (9.16)$$

Here, we have used the change of variable $m = n - k$. We can now invoke Euler's infinite product expansion for the sine function

$$\sin(\pi\tau) = (\pi\tau)\prod_{k=1}^{\infty}\left(1 - \frac{\tau^2}{k^2}\right)$$

(whose derivation is in the appendix) to finally obtain

$$\lim_{N\to\infty} L_n^{(N)}(t) = \text{sinc}\left(\frac{t - nT_s}{T_s}\right) \tag{9.17}$$

The convergence of the Lagrange polynomial $L_0^{(N)}(t)$ to the sinc function is illustrated in Figure 9.6. Note that, now, as the number of points becomes infinite, the Lagrange polynomials converge to shifts of the *same* prototype function, i.e. the sinc; therefore, the interpolation formula can be expressed as in (9.11) with $I(t) = \text{sinc}(t)$; indeed, if we consider an infinite sequence $x[n]$ and apply the Lagrange interpolation formula (9.15), we obtain:

$$x(t) = \sum_{n=-\infty}^{\infty} x[n]\,\text{sinc}\left(\frac{t - nT_s}{T_s}\right) \tag{9.18}$$

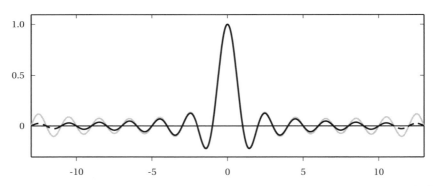

Figure 9.6 A portion of the sinc function and its Lagrange approximation $L_0^{(100)}(t)$ (light gray).

Spectral Properties of the Sinc Interpolation. The sinc interpolation of a discrete-time sequence gives rise to a strictly bandlimited continuous-time function. If the DTFT $X(e^{j\omega})$ of the discrete-time sequence exists, the spectrum of the interpolated function $X(j\Omega)$ can be obtained as follows:

$$X(j\Omega) = \int_{-\infty}^{\infty} \sum_{n=-\infty}^{\infty} x[n]\,\text{sinc}\left(\frac{t - nT_s}{T_s}\right) e^{-j\Omega t}\,dt$$

$$= \sum_{n=-\infty}^{\infty} x[n] \int_{-\infty}^{\infty} \text{sinc}\left(\frac{t - nT_s}{T_s}\right) e^{-j\Omega t}\,dt$$

Now we use (9.9) to obtain the Fourier Transform of the scaled and shifted sinc:

$$X(j\Omega) = \sum_{n=-\infty}^{\infty} x[n] \left(\frac{\pi}{\Omega_N}\right) \text{rect}\left(\frac{\Omega}{2\Omega_N}\right) e^{-jnT_s\Omega}$$

and use the fact that, as usual, $T_s = \pi/\Omega_N$:

$$X(j\Omega) = \left(\frac{\pi}{\Omega_N}\right) \text{rect}\left(\frac{\Omega}{2\Omega_N}\right) \sum_{n=-\infty}^{\infty} x[n] e^{-j\pi(\Omega/\Omega_N)n}$$

$$= \begin{cases} \dfrac{\pi}{\Omega_N} X(e^{j\pi\Omega/\Omega_N}) & \text{for } |\Omega| \leq \Omega_N \\ 0 & \text{otherwise} \end{cases}$$

In other words, the continuous-time spectrum is just a scaled and stretched version of the DTFT of the discrete-time sequence between $-\pi$ and π. The duration of the interpolation interval T_s is inversely proportional to the resulting bandwidth of the interpolated signal. Intuitively, a slow interpolation (T_s large) results in a spectrum concentrated around the low frequencies; conversely, a fast interpolation (T_s small) results in a spread-out spectrum (more high frequencies are present).[6]

9.5 The Sampling Theorem

We have seen in the previous Section that the "natural" polynomial interpolation scheme leads to the so-called sinc interpolation for infinite discrete time sequences. Another way to look at the previous result is that any square summable discrete-time signal can be interpolated into a continuous-time signal which is smooth in time and strictly bandlimited in frequency. This suggests that the class of bandlimited functions must play a special role in bridging the gap between discrete and continuous time and this deserves further investigation. In particular, since any discrete-time signal can be interpolated exactly into a bandlimited function, we now ask ourselves whether the converse is true: can any bandlimited signal be transformed into a discrete-time signal with no loss of information?

---

[6] To find a simple everyday (yesterday's?) analogy, think of a 45 rpm vinyl record played at either 33 rpm (slow interpolation) or at 78 rpm (fast interpolation) and remember the acoustic effect on the sounds.

The Space of Bandlimited Signals. The class of Ω_N-bandlimited functions of finite energy forms a Hilbert space, with the inner product defined in (9.1). An orthogonal basis for the space of Ω_N-bandlimited functions can easily be obtained from the prototypical bandlimited function, the sinc; indeed, consider the family:

$$\varphi^{(n)}(t) = \text{sinc}\left(\frac{t - nT_s}{T_s}\right), \qquad n \in \mathbb{Z} \tag{9.19}$$

where, once again, $T_s = \pi/\Omega_N$. Note that we have $\varphi^{(n)}(t) = \varphi^{(0)}(t - nT_s)$ so that each basis function is simply a translated version of the prototype basis function $\varphi^{(0)}$. Orthogonality can easily be proved as follows: first of all, because of the symmetry of the sinc function and the time-invariance of the convolution, we can write

$$\begin{aligned}\langle \varphi^{(n)}(t), \varphi^{(m)}(t)\rangle &= \langle \varphi^{(0)}(t - nT_s), \varphi^{(0)}(t - mT_s)\rangle \\ &= \langle \varphi^{(0)}(nT_s - t), \varphi^{(0)}(mT_s - t)\rangle \\ &= (\varphi^{(0)} * \varphi^{(0)})((n - m)T_s)\end{aligned}$$

We can now apply the convolution theorem and (9.9) to obtain:

$$\begin{aligned}\langle \varphi^{(n)}(t), \varphi^{(m)}(t)\rangle &= \frac{1}{2\pi}\int_{-\infty}^{\infty}\left(\frac{\pi}{\Omega_N}\text{rect}\left(\frac{\Omega}{\Omega_N}\right)\right)^2 e^{j\Omega(n-m)T_s}\,d\Omega \\ &= \frac{\pi}{2\Omega_N^2}\int_{-\Omega_N}^{\Omega_N} e^{j\Omega(n-m)T_s}\,d\Omega \\ &= \begin{cases}\dfrac{\pi}{\Omega_N} = T_s & \text{if } n = m \\ 0 & \text{if } n \neq m\end{cases}\end{aligned}$$

so that $\{\varphi^{(n)}(t)\}_{n\in\mathbb{Z}}$ is orthogonal with normalization factor Ω_N/π (or, equivalently, $1/T_s$).

In order to show that the space of Ω_N-bandlimited functions is indeed a Hilbert space, we should also prove that the space is complete. This is a more delicate notion to show[7] and here it will simply be assumed.

---

[7] Completeness of the sinc basis can be proven as a consequence of the completeness of the Fourier series in the continuous-time domain.

Sampling as a Basis Expansion. Now that we have an orthogonal basis, we can compute coefficients in the basis expansion of an arbitrary Ω_N-bandlimited function $x(t)$. We have

$$\langle \varphi^{(n)}(t), x(t) \rangle = \langle \varphi^{(0)}(t - nT_s), x(t) \rangle \qquad (9.20)$$

$$= (\varphi^{(0)} * x)(nT_s) \qquad (9.21)$$

$$= \frac{1}{2\pi} \int_{-\infty}^{\infty} \frac{\pi}{\Omega_N} \operatorname{rect}\left(\frac{\Omega}{\Omega_N}\right) X(j\Omega) e^{j\Omega nT_s} d\Omega \qquad (9.22)$$

$$= \frac{\pi}{\Omega_N} \frac{1}{2\pi} \int_{-\Omega_N}^{\Omega_N} X(j\Omega) e^{j\Omega nT_s} d\Omega \qquad (9.23)$$

$$= T_s x(nT_s) \qquad (9.24)$$

In the derivation, firstly we have rewritten the inner product as a convolution operation, after which we have applied the convolution theorem, and recognized the penultimate line as simply the inverse FT of $X(j\Omega)$ calculated in $t = nT_s$. We therefore have the remarkable result that the n-th basis expansion coefficient is *proportional to the sampled value of $x(t)$ at $t = nT_s$*. For this reason, the sinc basis expansion is also called *sinc sampling*.

Reconstruction of $x(t)$ from its projections can now be achieved via the orthonormal basis reconstruction formula (3.40); since the sinc basis is just orthogonal, rather than orthonormal, (3.40) needs to take into account the normalization factor and we have

$$x(t) = \frac{1}{T_s} \sum_{n=-\infty}^{\infty} \langle \varphi^{(n)}(t), x(t) \rangle \varphi^{(n)}(t)$$

$$= \sum_{n=-\infty}^{\infty} x(nT_s) \operatorname{sinc}\left(\frac{t - nT_s}{T_s}\right) \qquad (9.25)$$

which corresponds to the interpolation formula (9.18).

The Sampling: Theorem. *If $x(t)$ is a Ω_N-bandlimited continuous-time signal, a sufficient representation of $x(t)$ is given by the discrete-time signal $x[n] = x(nT_s)$, with $T_s = \pi/\Omega_N$. The continuous time signal $x(t)$ can be exactly reconstructed from the discrete-time signal $x[n]$ as*

$$x(t) = \sum_{n=-\infty}^{\infty} x[n] \operatorname{sinc}\left(\frac{t - nT_s}{T_s}\right)$$

The proof of the theorem is inherent to the properties of the Hilbert space of bandlimited functions, and is trivial once having proved the existence of an orthogonal basis. Now, call Ω_{max} the largest nonzero frequency of a bandlimited signal.[8] Such a signal is obviously Ω_N-bandlimited for all $\Omega_N > \Omega_{max}$. Therefore, a bandlimited signal $x(t)$ is uniquely represented by all sequences $x[n] = x(nT)$ for which $T \leq T_s = \pi/\Omega_{max}$; T_s is the largest sampling period which guarantees perfect reconstruction (i.e. we cannot take fewer than $1/T_s$ samples per second to perfectly capture the signal; we will see in the next Section what happens if we do). Another way to state the above point is to say that the *minimum* sampling frequency Ω_s for perfect reconstruction is exactly twice the signal's maximum frequency, or that the Nyquist frequency must coincide to the highest frequency of the bandlimited signal; the sampling frequency Ω must therefore satisfy the following relationship:

$$\Omega \geq \Omega_s = 2\Omega_{max}$$

or, in hertz,

$$F \geq F_s = 2F_{max}$$

9.6 Aliasing

The "naive" notion of sampling, as we have seen, is associated to the very practical idea of measuring the instantaneous value of a continuous-time signal at uniformly spaced instants in time. For bandlimited signals, we have seen that this is actually equivalent to an orthogonal decomposition in the space of bandlimited functions, which guarantees that the set of samples $x(nT_s)$ uniquely determines the signal and allows its perfect reconstruction. We now want to address the following question: what happens if we simply sample an *arbitrary* continuous time signal in the "naive" sense (i.e. in the sense of simply taking $x[n] = x(nT_s)$) and what can we reconstruct from the set of samples thus obtained?

9.6.1 Non-Bandlimited Signals

Given a sampling period of T_s seconds, the sampling theorem ensures that there is no loss of information by sampling the class of Ω_N-bandlimited signals, where as usual $\Omega_N = \pi/T_s$. If a signal $x(t)$ is not Ω_N-bandlimited (i.e. its spectrum is nonzero at least somewhere outside of $[-\Omega_N, \Omega_N]$) then the

---

[8] For real signals, whose spectrum is symmetric, $X(j\Omega) = 0$ for $|\Omega| > \Omega_{max}$ so that Ω_{max} is the largest *positive* nonzero frequency. For complex signals, Ω_{max} is the largest nonzero frequency in magnitude.

approximation properties of orthogonal bases state that its *best* approximation in terms of uniform samples T_s seconds apart is given by the samples *of its projection over the space of Ω_N-bandlimited signals* (Sect. 3.3.2). This is easily seen in (9.23), where the projection is easily recognizable as an ideal lowpass filtering operation on $x(t)$ (with gain T_s) which truncates its spectrum outside of the $[-\Omega_N, \Omega_N]$ interval.

Sampling as the result of a sinc basis expansion automatically includes this lowpass filtering operation; for a Ω_N-bandlimited signal, obviously, the filtering is just a scaling by T_s. For an arbitrary signal, however, we can now decompose the sinc sampling as in Figure 9.7, where the first block is a continuous-time lowpass filter with cutoff frequency Ω_N and gain $T_s = \pi/\Omega_N$. The discrete time sequence $x[n]$ thus obtained is the best discrete-time approximation of the original signal when the sampling is uniform.

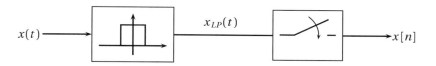

Figure 9.7 Bandlimited sampling (sinc basis expansion) as a combination of low-pass filtering (in the continuous-time domain) and sampling; $x_{LP}(t)$ is the projection of $x(t)$ over the space of Ω_N-bandlimited functions.

9.6.2 Aliasing: Intuition

Now let us go back to the naive sampling scheme in which simply $x[n] = x(nT_s)$, with $F_s = 1/T_s$, the sampling frequency of the system; what is the error we incur if $x(t)$ is not bandlimited or if the sampling frequency is less than twice the maximum frequency? We can develop the intuition by starting with the simple case of a single sinusoid before moving on to a formal demonstration of the aliasing phenomenon.

Sampling of Sinusoids. Consider a simple continuous-time signal[9] such as $x(t) = e^{j2\pi f_0 t}$ and its sampled version $x[n] = e^{j2\pi (f_0/F_s) n} = e^{j\omega_0 n}$ with

$$\omega_0 = 2\pi \frac{f_0}{F_s} \tag{9.26}$$

Clearly, since $x(t)$ contains only one frequency, it is Ω-bandlimited for all $\Omega > 2\pi |f_0|$. If the frequency of the sinusoid satisfies $|f_0| < F_s/2 = F_N$, then

---

[9] In the following examples we will express the frequencies of sinusoids in hertz, both out of practicality and to give an example of a different form of notation.

$\omega_0 \in (-\pi, \pi)$ and the frequency of the original sinusoid can be univocally determined from the sampled signal. Now assume that $f_0 = F_N = F_s/2$; we have

$$x[n] = e^{j\pi n} = e^{-j\pi n}$$

In other words, we encounter a first ambiguity with respect to the direction of rotation of the complex exponential: from the sampled signal we cannot determine whether the original frequency was $f_0 = F_N$ or $f_0 = -F_N$. If we increase the frequency further, say $f_0 = (1+\alpha)F_N$, we have

$$x[n] = e^{j(1+\alpha)\pi n} = e^{-j\alpha\pi n}$$

Now the ambiguity is both on the direction and on the frequency value: if we try to infer the original frequency from the sampled sinusoid from (9.26), we cannot discriminate between $f_0 = (1+\alpha)F_N$ or $f_0 = -\alpha F_N$. Matters get even worse if $f_0 > F_s$. Suppose we can write $f_0 = F_s + f_b$ with $f_b < F_s/2$; we have

$$x[n] = e^{j(2\pi F_s T_s + 2\pi f_b T_s)n} = e^{j(2\pi + \omega_b)n} = e^{j\omega_b n}$$

so that the sinusoid is completely indistinguishable from a sinusoid of frequency f_b sampled at F_s; the fact that two continuous-time frequencies are mapped to the same discrete-time frequency is called *aliasing*. An example of aliasing is depicted in Figure 9.8.

In general, because of the 2π-periodicity of the discrete-time complex exponential, we can always write

$$\omega_b = (2\pi f_0 T_s) + 2k\pi$$

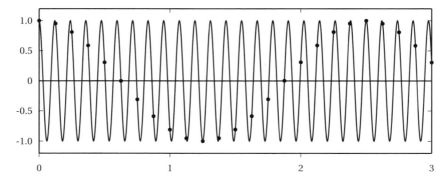

Figure 9.8 Example of aliasing: a sinusoid at 8400 Hz, $x(t) = \cos(2\pi \cdot 8400t)$ (solid line) is sampled at $F_s = 8000$ Hz. The sampled values (dots) are indistinguishable from those of at 400 Hz sinusoid sampled at F_s.

and choose $k \in \mathbb{Z}$ so that ω_b falls in the $[-\pi, \pi]$ interval. Seen the other way, all continuous-time frequencies of the form

$$f = f_b + kF_s$$

with $f_b < F_N$ are aliased to the same discrete-time frequency ω_b.

Consider now the signal $y(t) = Ae^{j2\pi f_b t} + Be^{j2\pi(f_b+F_s)t}$, with $f_b < F_N$. If we sample this signal with sampling frequency F_s we obtain:

$$\begin{aligned} x[n] &= A\,e^{j2\pi(f_b/F_s)n} + B\,e^{j2\pi(f_b/F_s+1)nT_s} \\ &= A\,e^{j\omega_b n} + B\,e^{j\omega_b n}e^{j2\pi n} \\ &= (A+B)\,e^{j\omega_b n} \end{aligned}$$

In other words, two continuous-time exponentials which are F_s Hz apart, give rise to a single discrete-time complex exponential, whose amplitude is equal to the sum of the amplitudes of both the original sinusoids.

Energy Folding of the Fourier Transform. To understand what happens to a general signal, consider the interpretation of the Fourier transform as a bank of (infinitely many) complex oscillators initialized with phase and amplitude, each contributing to the energy content of the signal at their respective frequency. Since, in the sampled version, any two frequencies F_s apart are indistinguishable, their contributions to the discrete-time Fourier transform of the sampled signal add up. This aliasing can be represented as a spectral *superposition*: the continuous-time spectrum above F_N is cut, shifted back to $-F_N$, summed over $[-F_N, F_N]$, and the process is repeated again and again; the same applies for the spectrum below $-F_N$. This process is nothing but the familiar periodization of a signal:

$$\sum_{k=-\infty}^{\infty} X(j2\pi f + j2k\pi F_s)$$

as we will prove formally in the next Section.

9.6.3 Aliasing: Proof

In the following, we consider the relationship between the DTFT of a sampled signal $x[n]$ and the FT of the originating continuous-time signal $x_c(t)$. For clarity, we add the subscript "c" to all continuous-time quantities so that, for instance, we write $x[n] = x_c(nT_s)$. Moreover, we use the usual tilde notation for periodic functions.

Consider $X(e^{j\omega})$, the DTFT of the sampled sequence (with, as usual, $T_s = (1/F_s) = (\pi/\Omega_N)$). The inversion formula states:

$$x[n] = \frac{1}{2\pi} \int_{-\pi}^{\pi} X(e^{j\omega}) e^{j\omega n} \, d\omega \qquad (9.27)$$

We will use this result later. We can also arrive at an expression for $x[n]$ from $X_c(j\Omega)$, the Fourier transform of the continuous-time function $x_c(t)$; indeed, by writing the formula for the inverse continuous-time Fourier transform computed in nT_s we can state that:

$$x[n] = x_c(nT_s) = \frac{1}{2\pi} \int_{-\infty}^{\infty} X_c(j\Omega) e^{j\Omega nT_s} \, d\Omega \qquad (9.28)$$

The idea is to split the integration interval in the above expression as the sum of non-overlapping intervals whose width is equal to the sampling bandwidth $\Omega_s = 2\Omega_N$; this stems from the realization that, in the inversion process, all frequencies Ω_s apart give indistinguishable contributions to the discrete-time spectrum. We have

$$x[n] = \frac{1}{2\pi} \sum_{k=-\infty}^{\infty} \int_{(2k-1)\Omega_N}^{(2k+1)\Omega_N} X_c(j\Omega) e^{j\Omega nT_s} \, d\Omega$$

which, by exploiting the Ω_s-periodicity of $e^{j\Omega nT_s}$ (i.e. $e^{j(\Omega + k\Omega_s)nT_s} = e^{j\Omega nT_s}$), becomes

$$x[n] = \frac{1}{2\pi} \sum_{k=-\infty}^{\infty} \int_{-\Omega_N}^{\Omega_N} X_c(j\Omega - jk\Omega_s) e^{j\Omega nT_s} \, d\Omega \qquad (9.29)$$

Now we interchange the order of integration and summation (this can be done under fairly broad conditions for $x_c(t)$):

$$x[n] = \frac{1}{2\pi} \int_{-\Omega_N}^{\Omega_N} \left\{ \sum_{k=-\infty}^{\infty} X_c(j\Omega - jk\Omega_s) \right\} e^{j\Omega nT_s} \, d\Omega \qquad (9.30)$$

and if we define the periodized function:

$$\tilde{X}_c(j\Omega) = \sum_{k=-\infty}^{\infty} X_c(j\Omega - jk\Omega_s)$$

we can write

$$x[n] = \frac{1}{2\pi} \int_{-\Omega_N}^{\Omega_N} \tilde{X}_c(j\Omega) e^{j\Omega nT_s} \, d\Omega \qquad (9.31)$$

after which, finally, we operate the change of variable $\theta = \Omega T_s$:

$$x[n] = \frac{1}{2\pi} \int_{-\pi}^{\pi} \frac{1}{T_s} \tilde{X}_c\left(j\frac{\theta}{T_s}\right) e^{j\theta n} \, d\theta \tag{9.32}$$

It is immediately verifiable that $\tilde{X}_c(j(\theta/T_s))$ is now 2π-periodic in θ. If we now compare (9.32) to (9.27) we can easily see that (9.32) is nothing but the DTFT inversion formula for the 2π-periodic function $(1/T_s)\tilde{X}(j\theta/T_s)$; since the inversion formulas (9.32) and (9.27) yield the same result (namely, $x[n]$) we can conclude that:

$$X(e^{j\omega}) = \frac{1}{T_s} \sum_{k=-\infty}^{\infty} X_c\left(j\frac{\omega}{T_s} - j\frac{2\pi k}{T_s}\right) \tag{9.33}$$

which is the relationship between the Fourier transform of a continuous-time function and the DTFT of its sampled version, with T_s being the sampling period. The above result is a particular version of a more general result in Fourier theory called the *Poisson sum formula*. In particular, when $x_c(t)$ is Ω_N-bandlimited, the copies in the periodized spectrum do not overlap and the (periodic) discrete-time spectrum between $-\pi$ and π is simply

$$X(e^{j\omega}) = \frac{1}{T_s} X_c\left(j\frac{\omega}{T_s}\right) \tag{9.34}$$

9.6.4 Aliasing: Examples

Figures 9.9 to 9.12 illustrate several examples of the relationship between the continuous-time spectrum and the discrete-time spectrum. For all figures, the top panel shows the continuous-time spectrum $X(j\Omega)$, with labels indicating the Nyquist frequency and the sampling frequency. The middle panel shows the periodized function $\tilde{X}_c(j\Omega)$; the single copies are plotted with a dashed line (they are not be visible if there is no overlap) and their sum is plotted in gray, with the main period highlighted in black. Finally, the last panel shows the DTFT after sampling over the $[-\pi, \pi]$ interval.

Oversampling. Figure 9.9 shows the result of sampling a bandlimited signal with a sampling frequency in excess of the minimum (in this case, $\Omega_N = 3\Omega_{\max}/2$); in this case we say that the signal has been *oversampled*. The result is that, in the periodized spectrum, the copies do not overlap and the discrete-time spectrum is just a scaled version of the original spectrum (with even a narrower support than the full $[-\pi, \pi]$ range because of the oversampling; in this case $\omega_{\max} = 2\pi/3$).

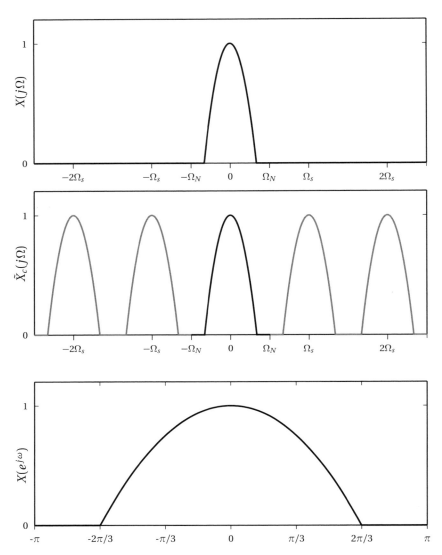

Figure 9.9 Sampling of a bandlimited signal with $\Omega_N > \Omega_{\max}$ (in this case $\Omega_{\max} = 2\Omega_N/3$). Original continuous-time spectrum $X_c(j\Omega)$ (top panel); periodized spectrum $\tilde{X}_c(j\Omega)$ (middle panel, with repetitions in gray); discrete-time spectrum $X(e^{j\omega})$ over the $[-\pi, \pi]$ interval (bottom panel).

Critical Sampling. Figure 9.10 shows the result of sampling a bandlimited signal with a sampling frequency exactly equal to twice the maximum frequency; in this case we say that the signal has been *critically sampled*. In the periodized spectrum, again the copies do not overlap and the discrete-time spectrum is a scaled version of the original spectrum occupying the whole $[-\pi, \pi]$ range.

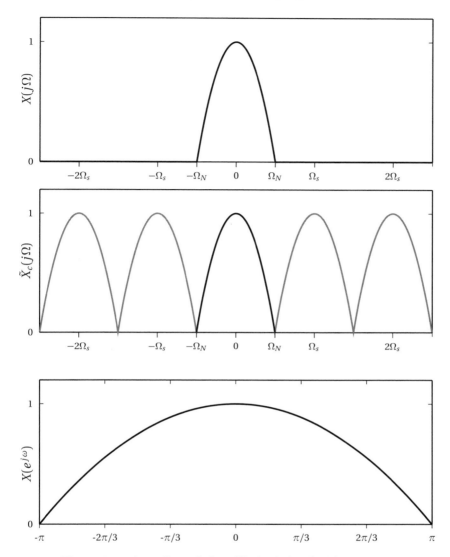

Figure 9.10 Sampling of a bandlimited signal with $\Omega_{\max} = \Omega_N$.

Undersampling (Aliasing). Figure 9.11 shows the result of sampling a bandlimited signal with a sampling frequency less than twice the maximum frequency. It this case, copies do overlap in the periodized spectrum and the resulting discrete-time spectrum is an aliased version of the original; the original spectrum can*not* be reconstructed from the sampled signal. Note, in particular, that the original lowpass shape is now a highpass shape in the sampled domain (energy at $\omega = \pi$ is larger than at $\omega = 0$).

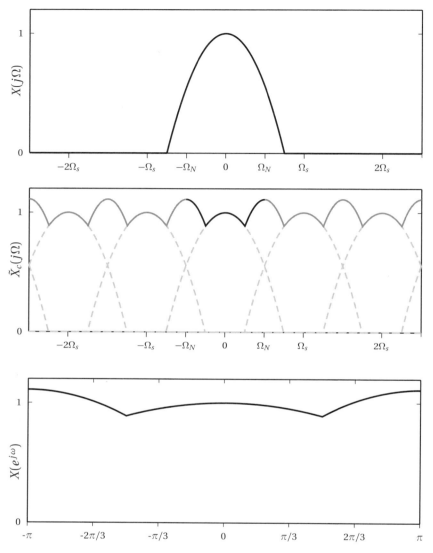

Figure 9.11 Sampling of a bandlimited signal with $\Omega_{max} > \Omega_N$ (in this case $\Omega_{max} = 3\Omega_N/2$). Original continuous-time spectrum $X_c(j\Omega)$ (top panel); periodized spectrum $\tilde{X}_c(j\Omega)$ (middle panel, with overlapping repetitions dashed gray); aliased discrete-time spectrum (bottom panel).

Sampling of Non-Bandlimited Signals. Finally, Figure 9.12 shows the result of sampling a non-bandlimited signal with a sampling frequency which is chosen as a tradeoff between alias and number of samples per second. The idea is to disregard the low-energy "tails" of the original spectrum so that their alias does not exceedingly corrupt the discrete-time spectrum.

In the periodized spectrum, the copies do overlap and the resulting discrete-time spectrum is an aliased version of the original, which is similar to the original; however, the original spectrum cannot be reconstructed from the sampled signal. In a practical sampling scenario, the correct design choice would have been to lowpass filter (in the continuous-time domain) the original signal so as to eliminate the spectral tails beyond $\pm\Omega_N$.

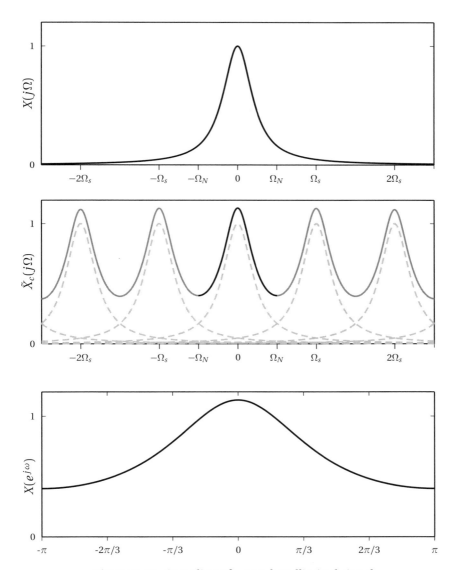

Figure 9.12 Sampling of a non-bandlimited signal.

9.7 Discrete-Time Processing of Analog Signals

Sampling and interpolation (or, more precisely, the A/D and D/A conversions strategies which we will see in the next Chapter) represent the entry and exit points of the powerful processing paradigm for which discrete-time signal processing is "famous". Samplers and interpolators are the only interfaces with the physical world, while all the processing and the analysis are performed in the abstract, dimensionless and timeless world of a general purpose microprocessor.

The generic setup of a real-world processing device is as shown in Figure 9.13. In most cases, the sampler's and the interpolator's frequencies are the same, and they are chosen as a function of the bandwidth of the class of signals for which the device is conceived; let us assume that the input is bandlimited to $\Omega_N = \pi/T_s$ (or to $F_s/2$, if we reason in hertz). For the case in which the processing block is a linear filter $H(z)$, the overall processing chain implements an analog transfer function; from the relations:

$$X(e^{j\omega}) = \frac{1}{T_s} X_c\left(j\frac{\omega}{T_s}\right) \tag{9.35}$$

$$Y(e^{j\omega}) = H(e^{j\omega})X(e^{j\omega}) \tag{9.36}$$

$$Y_c(j\Omega) = T_s Y(e^{j\Omega T_s}) \tag{9.37}$$

we have

$$Y_c(j\Omega) = H(e^{j\Omega T_s})X_c(j\Omega) \tag{9.38}$$

So, for instance, if $H(z)$ is a lowpass filter with cutoff frequency $\pi/3$, the processing chain in Figure 9.13 implements the transfer function of an analog lowpass filter with cutoff frequency $\Omega_N/3$ (or, in hertz, $F_s/6$).

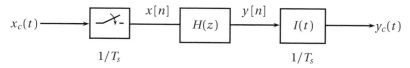

Figure 9.13 Discrete-time processing of analog signals.

9.7.1 A Digital Differentiator

In Section 2.1.4 we introduced an approximation to the differentiator operator in discrete time as the first-order difference between neighboring samples. The processing paradigm that we have just introduced, will now allow us to find the exact differentiator for a discrete-time signal. We start from a classical result of Fourier theory stating that, under broad conditions, if $x(t)$ is a differentiable function and if $x(t) \overset{\text{FT}}{\longleftrightarrow} X(j\Omega)$ then it is

$$x'(t) \overset{\text{FT}}{\longleftrightarrow} j\Omega X(j\Omega)$$

This is actually easy to prove by applying integration by part to the expression for the Fourier transform of $x'(t)$. Now, it is immediately verifiable that if we set $H(e^{j\omega}) = j\omega$ in (9.38) we obtain the followings: the system in Figure 9.13 exactly implements the transfer function of a continuous-time differentiator or, in other words, we can safely state that:

$$y[n] = x'[n]$$

where the derivative is clearly interpreted as the derivative of the underlying continuous-time signal. From the frequency response of the digital differentiator, it is easy to determine its impulse response via an inverse DTFT (and an integration by parts); we have

$$h[n] = \begin{cases} 0 & \text{for } n = 0 \\ \dfrac{(-1)^n}{n} & \text{otherwise} \end{cases}$$

From its infinite, two-sided impulse response, it is readily seen that the digital differentiator is an ideal filter; good FIR approximations can be obtained using the Parks-McClellan algorithm.

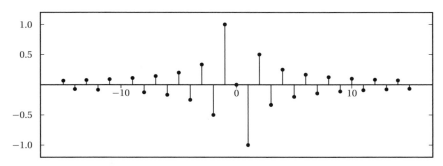

Figure 9.14 Impulse response (portion) of the ideal discrete-time differentiator.

9.7.2 Fractional Delays

We know that a discrete-time delay of D samples is a linear system with transfer function $H_d(z) = z^{-D}$ (or, alternatively, $H_d(e^{j\omega}) = e^{-j\omega D}$). If D is *not* an integer, the transfer function is still formally valid and we have stated that it implements a so-called *fractional delay*, i.e. a delay which lies "in between" two integer delays. The processing paradigm above will allow us to understand exactly the inner workings of a fractional delay. Assume $D \in \mathbb{R}$ and $0 < D < 1$; for all other values of D we can always split the delay into an integer part (which poses no problems) and a fractional part between zero and one. If we use $H(z) = z^{-D}$ in (9.38) we have

$$Y_c(j\Omega) = e^{-j\Omega DT_s} X_c(j\Omega)$$

which in the time domain becomes

$$y_c(t) = x_c(t - DT_s)$$

that is, the output continuous-time signal is just the input continuous-time signal delayed by a fraction D of the sampling interval T_s. In discrete time therefore we have

$$y[n] = x_c(nT_s - DT_s)$$

so that the action of a fractional delay is "resampling" the signal at any given point between the original sampling instants. We can appreciate this by also considering the impulse response; it is obvious that:

$$h_d[n] = \mathrm{sinc}(n - D)$$

Now, if we write out the convolution sum explicitly, we have

$$\begin{aligned} y[n] &= x[n] * h_d[n] \\ &= \sum_{k=-\infty}^{\infty} x[k] \mathrm{sinc}(n - D - k) \\ &= \sum_{k=-\infty}^{\infty} x[k] \mathrm{sinc}\left(\frac{(nT - DT) - kT}{T}\right) \end{aligned}$$

which is valid for any T and which clearly shows the convolution sum as the interpolation formula (9.18) sampled at the instants $t = nT - DT$.

Examples

Example 9.1: Another way to aliasing

Consider a real function $f(t)$ for which the Fourier transform is well defined:

$$F(j\Omega) = \int_{-\infty}^{\infty} f(t) e^{-j\Omega t} \, dt \tag{9.39}$$

Suppose that we only possess a sampled version of $f(t)$, that is, we only know the numeric value of $f(t)$ at times multiples of a sampling interval T_s and that we want to obtain an approximation of the Fourier transform above.

Assume we do not know about the DTFT; an intuitive (and standard) place to start is to write out the Fourier integral as a Riemann sum:

$$F(j\Omega) \approx \hat{F}(j\Omega) = \sum_{n=-\infty}^{\infty} T_s f(nT_s) e^{-jT_s n\Omega} \qquad (9.40)$$

indeed, this expression only uses the known sampled values of $f(t)$. In order to understand whether (9.40) is a good approximation consider the periodization of $F(j\Omega)$:

$$\tilde{F}(j\Omega) = \sum_{n=-\infty}^{\infty} F\left(j\left(\Omega + \frac{2\pi}{T_s}n\right)\right) \qquad (9.41)$$

in which $F(j\Omega)$ is repeated with overlap with period $2\pi/T_s$. We will show that:

$$\hat{F}(j\Omega) = \tilde{F}(j\Omega)$$

that is, the Riemann approximation is equivalent to a periodization of the original Fourier transform; in mathematics this is known as a particular form of the *Poisson sum formula*.

To see this, consider the periodic nature of $\tilde{F}(j\Omega)$ and remember that any periodic function $f(x)$ of period L admits a *Fourier series* expansion:

$$f(t) = \sum_{n=-\infty}^{\infty} A_n e^{j\frac{2\pi}{L}nt} \qquad (9.42)$$

where

$$A_n = \frac{1}{L} \int_{-L/2}^{L/2} f(t) e^{-j\frac{2\pi}{L}nt} dt \qquad (9.43)$$

Here's the trick: we regard $\hat{F}(j\Omega)$ as an anonymous periodic complex function and we compute its Fourier series expansion coefficients. If we replace L by $2\pi/T_s$ in (9.43) we can write

$$A_n = \frac{T_s}{2\pi} \int_{-\pi/T_s}^{\pi/T_s} \tilde{F}(j\Omega) e^{-jnT_s\Omega} d\Omega$$

$$= \frac{T_s}{2\pi} \int_{-\pi/T_s}^{\pi/T_s} \sum_{k=-\infty}^{+\infty} F\left(j\left(\Omega + \frac{2\pi}{T_s}k\right)\right) e^{-jnT_s\Omega} d\Omega$$

By inverting integral and summation, which we can do if the Fourier transform (9.40) is well defined:

$$A_n = \frac{T_s}{2\pi} \sum_k \int_{-\pi/T_s}^{\pi/T_s} F\left(j\left(\Omega + \frac{2\pi}{T_s}k\right)\right) e^{-jnT_s\Omega} d\Omega$$

and, with the change of variable $\Omega' = \Omega + (2\pi/T_s)k$,

$$A_n = \frac{T_s}{2\pi} \sum_k \int_{(2k-1)\pi/T_s}^{(2k+1)\pi/T_s} F(j\Omega') e^{-jnT_s\Omega'} e^{jT_s\frac{2\pi}{T_s}nk} d\Omega'$$

where $e^{jT_s\frac{2\pi}{T_s}nk} = 1$. The integrals in the sum are on contiguous and non-overlapping intervals, therefore:

$$A_n = T_s \frac{1}{2\pi} \int_{-\infty}^{+\infty} F(j\Omega') e^{-jnT_s\Omega'} d\Omega'$$
$$= T_s f(-nT_s)$$

so that by replacing the values for all the A_n in (9.42) we obtain $\tilde{F}(j\Omega) = \hat{F}(j\Omega)$.

What we just found is another derivation of the aliasing formula. Intuitively, there is a duality between the time domain and the frequency domain in that a discretization of the time domain leads to a periodization of the frequency domain; similarly, a discretization of the frequency domain leads to a periodization of the time domain (think of the DFS and see also Exercise 9.9).

Example 9.2: Time-limited vs. bandlimited functions

The trick of periodizing a function and then computing its Fourier series expansion comes very handy in proving that a function cannot be bandlimited and have a finite support at the same time. The proof is by contradiction and goes as follows: assume $f(t)$ has finite support

$$f(t) = 0, \quad \text{for } |t| > T_0$$

assume that $f(t)$ has a well-defined Fourier transform:

$$F(j\Omega) = \int_{-\infty}^{\infty} f(t) e^{-j\Omega t} dt = \int_{-T_0}^{T_0} f(t) e^{-j\Omega t} dt$$

and that it is *also* bandlimited so that:

$$F(j\Omega) = 0, \quad \text{for } |\Omega| > \Omega_0$$

Now consider the periodized version:

$$\tilde{f}(t) = \sum_{n=-\infty}^{\infty} f(t - 2nS)$$

since $f(t) = 0$ for $|t| > T_0$, if we choose $S > T_0$ the copies in the sum do not overlap, as shown in Figure 9.15. If we compute the Fourier series expansion (9.43) for the $2S$-periodic $\tilde{f}(t)$ we have

$$A_n = \frac{1}{2S} \int_{-S}^{S} \tilde{f}(t) e^{-j(\pi/S)nt} \, dt$$

$$= \frac{1}{2S} \int_{-T_0}^{T_0} f(t) e^{-j(\pi/S)nt} \, dt$$

$$= F\left(j\frac{\pi}{S}n\right)$$

Since we assumed that $f(t)$ is bandlimited, it is

$$A_n = 0, \qquad \text{for } |n| > \left\lfloor \frac{\Omega_0 S}{\pi} \right\rfloor = N_0$$

and therefore we can write the reconstruction formula (9.42):

$$\tilde{f}(t) = \sum_{n=-N_0}^{N_0} A_n e^{j(\pi/S)nt}$$

Now consider the complex-valued polynomial of degree $2N_0 + 1$

$$P(z) = \sum_{n=-N_0}^{N_0} A_n z^n$$

obviously $P(e^{j(\pi/S)t}) = \tilde{f}(t)$ but we also know that $\tilde{f}(t)$ is identically zero over the $[T_0, 2S - T_0]$ interval (Fig. 9.15). Now, a finite-degree polynomial $P(z)$ has only a finite number of roots and therefore it cannot be identically zero over an interval unless it is zero everywhere (see also Example 6.2). Hence, either $f(t) = 0$ everywhere or $f(t)$ cannot be both bandlimited and time-limited.

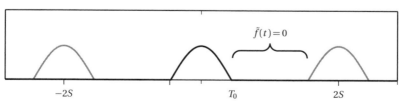

Figure 9.15 Finite support function $f(t)$ (black) and non-overlapping periodization (gray).

Appendix

The Sinc Product Expansion Formula

The goal is to prove the product expansion

$$\frac{\sin(\pi t)}{\pi t} = \prod_{n=1}^{\infty}\left(1 - \frac{t^2}{n^2}\right) \tag{9.44}$$

We present two proofs; the first was proposed by Euler in 1748 and, while it certainly lacks rigor by modern standards, it has the irresistible charm of elegance and simplicity in that it relies only on basic algebra. The second proof is more rigorous, and is based on the theory of Fourier series for periodic functions; relying on Fourier theory, however, hides most of the convergence issues.

Euler's Proof. Consider the N roots of unity for N odd. They comprise $z = 1$ plus $N-1$ complex conjugate roots of the form $z = e^{\pm j\omega_N k}$ for $k = 1, \ldots, (N-1)/2$ and $\omega_N = 2\pi/N$. If we group the complex conjugate roots pairwise we can factor the polynomial $z^N - 1$ as

$$z^N - 1 = (z-1)\prod_{k=1}^{(N-1)/2}\left(z^2 - 2z\cos(\omega_N k) + 1\right)$$

The above expression can immediately be generalized to

$$z^N - a^N = (z-a)\prod_{k=1}^{(N-1)/2}\left(z^2 - 2az\cos(\omega_N k) + a^2\right)$$

Now replace z and a in the above formula by $z = (1+x/N)$ and $a = (1-x/N)$; we obtain the following:

$$\left(1+\frac{x}{N}\right)^N - \left(1-\frac{x}{N}\right)^N =$$

$$= \frac{4x}{N}\prod_{k=1}^{(N-1)/2}\left(1 - \cos(\omega_N k) + \frac{x^2}{N^2}(1+\cos(\omega_N k))\right)$$

$$= \frac{4x}{N}\prod_{k=1}^{(N-1)/2}(1-\cos(\omega_N k))\left(1 + \frac{x^2}{N^2}\cdot\frac{1+\cos(\omega_N k)}{1-\cos(\omega_N k)}\right)$$

$$= Ax\prod_{k=1}^{(N-1)/2}\left(1 + \frac{x^2(1+\cos(\omega_N k))}{N^2(1-\cos(\omega_N k))}\right)$$

where A is just the finite product $(4/N)\prod_{k=1}^{(N-1)/2}(1-\cos(\omega_N k))$. The value A is also the coefficient for the degree-one term x in the right-hand side and it can be easily seen from the expansion of the left hand-side that $A = 2$ for all N; actually, this is an application of Pascal's triangle and it was proven by Pascal in the general case in 1654. As N grows larger we have that:

$$\left(1 \pm \frac{x}{N}\right)^N \approx e^{\pm x}$$

and at the same time, if N is large, then $\omega_N = 2\pi/N$ is small and, for small values of the angle, the cosine can be approximated as

$$\cos(\omega) \approx 1 - \frac{\omega^2}{2}$$

so that the denominator in the general product term can, in turn, be approximated as

$$N^2\left(1 - \cos\left(\frac{2\pi}{N}k\right)\right) \approx N^2 \cdot \frac{4k^2\pi^2}{2N^2} = 2k^2\pi^2$$

By the same token, for large N, the numerator can be approximated as $1 + \cos((2\pi/n)k) \approx 2$ and therefore (by bringing $A = 2$ over to the left-hand side) the above expansion becomes

$$\frac{e^x - e^{-x}}{2} = x\left(1 + \frac{x^2}{\pi^2}\right)\left(1 + \frac{x^2}{4\pi^2}\right)\left(1 + \frac{x^2}{9\pi^2}\right)\cdots$$

Finally, we replace x by $j\pi t$ to obtain:

$$\frac{\sin(\pi t)}{\pi t} = \prod_{n=1}^{\infty}\left(1 - \frac{t^2}{n^2}\right)$$

Rigorous Proof. Consider the Fourier series expansion of the *even* function $f(x) = \cos(\tau x)$ periodized over the interval $[-\pi, \pi]$. We have

$$f(x) = \frac{1}{2}a_0 + \sum_{n=1}^{\infty} a_n \cos(nx)$$

with

$$\begin{aligned}
a_n &= \frac{1}{\pi}\int_{-\pi}^{\pi} \cos(\tau x)\cos(nx)\,dx \\
&= \frac{2}{\pi}\int_0^{\pi} \frac{1}{2}\left(\cos((\tau + n)x) + \cos((\tau - n)x)\right)\,dx \\
&= \frac{1}{\pi}\left(\frac{\sin((\tau + n)\pi)}{\tau + n} + \frac{\sin((\tau - n)\pi)}{\tau - n}\right) \\
&= \frac{2\sin(\tau\pi)}{\pi}\frac{(-1)^n \tau}{\tau^2 - n^2}
\end{aligned}$$

so that

$$\cos(\tau x) = \frac{2\tau \sin(\tau \pi)}{\pi} \left(\frac{1}{2\tau^2} - \frac{\cos(x)}{\tau^2 - 1} + \frac{\cos(2x)}{\tau^2 - 2^2} - \frac{\cos(3x)}{\tau^2 - 3^2} + \cdots \right)$$

In particular, for $x = \pi$ we have

$$\cot(\pi \tau) = \frac{2\tau}{\pi} \left(\frac{1}{2\tau^2} + \frac{1}{\tau^2 - 1} + \frac{1}{\tau^2 - 2^2} + \frac{1}{\tau^2 - 3^2} + \cdots \right)$$

which we can rewrite as

$$\pi \left(\cot(\pi \tau) - \frac{1}{\pi \tau} \right) = \sum_{n=1}^{\infty} \frac{-2\tau}{n^2 - \tau^2}$$

If we now integrate between 0 and t both sides of the equation we have

$$\int_0^t \left(\cot(\pi \tau) - \frac{1}{\pi \tau} \right) d\pi\tau = \ln \left. \frac{\sin(\pi \tau)}{\pi \tau} \right|_0^t = \ln \left[\frac{\sin(\pi t)}{\pi t} \right]$$

and

$$\int_0^t \sum_{n=1}^{\infty} \frac{-2\tau}{n^2 - \tau^2} d\tau = \sum_{n=1}^{\infty} \ln \left[1 - \frac{t^2}{n^2} \right] = \ln \left[\prod_{n=1}^{\infty} \left(1 - \frac{t^2}{n^2} \right) \right]$$

from which, finally,

$$\frac{\sin(\pi t)}{\pi t} = \prod_{n=1}^{\infty} \left(1 - \frac{t^2}{n^2} \right)$$

Further Reading

The sampling theorem is often credited to C. Shannon, and indeed it appears with a sketchy proof in his foundational 1948 paper "A Mathematical Theory of Communication", *Bell System Technical Journal*, Vol. 27, 1948, pp. 379-423 and pp. 623-656. Contemporary treatments can be found in all signal processing books, but also in more mathematical texts, such as S. Mallat's *A Wavelet Tour of Signal Processing* (Academic Press, 1998), or the soon to be published *The World of Fourier and Wavelets* by M. Vetterli, J. Kovacevic and V. Goyal. These more modern treatments take a Hilbert space point of view, which allows the generalization of sampling theorems to more general spaces than just bandlimited functions.

Exercises

Exercise 9.1: Zero-order hold. Consider a discrete-time sequence $x[n]$ with DTFT $X(e^{j\omega})$. Next, consider the continuous-time interpolated signal

$$x_0(t) = \sum_{n=-\infty}^{\infty} x[n]\,\text{rect}(t-n)$$

i.e. the signal interpolated with a zero-centered zero-order hold and $T = 1$ sec.

(a) Express $X_0(j\Omega)$ (the spectrum of $x_0(t)$) in terms of $X(e^{j\omega})$.

(b) Compare $X_0(j\Omega)$ to $X(j\Omega)$. We can look at $X(j\Omega)$ as the Fourier transform of the signal obtained from the sinc interpolation of $x[n]$ (always with $T = 1$):

$$x(t) = \sum_{n \in \mathbb{Z}} x[n]\,\text{sinc}(t-n)$$

Comment on the result: you should point out two major problems.

So, as it appears, interpolating with a zero-order hold introduces a distortion in the interpolated signal with respect to the sinc interpolation in the region $-\pi \leq \Omega \leq \pi$. Furthermore, it makes the signal non-bandlimited outside the region $-\pi \leq \Omega \leq \pi$. The signal $x(t)$ can be obtained from the zero-order hold interpolation $x_0(t)$ as $x(t) = x_0(t) * g(t)$ for some filter $g(t)$.

(c) Sketch the frequency response of $g(t)$.

(d) Propose two solutions (one in the continuous-time domain, and another in the discrete-time domain) to eliminate or attenuate the distortion due to the zero-order hold. Discuss the advantages and disadvantages of each.

Exercise 9.2: A bizarre interpolator. Consider the local interpolation scheme of the previous exercise but assume that the characteristic of the interpolator is the following:

$$I(t) = \begin{cases} 1 - 2|t| & \text{for } |t| \leq 1/2 \\ 0 & \text{otherwise} \end{cases}$$

This is a triangular characteristic with the same support as the zero-order hold. If we pick an interpolation interval T_s and interpolate a given discrete-time signal $x[n]$ with $I(t)$, we obtain a continuous-time signal:

$$x(t) = \sum_{n} x[n] I\left(\frac{t - nT_s}{T_s}\right)$$

which looks like this:

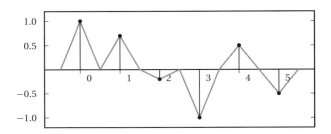

Assume that the spectrum of $x[n]$ between $-\pi$ and π is

$$X(e^{j\omega}) = \begin{cases} 1 & \text{for } |\omega| \leq 2\pi/3 \\ 0 & \text{otherwise} \end{cases}$$

(with the obvious 2π-periodicity over the entire frequency axis).

(a) Compute and sketch the Fourier transform $I(j\Omega)$ of the interpolating function $I(t)$. (Recall that the triangular function can be expressed as the convolution of $\text{rect}(t/2)$ with itself).

(b) Sketch the Fourier transform $X(j\Omega)$ of the interpolated signal $x(t)$; in particular, clearly mark the Nyquist frequency $\Omega_N = \pi/T_s$.

(c) The use of $I(t)$ instead of a sinc interpolator introduces two types of errors: briefly describe them.

(d) To eliminate the error *in the baseband* $[-\Omega_N, \Omega_N]$ we can pre-filter the signal $x[n]$ with a filter $h[n]$ *before* interpolating with $I(t)$. Write the frequency response of the discrete-time filter $H(e^{j\omega})$.

Exercise 9.3: Another view of sampling. One of the standard ways of describing the sampling operation relies on the concept of "modulation by a pulse train". Choose a sampling interval T_s and define a continuous-time pulse train $p(t)$ as

$$p(t) = \sum_{k=-\infty}^{\infty} \delta(t - kT_s)$$

The Fourier Transform of the pulse train is

$$P(j\Omega) = \frac{2\pi}{T_s} \sum_{k=-\infty}^{\infty} \delta\left(\Omega - k\frac{2\pi}{T_s}\right)$$

This is tricky to show, so just take the result as is. The "sampled" signal is simply the modulation of an arbitrary-continuous time signal $x(t)$ by the pulse train:

$$x_s(t) = p(t)x(t)$$

Note that, now, this sampled signal is still continuous time but, by the properties of the delta function, is non-zero only at multiples of T_s; in a sense, $x_s(t)$ is a discrete-time signal brutally embedded in the continuous time world. Here is the question: derive the Fourier transform of $x_s(t)$ and show that if $x(t)$ is bandlimited to π/T_s then we can reconstruct $x(t)$ from $x_s(t)$.

Exercise 9.4: Aliasing can be good! Consider a real, continuous-time signal $x_c(t)$ with the following spectrum $X_c(j\Omega)$:

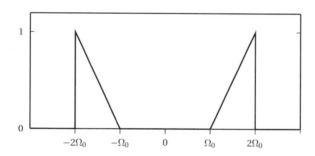

(a) What is the bandwidth of the signal? What is the minimum sampling period in order to satisfy the sampling theorem?

(b) Take a sampling period $T_s = \pi/\Omega_0$; clearly, with this sampling period, there will be aliasing. Plot the DTFT of the discrete-time signal $x_a[n] = x_c(nT_s)$.

(c) Suggest a block diagram to reconstruct $x_c(t)$ from $x_a[n]$.

(d) With such a scheme available, we can therefore exploit aliasing to reduce the sampling frequency necessary to sample a bandpass signal. In general, what is the minimum sampling frequency to be able to reconstruct, with the above strategy, a real signal whose frequency support on the positive axis is $[\Omega_0, \Omega_1]$ (with the usual symmetry around zero, of course)?

Exercise 9.5: Digital processing of continuous-time signals. For your birthday, you receive an unexpected present: a 4 MHz A/D converter, complete with anti-aliasing filter. This means you can safely sample signals up to a frequency of 2 MHz; since this frequency is above the AM radio frequency band, you decide to hook up the A/D to your favorite signal processing system and build an entirely digital radio receiver. In this exercise we will explore how to do so.

Simply, assume that the AM radio spectrum extends from 1 Mhz to 1.2 Mhz and that in this band you have ten channels side by side, each one of which occupies 20 KHz.

(a) Sketch the digital spectrum at the output of the A/D converter, and show the bands occupied by the channels, numbered from 1 to 10, with their beginning and end frequencies.

The first thing that you need to do is to find a way to isolate the channel you want to listen to and to eliminate the rest. For this, you need a bandpass filter centered on the band of interest. Of course, this filter must be *tunable* in the sense that you must be able to change its spectral location when you want to change station. An easy way to obtain a tunable bandpass filter is by modulating a lowpass filter with a sinusoidal oscillator whose frequency is controllable by the user:

(b) As an example of a tunable filter, assume $h[n]$ is an ideal lowpass filter with cutoff frequency $\pi/8$. Plot the magnitude response of the filter $h_m[n] = \cos(\omega_m n) h[n]$, where $\omega_m = \pi/2$; ω_m is called the *tuning frequency*.

(c) Specify the cutoff frequency of a lowpass filter which can be used to select one of the AM channels above.

(d) Specify the tuning frequencies for channel 1, 5 and 10.

Now that you know how to select a channel, all that is left to do is to demodulate the signal and feed it to a D/A converter and to a loudspeaker.

(e) Sketch the complete block diagram of the radio receiver, from the antenna going into the A/D converter to the final loudspeaker. Use only one sinusoidal oscillator. Do not forget the filter before the D/A (specify its bandwidth).

The whole receiver now works at a rate of 4 MHz; since it outputs audio signals, this is clearly a waste.

(f) Which is the minimum D/A frequency you can use? Modify the receiver's block diagram with the necessary elements to use a low frequency D/A.

Exercise 9.6: Acoustic aliasing. Assume $x(t)$ is a continuous-time pure sinusoid at 10 KHz. It is sampled with a sampler at 8 KHz and then interpolated back to a continuous-time signal with an interpolator at 8 KHz. What is the perceived frequency of the interpolated sinusoid?

Exercise 9.7: Interpolation subtleties. We have seen that any discrete-time sequence can be sinc-interpolated into a continuous-time signal which is Ω_N-bandlimited; Ω_N depends on the interpolation interval T_s via the relation $\Omega_N = \pi/T_s$.
Consider the continuous-time signal $x_c(t) = e^{-t/T_s}$ and the discrete-time sequence $x[n] = e^{-n}$. Clearly, $x_c(nT_s) = x[n]$; but, can we also say that $x_c(t)$ is the signal we obtain if we apply sinc interpolation to the sequence $x[n] = e^{-n}$ with interpolation interval T_s? Explain in detail.

Exercise 9.8: Time and frequency. Consider a real continuous-time signal $x(t)$. All you know about the signal is that $x(t) = 0$ for $|t| > t_0$. Can you determine a sampling frequency F_s so that when you sample $x(t)$, there is no aliasing? Explain.

Exercise 9.9: Aliasing in time? Consider an N-periodic discrete-time signal $\tilde{x}[n]$, with N an *even* number, and let $\tilde{X}[k]$ be its DFS:

$$\tilde{X}[k] = \sum_{n=0}^{N-1} \tilde{x}[n] e^{-j\frac{2\pi}{N}nk}, \qquad k \in \mathbb{Z}$$

Let $\tilde{Y}[m] = \tilde{X}[2m]$, i.e. a "subsampled" version of the DFS coefficients; clearly this defines a $(N/2)$-periodic sequence of DFS coefficients. Now consider the $(N/2)$-point inverse DFS of $\tilde{Y}[m]$ and call this $(N/2)$-periodic signal $\tilde{y}[n]$:

$$\tilde{y}[n] = \frac{2}{N} \sum_{k=0}^{N/2-1} \tilde{Y}[k] e^{j\frac{2\pi}{N/2}nk}, \qquad n \in \mathbb{Z}$$

Express $\tilde{y}[n]$ in terms of $\tilde{x}[n]$ and describe in a few words what has happened to $\tilde{x}[n]$ and why.

Chapter 10

A/D and D/A Conversions

The word "digital" in "digital signal processing" indicates that, in the representation of a signal, *both* time and amplitude are discrete quantities. The necessity to discretize the *amplitude* values of a discrete-time signal comes from the fact that, in the digital world, all variables are necessarily represented with a finite precision. Specifically, general-purpose signal processors are nothing but streamlined processing units which address memory locations whose granularity is an integer number of bits. The conversion from the "real world" analog value of a signal to its discretized digital counterpart is called analog-to-digital (A/D) conversion. Analogously, a transition in the opposite direction is shorthanded as a D/A conversion; in this case, we are associating a physical analog value to a digital internal representation of a signal sample. Note that, just as was the case with sampling, quantization and its inverse lie at the boundary between the analog and the digital world and, as such, they are performed by actual pieces of complex, dedicated hardware.

10.1 Quantization

The sampling theorem described in Chapter 9 allowed us to represent a bandlimited signal by means of a discrete-time sequence of samples taken at instants multiple of a sampling period T_s. In order to store or process this sequence of numbers $x[n] = x(nT_s)$, $n \in \mathbb{Z}$, we need to transform the real values $x[n]$ into a format which fits the memory model of a computer; two such formats are, for instance, finite-precision integers or finite-precision floating point numbers (which are nothing but integers associated to a scale factor). In both cases, we need to map the real line or an interval thereof (i.e. the range of $x(t)$) onto a *countable set* of values. Unfortunately, because of this loss of dimensionality, this mapping is irreversible, which leads to approximation errors. Consider a bandlimited input signal $x(t)$, whose amplitude is known to vary between the values A and B. After sampling, each sample $x[n]$ will have to be stored as a R-bit integer, i.e. as one out of $K = 2^R$ possible values. An intuitive solution is to divide the $[A, B]$ interval into K non-overlapping subintervals I_k such that

$$\bigcup_{k=0}^{K-1} I_k = [A, B]$$

The intervals are defined by $K+1$ points i_k so that for each interval we can write

$$I_k = [i_k, i_{k+1}], \qquad k = 0, 1, \ldots, K-1$$

So $i_0 = A$, $i_K = B$, and $i_0 < i_1 < \ldots < i_{K-1} < i_K$. An example of this subdivision for $R = 2$ is shown in Figure 10.1 in which, arbitrarily. In order to map the input samples to a set of integers, we introduce the following *quantization* function:

$$\mathcal{Q}\{x[n]\} = \{k \mid x[n] \in I_k\} \tag{10.1}$$

In other words, quantization associates to the sample value $x[n]$, the integer index of the interval onto which $x[n]$ "falls". One of the fundamental questions in the design of a good quantizer is how to choose the splitting points i_k for a given class of input signals as well as the reconstruction values.

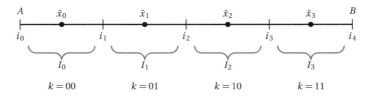

Figure 10.1 Example of quantization for an interval; quantization intervals $I_k = [i_k, i_{k+1}]$, representative values $\hat{x}_k$ and binary quantization levels k.

Quantization Error. Since all the (infinite) values falling onto the interval I_k are mapped to the same index, precious information is lost. This introduces an error in the quantized representation of a signal which we analyze in detail in the following and which we will strive to minimize. Note that quantization is a highly nonlinear operation since, in general,

$$\mathcal{Q}\{x[n] + y[n]\} \neq \mathcal{Q}\{x[n]\} + \mathcal{Q}\{y[n]\}$$

This is related to the quantization error since, if $a \in I_k$ and $a+b \in I_k$ then $\mathcal{Q}\{a+b\} = \mathcal{Q}\{a\}$; small perturbations in the signal are lost in quantization. On the other hand, consider a constant signal $x[n] = i_m$ for some $m < K$ (i.e. the value of the signal coincides with the lower limit of one of the quantization intervals); consider now a small "perturbation" noise sequence $\epsilon[n]$, with uniform distribution over $[-U, U]$ and U arbitrarily small. The quantized signal:

$$\mathcal{Q}\{x[n] + \epsilon[n]\} = \begin{cases} k-1 & \text{if } \epsilon[n] \leq 0 \\ k & \text{if } \epsilon[n] > 0 \end{cases}$$

In other words, the quantized signal will oscillate randomly with 50% chance between two neighboring quantization levels. This can create disruptive artifacts in a digital signal, and it is counteracted by special techniques called *dithering* for which we refer to the bibliography.

Reconstruction. The output of a quantizer is an abstract internal representation of a signal, where actual values are replaced by (binary) indices. In order to process or interpolate back such a signal, we need to somehow "undo" the quantization and, to achieve this, we need to associate back a physical value to each of the quantizer's indices. Reconstruction is entirely dependent on the original quantizer and, intuitively, it is clear that the value $\hat{x}_k$ which we associate to index k should belong to the interval I_k. For example, in the absence of any other information on $x[n]$, it is reasonable to choose the interval's midpoint as the representative value. In any case, the reconstructed signal will be

$$\hat{x}[n] = \{\hat{x}_k, \, k = \mathcal{Q}\{x[n]\}\} \tag{10.2}$$

and the second fundamental question in quantizer design is how to choose the representative values $\hat{x}_k$ so that the quantization error is minimized. The error introduced by the whole quantization/reconstruction chain can therefore be written out as

$$e[n] = \mathcal{E}(x[n]) = (\hat{x}[n] - x[n]) \tag{10.3}$$

For a graphical example see Figure 10.2. Note that, often, with an abuse of notation, we will use the term "quantized signal" to indicate the sequence

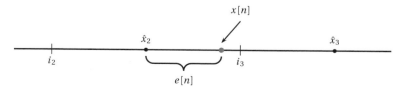

Figure 10.2 Example of quantization error.

of *reconstruction values* associated to the sequence of quantization indices as produced by the quantizer. In other words, we will often imply, for short,

$$\hat{x}[n] = \mathcal{Q}\{x[n]\}$$

10.1.1 Uniform Scalar Quantization

Quantization, as we mentioned, is a non-linear operation and it is therefore quite difficult to analyze in general; the goal is to obtain some statistical description of the quantization error for classes of stochastic input signals. In order to make the problem tractable and solvable, and thus gain some precious insight on the mathematics of quantization, simplified models are often used. The simplest such model is the uniform scalar quantization scenario. By *scalar* quantization, we indicate that each input sample $x[n]$ is quantized independently; more sophisticated techniques would take advantage of the correlation between neighboring samples to perform a joint quantization which goes under the name of "vector quantization". By uniform quantization, we indicate the key design choices for quantization and reconstruction: given a budget of R bits per sample (known as the *rate* of the digital signal), we will be able to quantize the input signal into $K = 2^R$ distinct levels; in order to do so we need to split the range of the signal into K intervals. It is immediately clear that such intervals should be disjoint (as to have a unique quantized value for each input value) and they should cover the whole range of the input. In the case of uniform quantization the following design is used:

- The range of the input $x[n]$ is assumed to be in the interval $[A, B]$, with $A, B \in \mathbb{R}$.

- The range $[A, B]$ is split into $K = 2^R$ contiguous intervals I_k of equal width $\Delta = (B - A)/K$.

- Each reconstruction point $\hat{x}_k$ is chosen to be the midpoint of the corresponding interval I_k.

An example of the input/output characteristic of a uniform quantizer is shown in Figure 10.3 for a rate of 3 bits/sample and a signal limited between -1 and 1.

Uniform Quantization of a Uniformly Distributed Input. In order to precisely evaluate the distortion introduced by a quantization and reconstruction chain we need to formulate some assumption on the statistical

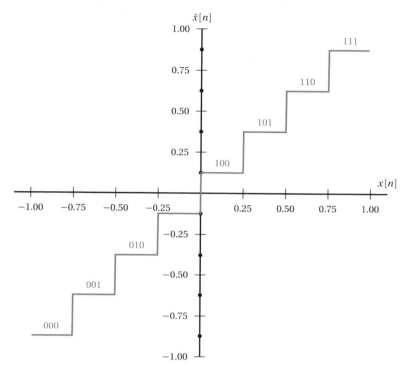

Figure 10.3 3-bit quantization of $x[n] \in [-1,1]$. Eight uniform intervals are defined between -1 and 1, and the approximation $\hat{x}[n]$ is chosen as the interval's mid-point. Nominal quantization indices (in binary) are shown over the reconstruction levels.

properties of the input signal. of the input signal. Let us start with a discrete-time signal $x[n]$ with the following characteristics:

- $x[n]$ is uniformly distributed over the $[A, B]$ interval;
- $x[n]$ is an i.i.d. process (independent and identically distributed).

While simple, this signal model manages to capture the essence of many real-world quantization problems. For a uniformly distributed input signal, the design choices of a uniform quantizer turn out to be optimal with respect to the minimization of the power of the quantization error P_e. If we consider the expression for the power

$$P_e = \mathrm{E}\left[|e[n]|^2\right] = \int_A^B \frac{1}{B-A} \mathcal{E}^2(x)\,dx \qquad (10.4)$$

and we remark that the error function localizes over the quantization intervals as

$$\mathcal{E}(x)\big|_{x \in I_k} = (x - \hat{x}_k)$$

then we can split the above integral as

$$P_e = \frac{1}{B-A} \sum_{k=0}^{K-1} \int_{i_k}^{i_{k+1}} (x - \hat{x}_k)^2 \, dx \qquad (10.5)$$

where $i_k = A + k\Delta$, $x_k = i_k + \Delta/2$ and $\Delta = (B-A)/K$. In order to show that the quantizer is optimal, we need to show that these values for i_k and $\hat{x}_k$ lie at a minimum for the error function P_e or, in other words, that they are a solution to the following system of equations:

$$\begin{cases} \dfrac{\partial P_e}{\partial \hat{x}_k} = 0 & k = 0, 1, \ldots, K-1 \\ \dfrac{\partial P_e}{\partial i_k} = 0 & k = 1, 2, \ldots, K-1 \end{cases} \qquad (10.6)$$

By plugging (10.5) into the first equation in (10.6) we have

$$\frac{\partial P_e}{\partial \hat{x}_k} \propto \frac{\partial}{\partial \hat{x}_k} \int_{i_k}^{i_{k+1}} (x - \hat{x}_k)^2 \, dx = (i_k - \hat{x}_k)^2 - (i_{k+1} - \hat{x}_k)^2$$

$$= \left(\frac{\Delta}{/2}\right)^2 - \left(-\frac{\Delta}{2}\right)^2 = 0$$

The same can easily be verified for the second equation in (10.6).

Finally, we can determine the power of the quantization error:

$$P_e = \sum_{k=0}^{K-1} \mathrm{E}\left[(x[n] - x_k)^2 \,\Big|\, x[n] \in I_k\right] \mathrm{P}[x[n] \in I_k]$$

$$= \int_{-\Delta/2}^{\Delta/2} \frac{1}{\Delta} x^2 \, dx$$

$$= \frac{\Delta^2}{12} \qquad (10.7)$$

where we have used the fact that, since the signal is uniformly distributed, $\mathrm{P}[x[n] \in I_k] = 1/K$ for all k. If we consider the average power (i.e. the variance) of the input signal:

$$\sigma_x^2 = \mathrm{E}[x^2[n]] = \frac{(B-A)^2}{12}$$

and we use the fact that $\Delta = (B-A)/K = (B-A)/2^R$, we can write

$$P_e = \sigma_x^2 2^{-2R} \tag{10.8}$$

This exponential decay of the error, as a function of the rate, is a key concept, not only with respect to quantization but, much more generally, with respect to data compression. Finally, if we divide by the power of the input signal, we can arrive at a convenient and compact expression for the signal to noise ratio of the digital signal:

$$\text{SNR} = 2^{2R} \tag{10.9}$$

If we express the SNR in dB, the above equation becomes:

$$\text{SNR}_{\text{dB}} = 10 \log_{10} 2^{2R} \approx 6R \text{ dB} \tag{10.10}$$

This provides us with an extremely practical rule of thumb for the distortion caused by quantization: *each additional bit per sample improves the SNR by 6 dB*. A compact disk, for instance, which uses a quantization of 16 bits per sample, has an SNR (or maximum dynamic range) of approximately 96 dB. Remember, however, that the above expression has been derived under very unrealistic assumptions for the input signal, the most limiting of which is that input samples are uncorrelated, which makes the quantization error uncorrelated as well. This, is of course, far from true in any non-noise signal so that the 6 dB/bit rule must be treated as a rough estimate.

Uniform Quantization of Normally Distributed Input. Consider now a more realistic distribution for the input signal $x[n]$, namely the Gaussian distribution; the input signal is now assumed to be white Gaussian noise of variance σ^2. Suppose we fix the size of the quantization interval Δ; since, in this case, the support of the probability density function is infinite, we either need an infinite number of bins (and therefore an infinite rate), or we need to "clip" $x[n]$ so that all values fall within a finite interval. In the purely theoretical case of an infinite number of quantization bins, it can be proven that the error power for a zero-mean Gaussian random variable of variance σ^2 quantized into bins of size Δ is

$$P_e = \frac{\sqrt{3}\pi}{2} \sigma^2 \Delta^2 \tag{10.11}$$

In a practical system, when a finite number of bins is used, $x[n]$ is clipped outside of an $[A, B]$ interval, which introduces an additional error, due to the loss of the tails of the distribution. It is customary to choose $B = -A = 2\sigma$; this is an instance of the so-called "4σ rule", stating that over 99.5% of the probability mass of a zero-mean Gaussian variable with variance σ^2 is comprised between -2σ and 2σ. With this choice, and with a rate of R bits per sample, we can build a uniform quantizer with $\Delta = 4\sigma/2^R$. The total error power increase because of clipping but not by much; essentially, the behavior is still given by an expression similar to (10.11) which, expressed as a function of the rate, can be written as

$$P_e = C\sigma^2 2^{-2R} \qquad (10.12)$$

where C is a constant larger but of the same order as $\sqrt{3}\pi/2 = 2.72$ in (10.11). Again, the signal to noise ratio turns out to be an exponentially increasing function of the rate

$$\text{SNR} = \frac{1}{C} 2^{2R}$$

with only a constant to reduce performance with respect to the uniformly distributed input. Again, the 6 dB/bit rule applies, with the usual caveats.

10.1.2 Advanced Quantizers

The Lloyd-Max algorithm is a procedure to design an optimal quantizer for an input signal with an arbitrary (but known) probability density function. The starting hypotheses for the Lloyd-Max procedure are the following:

- $x[n]$ is bounded over the $[A, B]$ interval;
- $x[n]$ is an i.i.d. process;
- $x[n]$ is distributed with pdf $f_x(x)$.

Under these assumptions the idea is to solve a system of equations similar to (10.6) where, in this case, the pdf for the input is explicit in the error integral. The solution defines the optimal quantizer for the given input distribution, i.e. the quantizer which minimizes the associated error.

The error can be expressed as

$$P_e = \sum_{k=0}^{K-1} \int_{i_k}^{i_{k+1}} (x - \hat{x}_k)^2 f_x(x)\,dx \qquad (10.13)$$

where now the only known parameters are K, $i_0 = A$ and $i_K = B$ and we must solve

$$\begin{cases} \dfrac{\partial P_e}{\partial \hat{x}_k} = 0 & k = 0, 1, \ldots, K-1 \\ \dfrac{\partial P_e}{\partial i_k} = 0 & k = 1, 2, \ldots, K-1 \end{cases} \qquad (10.14)$$

The first equation can be efficiently solved noting that only one term in (10.13) depends on $\hat{x}_k$ for a given value of k. Therefore:

$$\frac{\partial P_e}{\partial \hat{x}_k} = \frac{\partial}{\partial \hat{x}_k} \int_{i_k}^{i_{k+1}} (x - \hat{x}_k)^2 f_x(x) dx$$

$$= \int_{i_k}^{i_{k+1}} 2(x - \hat{x}_k) f_x(x) dx$$

which gives the optimal value for $\hat{x}_k$ as

$$\hat{x}_k = \frac{\int_{i_k}^{i_{k+1}} x f_x(x) dx}{\int_{i_k}^{i_{k+1}} f_x(x) dx} \tag{10.15}$$

Note that the optimal value is the center of mass of the input distribution over the quantization interval: $\hat{x}_k = E[x | i_k \leq x \leq i_{k+1}]$. Similarly, we can determine the boundaries of the quantization intervals as

$$\frac{\partial P_e}{\partial i_k} = \frac{\partial}{\partial i_k} \left(\int_{i_{k-1}}^{i_k} (x - \hat{x}_k)^2 f_x(x) dx + \int_{i_k}^{i_{k+1}} (x - \hat{x}_k)^2 f_x(x) dx \right)$$

$$= (i_k - \hat{x}_{k-1}) f_x(i_k) + (i_k - \hat{x}_k) f_x(i_k)$$

from which

$$i_k = \frac{\hat{x}_{k-1} + \hat{x}_k}{2} \tag{10.16}$$

i.e. the optimal boundaries are the midpoints between optimal quantization points. The system of Equations (10.15) and (10.16) can be solved (either exacly or, more often, iteratively) to find the optimal parameters. In practice, however, the SNR improvement introduced by a Lloyd-Max quantizer does not justify an ad-hoc hardware design effort and uniform quantizers are used almost exclusively in the case of scalar quantization.

10.2 A/D Conversion

The process which transforms an analog continuous-time signal into a digital discrete-time signal is called analog-to-digital (A/D) conversion. Again, it is important to remember that A/D conversion is the operation that lies at the interface between the analog and the digital world and, therefore, it is performed by specialized hardware which encode the instantaneous voltage values of an electrical signal into a binary representation suitable for use on a general-purpose processor. Once a suitable sampling frequency F_s has been chosen, the process is composed of four steps in cascade.

Analog Lowpass Filtering. An analog lowpass with cutoff $F_s/2$ is a necessary step even if the analog signal is virtually bandlimited because of the noise: we need to eliminate the high-frequency noise which, because of sampling, would alias back into the signal's bandwidth. Since sharp analog lowpass filters are "expensive", the design usually allows for a certain amount of slack by choosing a sampling frequency higher than the minimum necessary.

Sample and Hold. The input signal is sampled by a structure similar to that in Figure 10.4. The FET *T1* acts as a solid-state switch and it is driven by a train of pulses $k(t)$ generated by a very stable crystal oscillator. The pulses arrive F_s times per second and they cause *T1* to close briefly so that the capacitor *C1* is allowed to charge to the instantaneous value of the input signal $x_c(nT_s)$ (in Volts); the FET then opens immediately and the capacitor remains charged to $x_c(nT_s)$ over the time interval $[nT_s,(n+1)T_s]$. The key element of the sample-and-hold circuit is therefore the capacitor, acting as an instantaneous memory element for the input signal's voltage value, while the op-amps provide the necessary high-impedance interfaces to the input signal and to the capacitor. The continuous-time signal produced by this structure looks like the output of a zero-order hold interpolator (Fig. 9.1); availability of a piecewise-constant signal between sampling instants is necessary to allow the next steps in the chain to reach a stable output value.

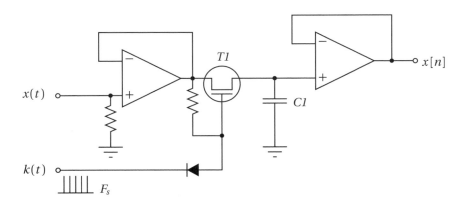

Figure 10.4 Sample and hold circuitry.

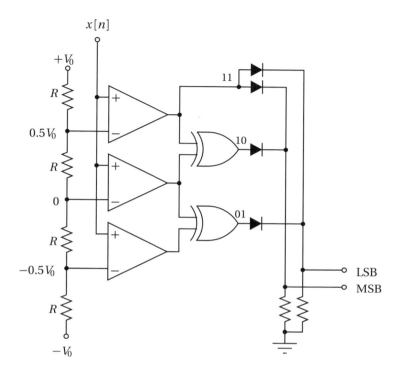

Figure 10.5 Simple two-bit quantization circuitry.

Clipping. Clipping limits the range of the voltages which enter the quantizer. The limiting function can be a hard threshold or, as in audio applications, a function called *compander* where the signal is leveled towards a maximum via a smooth function (usually a sigmoid).

Quantization. The quantizer, electrically connected (via the clipper) to the output of the sample-and-hold, is a circuit which follows the lines of the schematics in Figure 10.6. This is called a *flash* (or *parallel*) quantization scheme because the input sample is simultaneously compared to all of the quantization thresholds i_k. These are obtained via a voltage divider realized by the cascade of equally-valued resistors shown on the left of the circuit. In this simple example, the signal is quantized over the $[-V_0, V_0]$ interval with 2 bits per sample and therefore four voltage levels are necessary. To use the notation of Section 10.1 we have that $i_0 = -V_0$, $i_1 = -0.5 V_0$, $i_2 = 0$, $i_3 = 0.5 V_0$, and $i_4 = V_0$. These boundary voltages are fed to a parallel structure of op-amps acting as comparators; all the comparators, for which the reference voltage is less than the input voltage, will have a high output and the logic (XOR gates and diodes) will convert these high outputs into the proper binary value (in this case, a least significant and most significant bit (LSB and MSB)). Because of their parallel structure, flash quantizers exhibit a very short response time which allows their use with high sampling frequencies. Unfortunately, they require an exponentially increasing number of components per output bit (the number of comparators is on the order of 2^B where B is the rate of the signal in bits per sample). Other architectures are based on iterative conversion techniques; while they require fewer components, they are typically slower.

10.3 D/A Conversion

In the simplest case, digital-to-analog (D/A) conversion is performed by a circuit which translates the binary internal representation of the samples into a voltage output value; the voltage is kept constant between interpolation times, thereby producing a zero-order-hold interpolated signal. Further analog filtering may be employed to reduce the artifacts of the interpolation (Sect. 11.4.2).

A typical example of D/A circuitry is shown in Figure 10.6. The op-amp is configured as a voltage adder and is connected to a standard $R/2R$ ladder. The ladder has as many "steps" as the number of bits B used to encode each sample of the digital signal.[1] Each bit is connected to a non-inverting buffer which acts as a switch: for each "1" bit, the voltage V_0 is connected to the associated step of the ladder, while for each "0" bit the step is connected

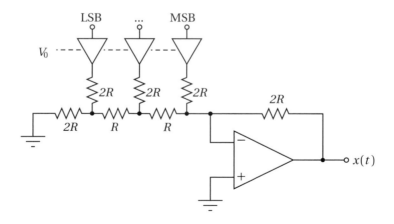

Figure 10.6 Simple three-bit D/A converter.

---

[1] We previously used the letter R to denote the rate of a quantized signal expressed in bits per sample; we use B here to avoid confusion with the resistor's value.

to the ground. By repeatedly applying Thevenin's theorem to each step in the ladder it is easy to show that a voltage of V_0 appearing at the k-th bit position (with $k = 0$ indicating the LSB and $k = B - 1$ indicating the MSB) is equivalent to a voltage of $V_0/2^{B-k}$ applied to the inverting input of the op-amp with an impedance of $2R$. By the property of superposition (applied here to the linear ladder circuit), the output voltage over the time interval $[nT_s, (n+1)T_s]$ is

$$x(t) = \sum_{k=0}^{B-1} b_k^n \frac{V_0}{2^{B-k}}$$

where $b_{B-1}^n b_{B-2}^n \cdots b_1^n b_0^n$ is the B-bit binary representation of the sample $x[n]$. Note that $0 \leq x(t) < V_0$; for a reconstruction interval between $-V_0$ and V_0, one can halve the value of the feedback resistor in the adder and add an offset of $-V_0$ Volts to the output.

Examples

Example 10.1: Nonuniform quantization

All signal processing systems have an intrinsic *range* of admissible input values; in analog systems, for instance, the nominal range corresponds to the voltage excursion for which the system behaves linearly. In a digital system the *internal* range is determined by the numerical precision used in representing both the samples and the filter coefficients, while the input range is fixed by the A/D converter, again in terms of maximum and minimum voltages. If the analog samples exceed the specifications, the system introduces some type of *nonlinear distortion*:

- In analog system the distortion depends on the saturation curves of the active components. Most of the audiophile debate on the superiority of tubes over solid-state electronics hinges upon the "natural" smoothness of the saturation curves sported by vacuum tubes. Also, note that sometimes the nonlinear distortion caused by saturation is a sought-after artifact, as in the case of guitar effects.

- In digital systems the distortion depends on the truncation characteristic of the A/D converter: however, quantization clipping is very disturbing acoustically and great care is taken to avoid it.

In order to avoid distortion, a possible approach is to *normalize* the input signal so that it never exceeds the nominal range. The *dynamic range* of a signal measures the ratio between its highest- and lowest-energy sections;

in the case of audio, the dynamic range measures how loud a signal gets with respect to silence. If an audio signal contains some amplitude peaks which exceed the nominal range but is otherwise not very loud, a normalization procedure would "squish" most of the signal to inaudible levels. This is for instance the case of speech, in which plosive sounds constitute high-energy bursts; if speech is normalized uniformly (i.e. if its amplitude is multiplied by a factor so that the peaks are within the range) softer sounds such as sibilants and some vowels end up being almost unintelligible. The solution is to use *nonuniform* dynamic range compression, also known as *companding*. The μ-law compander, commonly used in both analog and digital voice communication devices, performs an instantaneous nonlinear transformation of a signal according to the input-output relation:

$$\mathscr{C}\{x\} = \text{sgn}(x)\frac{\ln(1+\mu|x|)}{\ln(1+\mu)} \tag{10.17}$$

where it is assumed that $|x| < 1$; the parameter μ controls the shape of the resulting curve and it is usually $\mu = 255$. The transformation is shown in Figure 10.7 and we can see that it compresses the large-amplitude range while steeply expanding the low-amplitude range (hence the name "compander"); in so doing a previously peak-normalized signal regains power in the low-amplitude regions.

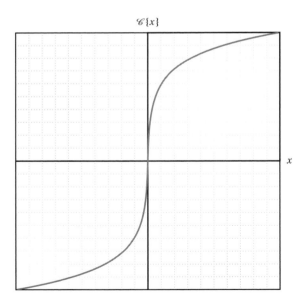

Figure 10.7 μ-law characteristic.

If the companded signal is needed in quantized form, instead of running the analog values through an analog compander and then quantize uniformly we can directly design a nonuniform quantizer which follows the μ-law characteristic. The method is shown in Figure 10.8 for a 3-bit quantizer with only the positive x-axis displayed. The uniform subdivision of the compander's output defines four unequal quantization intervals; the splitting points are obtained using the inverse μ-law transformation as

$$i_k = \text{sgn}(2^{-k}-1)\frac{1}{\mu}[(1+\mu)^{|2^{-k}-1|} - 1]$$

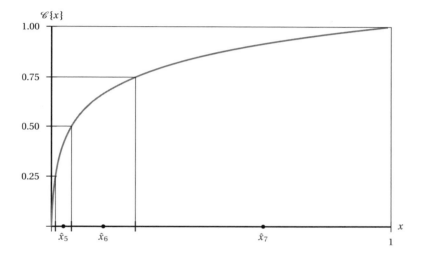

Figure 10.8 Quantized μ-law characteristic (axes not in scale and x_4 not shown for lack of space).

Usually, μ-law companding is used for voice signals in conjunction with 8-bit quantization. The dynamic range compression allows to represent a wider dynamic range at a lower SNR than with uniform quantization. To obtain a quantitative measure of the gain, consider that over 8 bits the smallest quantization interval is

$$\Delta = \frac{1}{\mu}[(1+\mu)^{1/128} - 1] \approx 1.7 \cdot 10^{-4}$$

To obtain the same resolution with a uniform quantizer we should use

$$\lceil -\log_2 \Delta \rceil = 13 \text{ bits}$$

so that the SNR gain is approximately 30 dBs.

Further Reading

Quantization is a key topic both in A/D and D/A conversion and in signal compression. Yet it is often overlooked in standard texts. The book by A. Gersho and R. M. Gray, *Vector Quantization and Signal Compression* (Springer, 1991) provides a good discussion of the topic. An excellent overview of quantization is given in R. M. Gray and D. L. Neuhof's article "Quantization", in *IEEE Transactions on Information Theory* (October 1998).

Exercises

Exercise 10.1: Quantization error – I. Consider a stationary i.i.d. random process $x[n]$ whose samples are uniformly distributed over the $[-1,1]$ interval. Consider a quantizer $\mathcal{Q}\{\cdot\}$ with the following characteristic:

$$\mathcal{Q}\{x\} = \begin{cases} -1 & \text{if } -1 \leq x < -0.5 \\ 0 & \text{if } -0.5 \leq x \leq 0.5 \\ 1 & \text{if } 0.5 < x \leq 1 \end{cases}$$

The quantized process $y[n] = \mathcal{Q}\{x[n]\}$ is still i.i.d.; compute its power spectral density $P_y(e^{j\omega})$

Exercise 10.2: Quantization error – II. Consider a stationary i.i.d. random process $x[n]$ whose samples are uniformly distributed over the $[-1,2]$ interval. The process is quantized with a 1-bit quantizer $\mathcal{Q}\{\cdot\}$ with the following characteristic:

$$\mathcal{Q}\{x\} = \begin{cases} -1 & \text{if } x < 0 \\ +1 & \text{if } x \geq 0 \end{cases}$$

Compute the signal to noise ratio at the output of the quantizer.

Exercise 10.3: More samples or more bits? You have a continuous-time signal (for example, a music source), which you want to store on a digital medium such as a memory card. Assume, for simplicity, that the signal has already been sampled (but *not* quantized) with a sampling frequency $F_s = 32{,}000$ Hz with no aliasing. Assume further that the sampled signal can be modeled as a white process $x[n]$ with power spectral density

$$P_x(e^{j\omega}) = \sigma_x^2$$

and that the pdf of each sample is uniform on the $[-1,1]$ interval.

Now you need to quantize and store the signal. Your constraints are the following:

(a) You want to store exactly one second's worth of the input signal.

(b) The capacity of the memory card is 32,000 bytes.

(c) You can either use a 8-bit quantizer ("Quantizer A") or a 16-bit quantizer ("Quantizer B"). Both quantizers are uniform over the $[-1, 1]$ interval.

You come up with two possible schemes:

I You quantize the samples with quantizer A and store them on the card.

II You first downsample the signal by 2 (with lowpass filtering) and then use quantizer B.

Clearly both schemes fulfill your constraints. Question: which is the configuration which minimizes the overall mean square error between the original signal and the digitized signal? Show why. As a guideline, note that the MSE will be composed of two independent parts: the one introduced by the quantizers and, for the second scheme, the one which is introduced by the lowpass filter before the downsampler. For the quantizer error, you can assume that the downsampled process still remains a uniform, i.i.d. process.

Chapter 11

Multirate Signal Processing

The sampling theorem in Chapter 9 provided us with a tool to map a continuous-time signal to a sequence of discrete-time samples taken at a given sampling rate. By choosing a different sampling rate, the same continuous-time signal can be mapped to an arbitrary number of different discrete-time signals. What is the relationship between these different discrete-time sequences? Can they be transformed into each other entirely from within the discrete-time world? These are the questions that multirate theory sets out to answer.

The conversion from one sampling rate to another can always take the "obvious" route via continuous time, i.e. via interpolation and resampling. This is clearly disadvantageous, both from the point of view of the needed equipment and from the point of view of the quality loss which always takes place upon quitting the digital discrete-time domain. That was the rationale, for instance, of an infamous engineering decision taken by the audio industry in the early 90's. In those years, after compact disk players had been around for about a decade, digital cassette players started to appear in the market under the name of DAT. The decision was to use a different and highly incompatible sampling rate for the DAT with respect to the CD (48 Khz vs. 44.1 Khz) so as to make it difficult to obtain perfect digital copies of existing CDs.[1] Multirate signal processing rendered that strategy moot, as we will see.

More generally, multirate signal processing not only comes to help whenever a conversion between different standards is needed, but it is also a full-fledged signal processing tool in its own right with many fruitful applications in the design of efficient filtering schemes and of telecommunication

---

[1] While DATs are all but extinct today, the problem remains of actuality since DVDs sport a sampling rate of 48 Khz as well.

systems. Finally, multirate theory is at the cornerstone of advanced processing techniques which go under the name of time-frequency analysis.

11.1 Downsampling

Downsampling by N (also called subsampling or decimation[2]) creates a lower-rate sequence by keeping only one out of N samples in the original signal. If we call $\mathcal{D}_N$ the downsampling operator, we have

$$x_{ND}[n] = \mathcal{D}_N\{x[n]\} = x[nN] \tag{11.1}$$

Downsampling effectively *discards* $N-1$ out of N samples and, as such, may cause a loss of information in the original sequence; to understand when and how this happens, we need to arrive at a frequency domain representation of the downsampled sequence.

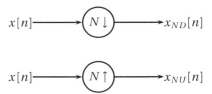

Figure 11.1 Downsampling and upsampling operators.

11.1.1 Properties of the Downsampling Operator

Let us consider, as an example, the downsampling by 2 operator $\mathcal{D}_2$ and let us write out explicitly its effect; if $x_{2D}[n] = \mathcal{D}_2\{x[n]\}$ we have

$$x[n] = \ldots, x[-2], x[-1], x[0], x[1], x[2], \ldots$$
$$x_{2D}[n] = \ldots, x[-4], x[-2], x[0], x[2], x[4], \ldots$$

Note that the time origin is extremely important, since:

$$\mathcal{D}_2\{x[n+1]\} = \ldots, x[-5], x[-3], x[1], x[3], x[5], \ldots$$

as, according to the definition, $\mathcal{D}_2\{x[n+1]\} = x[2n+1]$. We have just shown that *the downsampling operator is not time-invariant*. More precisely, the

---

[2] Technically, decimation means 9 out of 10 and refers to a roman custom of killing every 10th soldier of a defeated army...

downsampling operator is defined as *periodically time-varying* since, if $x_{ND}[n] = \mathscr{D}_N\{x[n]\}$, then:

$$\mathscr{D}_N\{x[n-kN]\} = x_{ND}[n-k] \tag{11.2}$$

It is trivial to show that the downsampling operator is indeed a linear operator.

One of the fundamental consequences of the lack of time-invariance is that, now, one of the key properties of LTI systems no longer holds for the downsampling operator; indeed, *complex sinusoids are no longer eigensequences*. As an example, consider $x[n] = (-1)^n = e^{j\pi n}$, which is the highest-frequency discrete-time sinusoid. After downsampling by 2, we obtain:

$$x_{2D}[n] = x[2n] = (-1)^{2n} = 1 \tag{11.3}$$

which is the lowest-frequency sinusoid in discrete time. This is one instance of the information loss inherent to downsampling and to understand how it operates we need to move to the frequency domain.

11.1.2 Frequency-Domain Representation

In order to obtain a frequency-domain representation of a downsampling by N, first consider the z-transform of the downsampled signal:

$$X_{ND}(z) = \sum_{n=-\infty}^{\infty} x[nN] z^{-n} \tag{11.4}$$

Now consider an "auxiliary" z-transform $X_a(z)$ defined as

$$X_a(z) = \sum_{n=-\infty}^{\infty} x[nN] z^{-nN} \tag{11.5}$$

The interest of $X_a(z)$ lies with the fact that, if we can obtain a closed-form expression for it, we can then write out $X_{ND}(z)$ simply as

$$X_{ND}(z) = X_a(z^{1/N}) \tag{11.6}$$

Clearly, $X_a(z)$ can be derived from $X(z)$, the z-transform of the original signal, by "killing off" all the terms in the z-transform sum whose index is not a multiple of N; in other words we can write

$$X_a(z) = \sum_{n=-\infty}^{\infty} \xi_N[n] x[n] z^{-n} \tag{11.7}$$

where $\xi_N[n]$ is a "selector" sequence defined as

$$\xi_N[n] = \begin{cases} 1 & \text{for } n \text{ multiple of } N \\ 0 & \text{otherwise} \end{cases}$$

The question now is to find an expression for such a sequence; to this end, let us recall a very early result about the orthogonality of the roots of unity (see Equation (4.4)), which we can rewrite as follows:

$$\sum_{k=0}^{N-1} W_N^{kn} = \begin{cases} N & \text{for } n \text{ multiple of } N \\ 0 & \text{otherwise} \end{cases} \quad (11.8)$$

where, as per usual, $W_N = e^{-j\frac{2\pi}{N}}$. Clearly, we can define our desired selector sequence as

$$\xi_N[n] = \frac{1}{N} \sum_{k=0}^{N-1} W_N^{kn}$$

and we can therefore write

$$X_a(z) = \frac{1}{N} \sum_{n=-\infty}^{\infty} \sum_{k=0}^{N-1} W_N^{kn} x[n] z^{-n}$$

$$= \frac{1}{N} \sum_{k=0}^{N-1} \sum_{n=-\infty}^{\infty} x[n] (W_N^k z^{-1})^n$$

$$= \frac{1}{N} \sum_{k=0}^{N-1} X(W_N^k z) \quad (11.9)$$

so that finally:

$$X_{ND}(z) = \frac{1}{N} \sum_{k=0}^{N-1} X(W_N^k z^{1/N}) \quad (11.10)$$

The Fourier transform of the downsampled signal is obtained by evaluating $X_{ND}(z)$ on the unit circle; explicitly, we have

$$X_{ND}(e^{j\omega}) = \frac{1}{N} \sum_{k=0}^{N-1} X(e^{j(\frac{\omega}{N} - \frac{2\pi}{N}k)}) \quad (11.11)$$

The resulting spectrum is, therefore, the scaled sum of N superimposed copies of the original spectrum $X(e^{j\omega})$; each copy is shifted in frequency by a multiple of $2\pi/N$ and the result is stretched by a factor of N. We are, in many ways, in a situation similar to that of equation (9.33) where sampling created a periodization of the underlying spectrum; here the spectra are already inherently 2π-periodic, and downsampling creates $N-1$ additional interleaved copies. Because of the superposition, aliasing can take place; this is a consequence of the potential loss of information that occurs when samples are discarded. It is easy to verify that in order for the spectral copies in (11.10) not to overlap, the maximum (positive) frequency ω_M of the original spectrum[3] must be less than π/N; this is the *non-aliasing condition* for the downsampling operator. Conceptually, fulfillment of the non-aliasing condition indicates that the discrete-time representation of the original signal is intrinsically redundant; $(N-1)/N$ of the information can be safely discarded and this is mirrored by the fact that only $1/N$ of the spectral frequency support is nonzero. We will see shortly that, in this case, the original signal can be perfectly reconstructed with an upsampling and filtering operation.

11.1.3 Examples

In the following graphical examples (Figs 11.2 to 11.6) the top panel shows the original spectrum $X(e^{j\omega})$; the second panel shows the same spectrum but plotted over a larger frequency interval so as to make its periodic nature explicit; the third panel shows (in different shades of gray) the individual components of the sum in (11.10) *before* scaling and stretching by N, i.e. the N copies $X(W_N^k e^{j\omega})$ for $k=0,1,\ldots,N-1$; the fourth panel shows the final $X_{ND}(e^{j\omega})$, with the individual components of the sum plotted with a dashed line; finally, the last panel shows $X_{ND}(e^{j\omega})$ over the usual $[-\pi,\pi]$ interval.

Downsampling by 2. If the downsampling factor is 2, the corresponding two roots of unity are just ± 1 and we have

$$X_{2D}(z) = \frac{1}{2}\left[X(z) + X(-z)\right]$$

$$X_{2D}(e^{j\omega}) = \frac{1}{2}\left[X(e^{j\frac{\omega}{2}}) + X(e^{j(\frac{\omega}{2}-\pi)})\right]$$

Figure 11.2 shows an example of downsampling by 2 for a lowpass signal whose maximum frequency is $\omega_M = \pi/2$ (i.e. a half-band signal). The non-aliasing condition is fulfilled and, in the superposition, the two shifted versions of the spectrum do not overlap. As the frequency axis expands by a factor of 2, the original half-band signal becomes full band.

---

[3] Here, for simplicity, we are imagining a lowpass real signal whose spectral magnitude is symmetric. More complex cases exist and some examples will be described next.

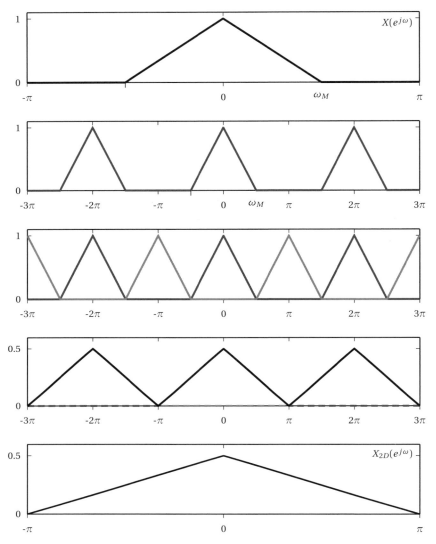

Figure 11.2 Downsampling by 2; the discrete-time signal's highest frequency is $\pi/2$ and no aliasing occurs.

Figure 11.3 shows an example in which the non-aliasing condition is violated. In this case, $\omega_M = 2\pi/3 > \pi/2$ and the spectral copies do overlap. We can see that, as a consequence, the downsampled signal loses its lowpass characteristics. Information is irretrievably lost and the original signal cannot be reconstructed. We will see in the next Section the customary way of dealing with this situation.

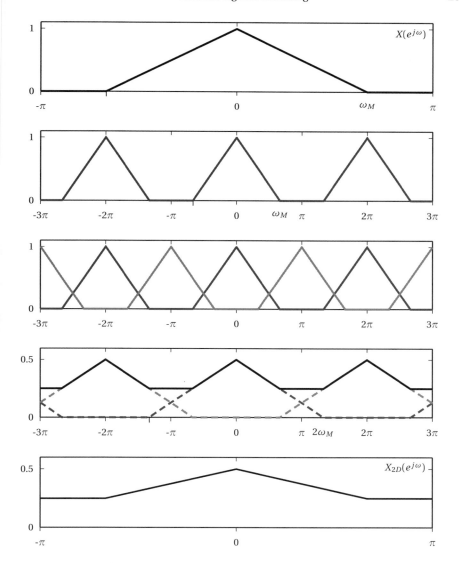

Figure 11.3 Downsampling by 2; the discrete-time signal's highest frequency is larger than $\pi/2$ (here, $\omega_M = 2\pi/3$) and aliasing corrupts the downsampled signal.

Downsampling by 3.

If the downsampling factor is 3 we have

$$X_{3D}(e^{j\omega}) = \frac{1}{3}\left[X(e^{j\frac{\omega}{3}}) + X(e^{j(\frac{\omega}{3} - \frac{2\pi}{3})}) + X(e^{j(\frac{\omega}{3} - \frac{4\pi}{3})})\right]$$

Figure 11.4 shows an example in which the non-aliasing condition is violated ($\omega_M = 2\pi/3 > \pi/3$). In particular, the superposition of the three spectral copies is such that the resulting spectrum is flat.

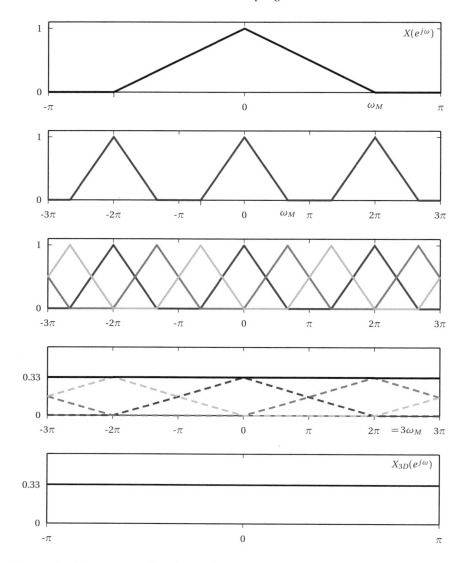

Figure 11.4 Downsampling by 3; the discrete-time signal's highest frequency is larger than $\pi/3$ (here, $\omega_M = 2\pi/3$) and aliasing corrupts the downsampled signal. Note the three replicas which contribute to the final spectrum.

Downsampling of a Highpass Signal. Figure 11.5 shows an example of downsampling by 2 of a half-band *highpass* signal. Since the signal occupies only the upper half of the $[0, \pi]$ frequency band (and, symmetrically, only the lower half of the $[-\pi, 0]$ interval), the interleaved copies do not overlap and, technically, there is no aliasing. The shape of the signal, however, is changed by the downsampling operation and what started out as a highpass signal is transformed into a lowpass signal. The details of the transformation are clearer if, for the sake of example, we consider a *complex* half-band highpass signal in which the positive and negative parts of the spectrum are different. The steps involved in the downsampling of such a signal are detailed in Figure 11.6 and it is apparent how the low and high parts of the spectrum are interchanged. In both cases the original signal can be exactly reconstructed (since there is no destructive overlap between spectral copies) but the required procedure (which we will study in the exercises) is more complex than a simple upsampling.

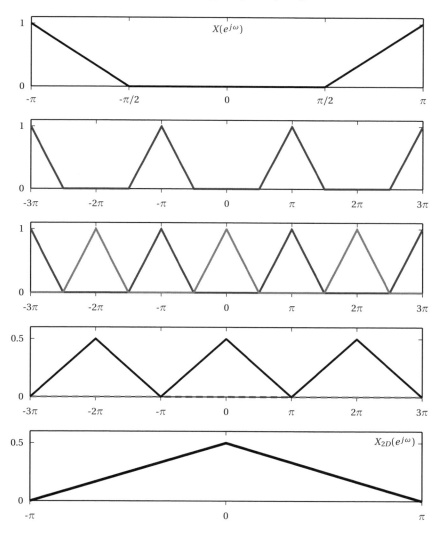

Figure 11.5 Downsampling by 2 of a highpass signal; note how aliasing changes the nature of the spectrum.

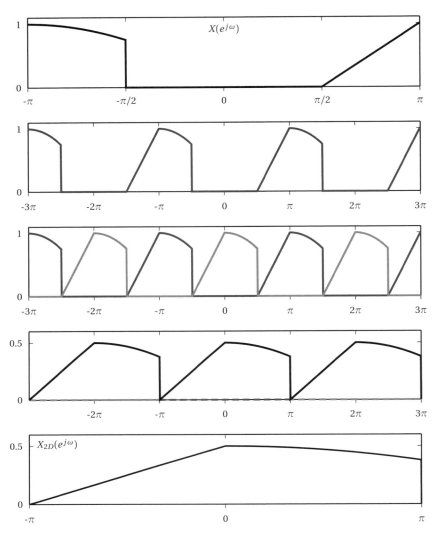

Figure 11.6 Downsampling by 2 of a *complex* highpass signal; the asymmetric spectrum helps to understand how aliasing works.

11.1.4 Downsampling and Filtering

Because of aliasing, it is customary to filter a signal prior to downsampling. The filter should be designed to eliminate aliasing by removing the high frequency components which fold back onto the lower frequencies (remember how the $(-1)^n$ signal ended up as the constant 1). For a downsampling by N, this is accomplished by a lowpass filter with cutoff frequency $\omega_c = \pi/N$, and the resulting structure is depicted in Figure 11.7.

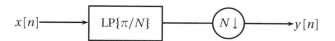

Figure 11.7 Anti-aliasing filter before downsampling. The filter is typically a low-pass filter with cut-off frequency π/N.

An example of the processing chain is shown in Figure 11.8 for a down-sampling factor of 2; a half-band lowpass filter is used to truncate the signal's spectrum outside of the $[-\pi/2, \pi/2]$ interval and then downsampling

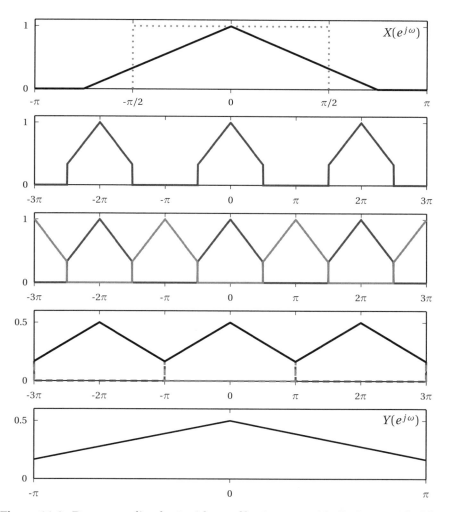

Figure 11.8 Downsampling by 2 with pre-filtering to avoid aliasing; an ideal low-pass with cutoff frequency of $\pi/2$ is used.

proceeds as usual with non-overlapping spectral copies. Clearly, some information is lost and the original signal cannot be recovered exactly but the distortion is controlled and less disruptive than foldover aliasing.

11.2 Upsampling

Upsampling by N produces a higher-rate sequence by creating N samples out of every sample in the original signal. The upsampling operation consists simply in inserting $N-1$ zeros between every two input samples; if we call $\mathcal{U}_N$ the upsampling operator, we have

$$x_{NU}[n] = \mathcal{U}_N\{x[n]\} = \begin{cases} x[k] & \text{for } n = kN, \ k \in \mathbb{Z} \\ 0 & \text{otherwise} \end{cases} \quad (11.12)$$

Upsampling is a much "nicer" operation than downsampling since no information is lost and the original signal can always be exactly recovered by downsampling:

$$\mathcal{D}_N\{\mathcal{U}_N\{x[n]\}\} = x[n] \quad (11.13)$$

Furthermore, the spectral description of upsampling is extremely simple; in the z-transform domain we have

$$\begin{aligned} X_{NU}(z) &= \sum_{n=-\infty}^{\infty} x_{NU}[n] z^{-n} \\ &= \sum_{k=-\infty}^{\infty} x[k] z^{-kN} = X(z^N) \end{aligned} \quad (11.14)$$

and therefore

$$X_{NU}(e^{j\omega}) = X(e^{j\omega N}) \quad (11.15)$$

so that upsampling is simply a contraction of the frequency axis by a factor of N. The inherent 2π-periodicity of the spectrum must be taken into account so that, in this contraction, the periodic repetitions of the base spectrum are "drawn in" the $[-\pi, \pi]$ interval. The effects of upsampling are shown graphically for a simple signal in Figures 11.9 to 11.11; in all figures the top panel shows the original spectrum $X(e^{j\omega})$ over $[-\pi, \pi]$; the middle panel shows the same spectrum over a wider range to make the 2π- periodicity explicitly; the last panel shows the upsampled spectrum $X_{NU}(e^{j\omega})$, highlighting the rescaling of the $[-N\pi, N\pi]$ interval.

Multirate Signal Processing

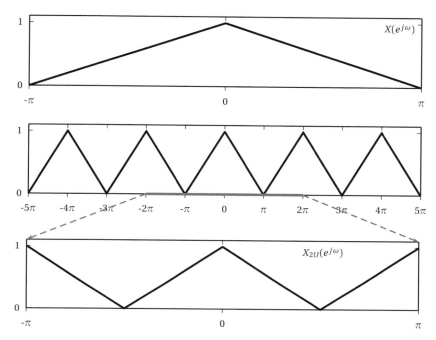

Figure 11.9 Upsampling by 2.

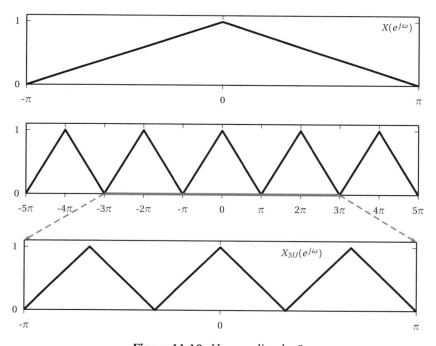

Figure 11.10 Upsampling by 3.

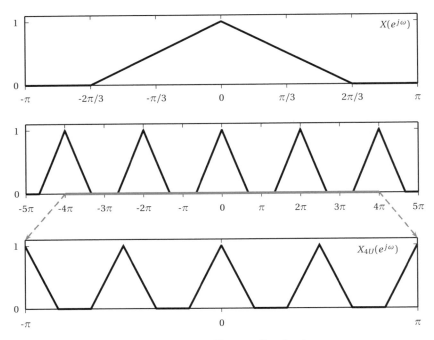

Figure 11.11 Upsampling by 4.

11.2.1 Upsampling and Interpolation

However simple, an upsampled signal suffers from two drawbacks. In the time domain, the upsampled signal does not look "natural" since there are $N-1$ zeros between every sample drawn from the input. Thus, a "smooth"[4] input signal no longer looks smooth after upsampling, as shown in the top two panels of Figure 11.13. A solution would be to try to interpolate the original samples in order to "fill in" the gaps. In the frequency domain, on the other hand, the repetitions of the base spectrum, which are drawn in by the upsampling, do not look as if they belong to the $[-\pi, \pi]$ interval and it seems natural to try to remove them. These two problems are actually one and the same and they can be solved by an appropriate filter.

Figure 11.12 Interpolation after upsampling. The interpolation filter is equivalent to a lowpass filter with cut-off frequency π/N.

---

[4] Informally, by "smooth" we refer to discrete-time signals which do not exhibit wide amplitude jumps between samples.

The problem of filling the gaps between nonzero samples in an upsampled sequence is, in many ways, similar to the discrete- to continuous-time interpolation problem of Section 9.4, except that now we are operating entirely in discrete-time. If we adapt the interpolation schemes that we have already studied, we can describe the following cases:

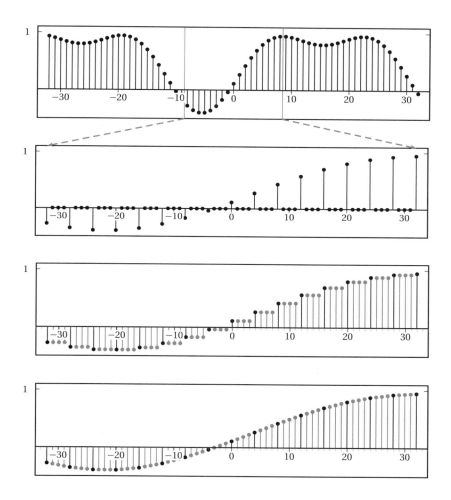

Figure 11.13 Upsampling by 4 in the time domain; original signal (top panel); portion of the upsampled signal (second panel); interpolated signal with zero- and first-order interpolators (third and fourth panels).

Zero-Order Hold. In this discrete-time interpolation scheme, also known as *piecewise-constant interpolation*, after upsampling by N, we use a filter with impulse response:

$$h_0[n] = \begin{cases} 1 & n = 0, 1, \ldots, N-1 \\ 0 & \text{otherwise} \end{cases} \qquad (11.16)$$

which is shown in Figure 11.14. This interpolation filter simply repeats the original input samples N times, giving a staircase approximation as shown for example in the third panel of Figure 11.13.

First-Order Hold. In this discrete-time interpolation scheme, we obtain a piecewise linear interpolation after upsampling by N by using

$$h_1[n] = \begin{cases} 1 - \dfrac{|n|}{N} & |n| < N \\ 0 & \text{otherwise} \end{cases} \qquad (11.17)$$

The impulse response is the familiar triangular function[5] shown in Figure 11.14. An example of the resulting interpolation is shown in Figure 11.13.

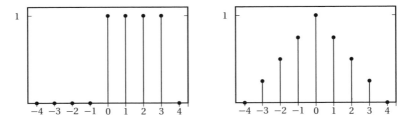

Figure 11.14 Discrete-time zero-order and first-order interpolators for $N=4$.

Sinc Interpolation. We know that, in continuous time, the smoothest interpolation is obtained by using a sinc function. This holds in discrete-time as well, and the resulting interpolation filter is a discrete-time sinc:

$$h[n] = \operatorname{sinc}\left(\frac{n}{N}\right) \qquad (11.18)$$

Note that the sinc above is equal to one for $n = 0$ and is equal to zero at all integer multiples of N, $n = kN$; this fulfills the interpolation condition that, after interpolation, the output equals the input at multiples of N (i.e. $(h * x_{NU})[n] = x[n]$ for $n = kN$).

---

[5] Once again, let us note that the triangle is the convolution of two rects, $h_1[n] = (1/N)(h_0[n] * h_0[n])$.

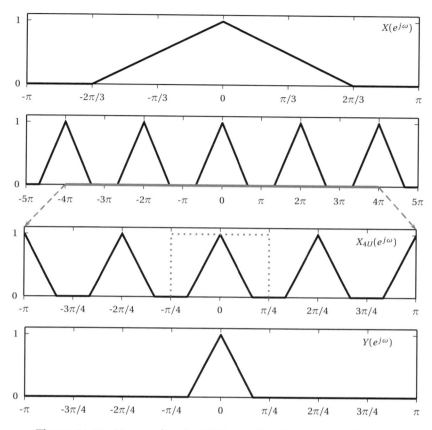

Figure 11.15 Upsampling by 4 followed by ideal lowpass filtering.

The three impulse responses above are all lowpass filters; in particular, the sinc interpolator is an ideal lowpass with cutoff frequency $\omega_c = \pi/N$ while the others are approximations of the same. As a consequence, the effect of the interpolator in the frequency domain is the removal of the $N-1$ repeat spectra which have been drawn in the $[-\pi, \pi]$ interval. An example is shown in Figure 11.15 where the signal in Figure 11.11 is filtered by an ideal lowpass filter with cutoff $\pi/4$. It turns out that the smoothest possible interpolation in the time domain corresponds to the removal of the spectral repetitions in the frequency domain. An interpolation by the zero-order, or first-order holds, only attenuates the replicas instead of performing a full removal, as we can readily see by considering their frequency responses. Since we are in discrete-time, however, there are no difficulties associated to the design of a digital lowpass filter which performs extremely well. This is in contrast to the design of discrete—to continuous—time interpolators, which are analog designs. That is why sampling rate changes are much more attractive in the discrete-time domain.

11.3 Rational Sampling Rate Changes

So far we have examined methods which change (multiply or divide) the implicit rate of a discrete-time signal by an integer factor. By combining upsampling and downsampling, we can achieve arbitrary rational sampling rate changes. Typically, a rate change by N/M is obtained by cascading an upsampler by N, a lowpass filter and a downsampler by M. The filter's cutoff frequency is the minimum of $\{\pi/N, \pi/M\}$; this follows from the fact that upsampling and downsampling require lowpass filters with cutoff frequencies of π/N and π/M respectively, and the minimum cutoff frequency dominates in the cascade. A block diagram of this system is shown in Figure 11.16.

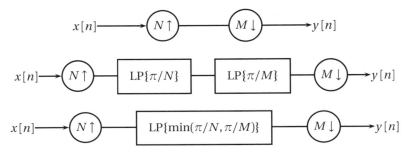

Figure 11.16 Sampling rate change by a rational factor N/M. Cascade of upsampling and downsampling (top diagram); cascade with interpolation after upsampling, and filtering before subsampling (middle diagram); the same cascade where the two filters are replaced by the narrower of the two (bottom diagram).

The order of the upsampling and downsampling operators is crucial since, in general, the operators are not commutative. It is easy to appreciate this fact by means of a simple example; for a given sequence $x[n]$ it is

$$\mathscr{D}_2\left(\mathscr{U}_2(x[n])\right) = x[n]$$

$$\mathscr{U}_2\left(\mathscr{D}_2(x[n])\right) = \begin{cases} x[n] & \text{for } n \text{ even} \\ 0 & \text{for } n \text{ odd} \end{cases}$$

Conceptually, using an upsampler first is the logical thing to do since no information is lost in a sample rate increase. Interestingly enough, however, if the downsampling and upsampling factors N and M are coprime, the operators do commute:

$$\mathscr{D}_N\left(\mathscr{U}_M(x[n])\right) = \mathscr{U}_M\left(\mathscr{D}_N(x[n])\right) \tag{11.19}$$

The proof of this property is left as an exercise. This property can be put to use in a rational sampling rate converter to minimize the number of operations, per sample in the middle filter.

As an example, we are now ready to solve the audio conversion problem which was quoted at the beginning of the Chapter. To convert an audio file sampled at 44 Khz ("CD-quality") into an audio file which can be played back at 48 Khz ("DVD-quality") a rate change of 12/11 is necessary; this can be achieved with the system shown at the top of Figure 11.17. Conversely, DVD to CD conversion can be performed with a 11/12 rate changer, shown at the bottom of Figure 11.17.[6]

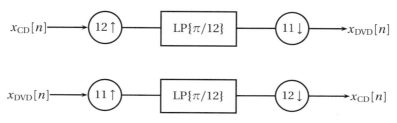

Figure 11.17 Conversion from CD to DVD and vice-versa conversely using rational sampling rate changes.

11.4 Oversampling

Manipulating the sampling rate is useful a many more ways beyond simple conversions between audio standards: oversampling is a case in point. The term "oversampling" describes a situation in which a signal's sampling rate is made to be deliberately higher than the minimum required by the sampling theorem. Oversampling is used to improve the performance of A/D and D/A converters.

11.4.1 Oversampled A/D Conversion

If a continuous-time signal $x(t)$ is bandlimited, the sampling theorem guarantees that we can choose a sampling period T_s such that no error is introduced by the sampling operation. The only source of error in A/D conversion remains the distortion due to quantization; oversampling, in this case, allows us to reduce this error by increasing the underlying sampling rate.

---

[6] In reality, the compact disk sampling rate is 44.1 Khz, so that the exact factor for the rational rate change should be 160/147. This is usually less practical so that other strategies are usually put in place. See for instance Exercise 12.2.

Under certain assumptions on the statistical properties of the input signal, the quantization error associated to A/D conversion has been modeled in Section 10.1.1 as an additive noise source. If $x(t)$ is a Ω_N-bandlimited signal and $T_s = \pi/\Omega_N$, we can write:

$$\hat{x}[n] = x(nT_s) + e[n]$$

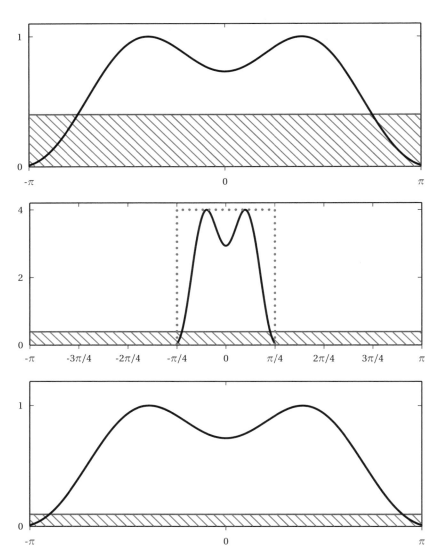

Figure 11.18 Oversampling for A/D conversion: signal's PSD (black) and quantization error's PSD (gray). Critically sampled signal (top panel); oversampled signal (middle panel); filtered and downsampled signal (bottom panel).

with $e[n]$ a white process of variance

$$P_e = \frac{\Delta^2}{12}$$

where Δ is the quantization interval. This is represented pictorially in the top panel of Figure 11.18 which shows the power spectral densities for an arbitrary critically sampled signal and for the associated quantization noise.[7] The bottom panel of Figure 11.18 shows the same quantities for the case in which the input signal has been oversampled by a factor of four, i.e. for the signal

$$x_u[n] = x(nT_u), \qquad T_u = T_s/4$$

The scale change between signal and noise comes from equation (9.34) but note that the signal-to-noise ratio of the oversampled signal is still the same. However, now we are in the digital domain and it is easy to build a discrete-time filter which removes the quantization error outside of the support of the signal (i.e. outside of the $[-\pi/4, \pi/4]$ interval) and this improves the SNR. Once the out-of-band noise is removed, we can use a downsampler by 4 to obtain a critically sampled signal for which the signal to noise ratio has improved by a factor of 4 (or, alternatively, by 6 dB). The processing chain is shown in Figure 11.19 for a generic oversampling factor N; as a rule of thumb, the signal-to-noise ratio is improved by about 3 dB per octave of oversampling, that is, each doubling of the sampling rate reduces the noise variance by a factor of two, which is $20 \log_{10}(2) \simeq 3$ dB.

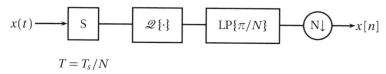

$T = T_s/N$

Figure 11.19 Oversampled A/D conversion chain.

The above example is deliberately lacking rigor in the derivations since it turns out that a precise analysis of A/D oversampling is very difficult. It is intuitively clear that some of the quantization noise will be rejected by this procedure, but the fundamental assumption that the input signal is white (and therefore that the quantization noise is uncorrelated) does not hold in reality. In fact, as the sampling rate increases, successive samples exhibit a higher and higher degree of correlation and most of the quantization noise power ends up falling within the band of the signal.

---

[7] In our previous analysis of the quantization error we have assumed that the input is uncorrelated; therefore the PSD in the figure should be flat for both the error *and* the signal. We, nevertheless, use a nonflat shape for the signal both for clarity and to stress the fact that, obviously such an assumption for the input is clearly flawed.

11.4.2 Oversampled D/A Conversion

The sampling theorem states that, under the hypothesis of a bandlimited input, sampling is invertible via a sinc interpolation. The sinc filter is an ideal filter and therefore it is not realizable either in the digital or in the analog domain. The analog sinc, therefore, must be approximated by some

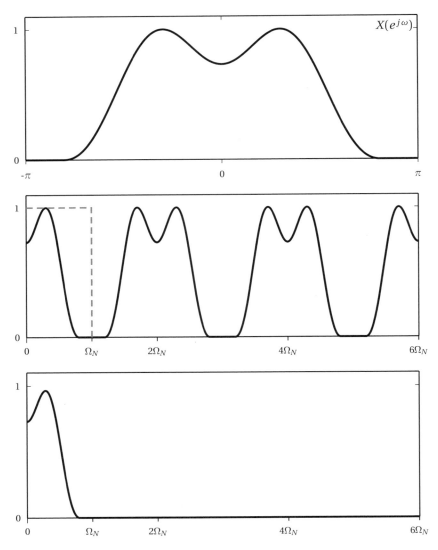

Figure 11.20 Ideal (sinc) D/A interpolation. Original spectrum (top panel); periodized analog spectrum and, in gray, frequency response of the sinc interpolator (middle panel); analog spectrum (bottom panel).

realizable interpolation filter. Recall that, once the interpolation period T_s is chosen, the continuous-time signal created by the interpolator is the mixed-domain convolution (9.11), which we rewite here:

$$x_c(t) = \sum_{n=-\infty}^{\infty} x[n] I\left(\frac{t-nT_s}{T_s}\right)$$

In the frequency domain this becomes

$$X_c(j\Omega) = \frac{\pi}{\Omega_N} I\left(j\frac{\Omega}{2\Omega_N}\right) X(e^{j\pi\Omega/\Omega_N}) \qquad (11.20)$$

with, as usual, $\Omega_N = \pi/T_s$. The above expression is the product of two terms; the last is the periodic digital spectrum, stretched so as to be $2\Omega_N$-periodic and the first is the frequency response of the analog interpolation filter, again stretched by $2\Omega_N$. In the case of sinc interpolation, the frequency response is a rect with cutoff frequency Ω_N, which "kills off" all the repetitions except for the baseband period of the periodic spectrum. The result of sinc interpolation is represented in Figure 11.20; the top panel shows the spectrum of an arbitrary discrete-time signal, the middle panel shows the two terms of Equation (11.20) with the sinc response dashed in gray, and the bottom panel shows the resulting analog spectrum. In both the middle and bottom panels only the positive frequency axis is shown since all signals are assumed to be real and, consequently, the magnitude spectra are symmetric.

With a realizable interpolator, the stopband of the interpolation filter cannot be uniformly zero and its transition band cannot be infinitely sharp. As a consequence, the spectral copies to the left and right of the baseband will "leak through" in the reconstructed analog signal. It is important to remark at this point that the interpolator filter is an analog filter and, as such, quite delicate to design. Without delving into too many details, there are no FIR filters in the continuous-time domain so that all analog filters are affected by stability problems and by design complexities associated to the passive and active electronic components. In short, a good interpolator is difficult to design and expensive to produce; so much so, in fact, that most of the interpolators used in practical circuitry are just zero-order holds. Unfortunately, the frequency response of the zero-order hold is quite poor; it is indeed easy to show that:

$$I_0(j\Omega) = \text{sinc}\left(\frac{\Omega}{2\pi}\right)$$

and that this response, while lowpass in nature, decays only as $1/\Omega$. The results of zero-order hold D/A conversion are shown in Figure 11.21; the top panel shows the original digital spectrum and the middle panel shows the two terms of Equation (11.20) with the magnitude response of the interpolator dashed in gray. The spectrum of the interpolated signal (shown in the bottom panel) exhibits several non-negligible instances of high-frequency

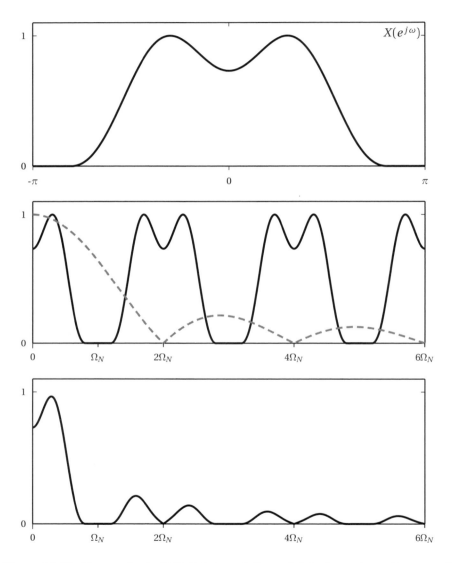

Figure 11.21 Zero-order hold D/A interpolation. Original spectrum (top panel); periodized analog spectrum and, in gray, frequency response of the zero-order hold interpolator (middle panel); resulting analog spectrum (bottom panel).

leakage centered around the multiples of twice the Nyquist frequency.[8] These are particularly undesirable in audio applications (such as in a CD player). Rather than using expensive and complex analog filters, the performance of the D/A converter can be dramatically improved if we are willing to perform the conversion at a higher rate than the strict minimum. This is achieved by oversampling the signal in the digital domain and the block diagram of the operation is shown in Figure 11.22. Note that this is a paradigmatic instance of cheap and easy discrete-time processing solving an otherwise difficult analog design: the lowpass filter used in discrete-time oversampling is an FIR with arbitrarily high performance, a filter which is much easier to design than an analog lowpass and has no stability problems. The only price paid is an increase in the working frequency of the converter.

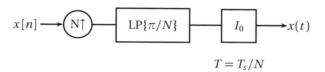

$T = T_s/N$

Figure 11.22 Oversampled D/A conversion chain.

Figure 11.23 details an example of D/A conversion with an oversampling factor of two. The top panel shows the spectrum of the oversampled discrete-time signal, together with the associated repetitions in the $[-\pi, \pi]$ interval which are going to be filtered out by a lowpass filter with cutoff $\pi/2$. The discrete-time filter response is dashed in gray in the top panel and, while the displayed characteristic is that of an ideal lowpass, note that in the discrete-time domain, we can approximate a very sharp filter rather easily. The two terms of Equation (11.20) (with the magnitude response of the interpolator dashed in gray) are shown in the middle panel; now the interpolation frequency is $\Omega_{N,O} = 2\Omega_N$, i.e. twice the frequency used in the previous example, in which the signal was critically sampled. Shrinking the spectrum in the digital domain and stretching in the analog makes sure that the analog spectrum is unchanged around the baseband. The final spectrum of the interpolated signal is shown in the bottom panel and we can notice how the first high frequency leakage occurs at twice the frequency of the previous example and is smaller in amplitude. An oversampling of N with $N > 2$ will

---

[8] Another distortion introduced by the zero-order hold interpolator is due to the nonflat response around zero in the passband; here, we will simply ignore this additional deviation from the ideal case, noting that this distortion can be easily compensated for either in the analog domain by an inverse filter or in the digital domain by an appropriate prefilter.

push the leakage even higher up in frequency; at this point a very simple analog lowpass (with a very large transition band) will suffice to remove all undesired frequency components.

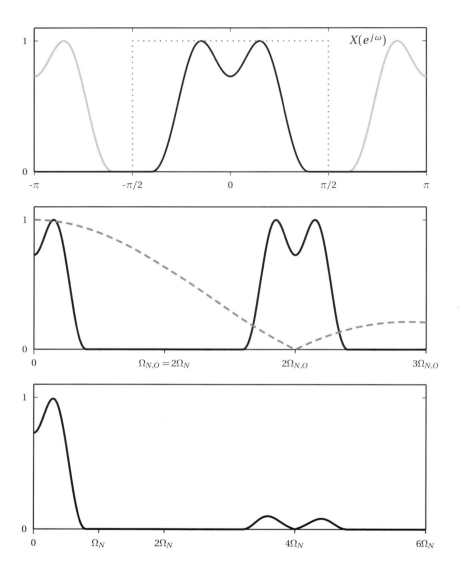

Figure 11.23 Oversampled zero-order hold D/A interpolation, oversampling by 2. Oversampled spectrum with digital interpolation filter response in gray (top panel); periodized analog spectrum and, in gray, frequency response of the zero-order hold interpolator at twice the minimum Nyquist frequency (middle panel); resulting analog spectrum (bottom panel).

Examples

Example 11.1: Radio over the phone

In the heyday of radio (up to the 50's), a main station would often have the necessity of providing audio content ahead of time to the ancillary broadcasters in its geographically distributed network. Before digital signal processing even existed, and before high-bandwidth communication lines became a possibility, most point-to-point real-time communications were over a standard telephone line and audio distribution was no exception. However, since telephone lines have a much smaller bandwidth than good quality audio, the idea was to play the content at lower speed so that the resulting bandwidth could be made to fit into the telephone band. At the other end, a tape would record the signal and then the signal could be sped up to normal pitch. In the continuous-time world, we know that (see (9.8)):

$$\text{FT}\{s(\alpha t)\} = \frac{1}{\alpha} S\left(j\frac{\Omega}{\alpha}\right)$$

so that a slowing down factor of two ($\alpha = 1/2$) would halve the spectral occupancy of the signal.

Today, with digital signal processing at our disposal, we have many more choices and here we will explore the difference between a discrete-time version of the analog scheme of yore and a full-fledged digital communication system such as the one we will study in detail in Chapter 12. Assume we have a DVD-quality audio signal $s[n]$; the signal is finite-length and it corresponds to 30 minutes of playback time. Recall that "DVD-quality" means that the audio is sampled at 48 KHz with 24 bits per sample and using 24 bits means that practically we can neglect the SNR introduced by the quantization. We want to send this signal over a telephone line knowing that the line is bandlimited to 3840 Hz and that the impairment introduced by the transmission can be modeled as a source of noise which brings down the SNR of the received signal to 40 dB.

Consider the transmission scheme in Figure 11.24; since the D/A is fixed by design (it is difficult to tune the frequency of a converter), we need to shrink the spectrum of the audio signal using multirate processing. The (positive) bandwidth of the DVD-audio signal is 24 KHz, while the telephone channel is limited to 3840 Hz. We have that

$$\frac{24{,}000}{3{,}840} = 6.25$$

and this is the factor by which we need to upsample; this can be achieved with a combination of a 25-times upsampler followed by a 4-times down-

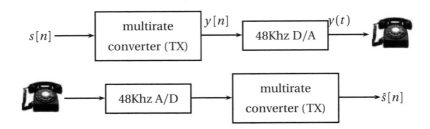

Figure 11.24 Transmission scheme for high-quality audio over a phone line.

sampler as in Figure 11.25 where $L_{TX}(z)$ is a lowpass filter with cutoff frequency $\pi/25$ and gain $L_0 = 4/25$. At the receiver the chain is inverted, with an upsampler by four, followed by a lowpass filter with cutoff frequency $\pi/4$ and gain $L_0 = 25/4$ followed by a 25-times downsampler.

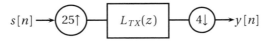

Figure 11.25 A 6.25-times upsampler.

Because of the upsampling (which translates to a slowed-down signal) it will take a little over three hours to send the audio ($6.25 \times 30 = 187.5$ minutes). The quality of the received signal is determined by the SNR of the telephone line; the in-band noise is unaffected by the multirate processing and so the final audio will have an overall SNR of 40 dBs.

Now let us compare the above solution to a fully digital communication scheme. For a telephone line with the bandwidth and the SNR specified above a commercially available digital modem can reliably achieve a throughput of 32 kbits per second. The 30-minute DVD-audio file contains ($30 \times 60 \times 48,000 \times 24$) bits. At 32 kbps, we will need approximately 18 hours to transmit the signal! The upside, however, is that the received audio will indeed be identical to the source, i.e. DVD-quality. Alternatively, we can sacrifice quality for time: if we quantize the original signal at 8 bits per sample, so that the SRN is approximately 48 dB, the transmission time reduces to 6 hours. Clearly, a modern audio transmission system would employ some advanced data compression scheme to reduce the necessary throughput.

Example 11.2: Spectral cut and paste

By using a suitable combination of upsampling and downsampling we can implement some nice tricks, such as swapping the upper and lower parts of a signal's spectrum. Consider a discrete-time signal $x[n]$ with the spectrum as in Figure 11.26. If we process the signal with the network in Figure 11.27 where the filters $L(z)$ and $H(z)$ are half-band lowpass and highpass respectively, the output spectrum will be swapped as in Figure 11.28.

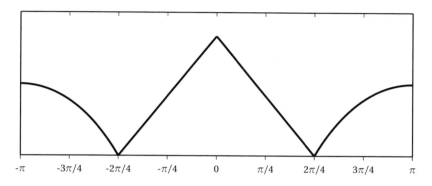

Figure 11.26 Original spectrum.

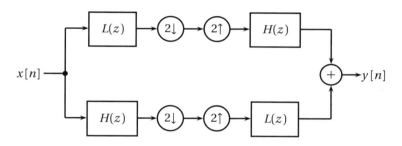

Figure 11.27 Spectral "swapper".

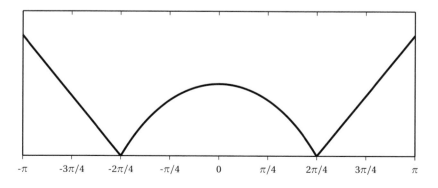

Figure 11.28 "Swapped" spectrum.

Further Reading

Historically, the topic of different sampling rates in signal processing was first treated in detail in R. E. Crochiere and L. R. Rabiner's, *Multirate Digital Signal Processing* (Prentice-Hall, 1983). With the advent of filter banks and wavelets, more recent books give a detailed treatment as well, such as P. P. Vaidyanathan, *Multirate Systems and Filter Banks* (Prentice Hall, 1992), and M. Vetterli and J. Kovacevic's, *Wavelets and Subband Coding* (Prentice Hall, 1995). Please note that the latter is now available in open access, see http://www.waveletsandsubbandcoding.org.

Exercises

Exercise 11.1: Multirate identities. Prove the following two identities:

(a) Downsampling by 2 followed by filtering by $H(z)$ is equivalent to filtering by $H(z^2)$ followed by downsampling by 2.

(b) Filtering by $H(z)$ followed by upsampling by 2 is equivalent to upsampling by 2 followed by filtering by $H(z^2)$.

Exercise 11.2: Multirate systems. Consider the input-output relations of the following multirate systems. Remember that, technically, one cannot talk of "transfer functions" in the case of multirate systems since sampling rate changes are not time invariant. It may happen, though, that by carefully designing the processing chain, this said relation does indeed implement a transfer function.

(a) Find the overall transformation operated by the following system:

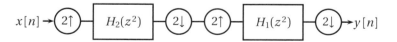

(b) In the system below, if $H(z) = E_0(z^2) + z^{-1} E_1(z^2)$ for some $E_{0,1}(z)$, prove that $Y(z) = X(z) E_0(z)$.

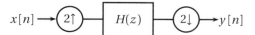

(c) Let $H(z)$, $F(z)$ and $G(z)$ be filters satisfying

$$H(z)G(z) + H(-z)G(-z) = 2$$
$$H(z)F(z) + H(-z)F(-z) = 0$$

Prove that one of the following systems is unity and the other zero:

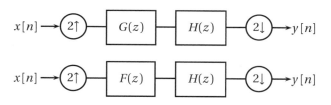

Exercise 11.3: Multirate Signal Processing. Consider a discrete-time signal $x[n]$ with the following spectrum:

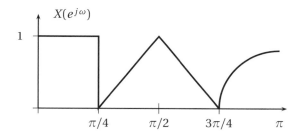

Now consider the following multirate processing scheme in which $L(z)$ is an ideal *lowpass* filter with cutoff frequency $\pi/2$ and $H(z)$ is an ideal *highpass* filter with cutoff frequency $\pi/2$:

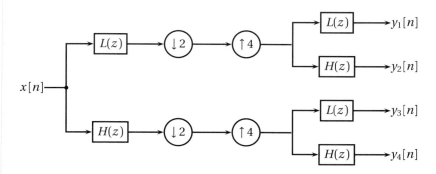

Plot the four spectra $Y_1(e^{j\omega})$, $Y_2(e^{j\omega})$, $Y_3(e^{j\omega})$, $Y_4(e^{j\omega})$.

Exercise 11.4: Digital processing of continuous-time signals. In your grandmother's attic you just found a treasure: a collection of super-rare 78 rpm vinyl jazz records. The first thing you want to do is to transfer the recordings to compact discs, so you can listen to them without wearing out the originals. Your idea is obviously to play the record on a turntable and use an A/D converter to convert the line-out signal into a discrete-time sequence, which you can then burn onto a CD. The problem is, you only have a "modern" turntable, which plays records at 33 rpm. Since you're a DSP wizard, you know you can just go ahead, play the 78 rpm record at 33 rpm and sample the output of the turntable at 44.1 KHz. You can then manipulate the signal in the discrete-time domain so that, when the signal is recorded on a CD and played back, it will sound right.

Design a system which performs the above conversion. If you need to get on the right track, consider the following:

- Call $s(t)$ the continuous-time signal encoded on the 78 rpm vinyl (the jazz music).

- Call $x(t)$ the continuous-time signal you obtain when you play the record at 33 rpm on the modern turntable.

- Let $x[n] = x(nT_s)$, with $T_s = 1/44,100$.

Answer the following questions:

(a) Express $x(t)$ in terms of $s(t)$.

(b) Sketch the Fourier transform $X(j\Omega)$ when $S(j\Omega)$ is as in the following figure. The highest nonzero frequency of $S(j\Omega)$ is $\Omega_{max} = (2\pi) \cdot 16,000$ Hz (old records have a smaller bandwidth than modern ones).

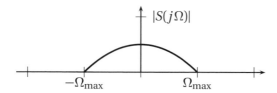

(c) Design a system to convert $x[n]$ into a sequence $y[n]$ so that, when you interpolate $y[n]$ to a continuous-time signal $y(t)$ with interpolation period T_s, you obtain $Y(j\Omega) = S(j\Omega)$.

(d) What if you had a turntable which plays records at 45 rpm? Would your system be different? Would it be better?

Exercise 11.5: Multirate is so useful! Consider the following block diagram:

$$x[n] \longrightarrow M\uparrow \longrightarrow \text{LP}\{\pi/M\} \longrightarrow z^{-L} \longrightarrow M\downarrow \longrightarrow y[n]$$

and show that this system implements a fractional delay (i.e. show that the transfer function of the system is that of a pure delay, where the delay is not necessarily an integer).

To see a practical use of this structure, consider now a data transmission system over an analog channel. The transmitter builds a discrete-time signal $s[n]$; this is converted to an analog signal $s_c(t)$ via an interpolator with period T_s, and finally $s_c(t)$ is transmitted over the channel. The signal takes a finite amount of time to travel all the way to the receiver; say that the transmission time over the channel is t_0 seconds: the received signal $\hat{s}_c(t)$ is therefore just a delayed version of the transmitted signal,

$$\hat{s}_c(t) = s_c(t - t_0)$$

At the receiver, $\hat{s}_c(t)$ is sampled with a sampler with period T_s so that no aliasing occurs to obtain $\hat{s}[n]$.

(a) Write out the Fourier Transform of $\hat{s}_c(t)$ as a function of $S_c(j\Omega)$.

(b) Write out the DTFT of the received signal sampled with rate T_s, $\hat{s}[n]$.

(c) Now we want to use the above multirate structure to compensate for the transmission delay. Assume $t_0 = 4.6\, T_s$; determine the values for M and L in the above block diagram so that $\hat{s}[n] = s[n - D]$, where $D \in \mathbb{N}$ has the smallest possible value (assume an ideal lowpass filter in the multirate structure).

Exercise 11.6: Multirate filtering. Assume $H(z)$ is an ideal lowpass filter with cutoff frequency $\pi/10$. Consider the system described by the following block diagram:

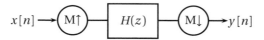

(a) Compute the transfer function of the system for $M = 2$.

(b) Compute the transfer function of the system for $M = 5$.

(c) Compute the transfer function of the system for $M=9$.

(d) Compute the transfer function of the system for $M=10$.

Exercise 11.7: Oversampled sequences. Consider a real-value sequence $x[n]$ for which:

$$X(e^{j\omega}) = 0 \qquad \frac{\pi}{3} \leq |\omega| \leq \pi$$

One sample of $x[n]$ may have been corrupted and we would like to approximately or exactly recover it. We denote n_0 the time index of the corrupted sample and $\hat{x}[n]$ the corresponding corrupted sequence.

(a) Specify a practical algorithm for exactly or approximately recovering $x[n]$ from $\hat{x}[n]$ if n_0 is known.

(b) What would you do if the value of n_0 is not known?

(c) Now suppose we have k corrupted samples at either known or unknown locations.

What is the condition that $X(e^{j\omega})$ must satisfy to be able to exactly recover $x[n]$? Specify the algorithm.

Chapter 12

Design of a Digital Communication System

The power of digital signal processing can probably be best appreciated in the enormous progresses which have been made in the field of telecommunications. These progresses stem from three main properties of digital processing:

- The flexibility and power of discrete-time processing techniques, which allow for the low-cost deployment of sophisticated and, more importantly, *adaptive* equalization and filtering modules.

- The ease of integration between low-level digital processing and high-level information-theoretical techniques which counteract transmission errors.

- The *regenerability* of a digital signal: in the necessary amplification of analog signals after transmission, the noise floor is amplified as well, thereby limiting the processing gain. Digital signals, on the other hand, can be *exactly* regenerated under reasonable SNR conditions (Fig. 1.10).

The fruits of such powerful communication systems are readily enjoyable in everyday life and it suffices here to mention the fast ADSL connections which take the power of high data rates into the home. ADSL is actually a quantitative evolution of a humbler, yet extraordinarily useful device: the voiceband modem. Voiceband modems, transmitting data at a rate of up to 56 Kbit/sec over standard telephone lines, are arguably the crown achievement of discrete-time signal processing in the late 90's and are still the cornerstone of most wired telecommunication devices such as laptops and fax machines.

In this Chapter, we explore the design and implementation of a voiceband modem as a paradigmatic example of applied digital signal processing. In principle, the development of a fully-functional device would require the use of concepts which are beyond the scope of this book, such as adaptive signal processing and information theory. Yet we will see that, if we neglect some of the impairments that are introduced by real-world telephone lines, we are able to design a working system which will flawlessly modulates and demodulates a data sequence.

12.1 The Communication Channel

A telecommunication system works by exploiting the propagation of electromagnetic waves in a medium. In the case of radio transmission, the medium is the electromagnetic spectrum; in the case of land-line communications such as those in voiceband or ADSL modems, the medium is a copper wire. In all cases, the properties of the medium determine two fundamental constraints around which any communication system is designed:

- **Bandwith constraint:** data transmission systems work best in the frequency range over which the medium behaves linearly; over this *passband* we can rely on the fact that a signal will be received with only phase and amplitude distortions, and these are "good" types of distortion since they amount to a linear filter. Further limitations on the available bandwidth can be imposed by law or by technical requirements and the transmitter must limit its spectral occupancy to the prescribed frequency region.

- **Power constraint:** the power of a transmitted signal is inherently limited by various factors, including the range over which the medium and the transmission circuitry behaves linearly. In many other cases, such as in telephone or radio communications, the maximum power is strictly regulated by law. Also, power could be limited by the effort to maximize the operating time of battery-powered mobile devices. At the same time, all analog media are affected by noise, which can come in the form of interference from neighboring transmission bands (as in the case of radio channels) or of parasitic noise due to electrical interference (as in the case of AC hum over audio lines). The *noise floor* is the noise level which cannot be removed and must be reckoned with in the transmission scheme. Power constraints limit the achievable *signal to noise ratio* (SNR) with respect to the channel's noise floor; in turn, the SNR determines the reliability of the data transmission scheme.

These constraints define a communication channel and the goal, in the design of a communication system, is to maximize the amount of information which can be reliably transmitted across a given channel. In the design of a *digital* communication system, the additional goal is to operate entirely in the discrete-time domain up to the interface with the physical channel; this means that:

- at the transmitter, the signal is synthesized, shaped and modulated in the discrete-time domain and is converted to a continuous-time signal just prior to transmission;

- at the receiver, the incoming signal is sampled from the channel and demodulation, processing and decoding is performed in the digital domain.

12.1.1 The AM Radio Channel

A classic example of a regulated electromagnetic channel is commercial radio. Bandwidth constraints in the case of the electromagnetic spectrum are rigorously put in place because the spectrum is a scarce resource which needs to be shared amongst a multitude of users (commercial radio, amateur radio, cellular telephony, emergency services, military use, etc). Power constraints on radio emissions are imposed for human safety concerns. The AM band, for instance, extends from 530 kHz to 1700 kHz; each radio station is allotted an 8 kHz frequency slot in this range. Suppose that a speech signal $x(t)$, obtained with a microphone, is to be transmitted over a slot extending from $f_{\min} = 650$ kHz to $f_{\max} = 658$ kHz. Human speech can be modeled as a bandlimited signal with a frequency support of approximately 12 kHz; speech can, however, be filtered through a lowpass filter with cutoff frequency 4 kHz with little loss of intelligibility so that its bandwidth can be made to match the 8 kHz bandwidth of the AM channel. The filtered signal now has a spectrum extending from −4 kHz to 4 kHz; multiplication by a sinusoid at frequency $f_c = (f_{\max} + f_{\min})/2 = 654$ KHz shifts its support according to the continuous-time version of the *modulation theorem*: if $x(t) \xleftrightarrow{FT} X(j\Omega)$ then:

$$x(t)\cos(\Omega_c t) \xleftrightarrow{FT} \frac{1}{2}[X(j\Omega - j\Omega_c) + X(j\Omega + j\Omega_c)] \quad (12.1)$$

where $\Omega_c = 2\pi f_c$. This is, of course, a completely analog transmission system, which is schematically displayed in Figure 12.1.

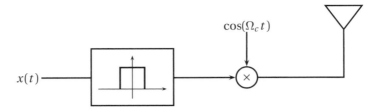

Figure 12.1 A simple AM radio transmitter.

12.1.2 The Telephone Channel

The telephone channel is basically a copper wire connecting two users. Because of the enormous number of telephone posts in the world, only a relatively small number of wires is used and the wires are *switched* between users when a call is made. The telephone network (also known as POTS, an acronym for "Plain Old Telephone System") is represented schematically in Figure 12.2. Each physical telephone is connected via a *twisted pair* (i.e. a pair of plain copper wires) to the nearest central office (CO); there are a lot of central offices in the network so that each telephone is usually no more than a few kilometers away. Central offices are connected to each other via the main lines in the network and the digits dialed by a caller are interpreted by the CO as connection instruction to the CO associated to the called number.

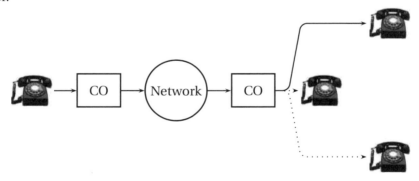

Figure 12.2 The Plain Old Telephone System (POTS).

To understand the limitations of the telephone channel we have to step back to the old analog times when COs were made of electromechanical switches and the voice signals traveling inside the network were boosted with simple operational amplifiers. The first link of the chain, the twisted pair to the central office, actually has a bandwidth of several MHz since it is just a copper wire (this is the main technical fact behind ADSL, by the way). Telephone companies, however, used to introduce what are called *loading coils* in the line to compensate for the attenuation introduced by the capacitive effects of longer wires in the network. A side effect of these coils was to turn the first link into a lowpass filter with a cutoff frequency of approximately 4 kHz so that, in practice, the *official* passband of the telephone channel is limited between $f_{min} = 300$ Hz and $f_{max} = 3000$ Hz, for a total usable positive bandwidth $W = 2700$ Hz. While today most of the network is actually digital, the official bandwidth remains in the order of 8 KHz (i.e. a positive bandwidth of 4 KHz); this is so that many more conversations can be multiplexed over the same cable or satellite link. The standard sampling rate for a telephone channel is nowadays 8 KHz and the bandwidth limitations are imposed only by the antialiasing filters at the CO, for a maximum bandwidth in excess of $W = 3400$ Hz. The upper and lower ends of the band are not usable due to possible great attenuations which may take place in the transmission. In particular, telephone lines exhibit a sharp notch at $f = 0$ (also known as *DC level*) so that any transmission scheme *will have to use bandpass signals* exclusively.

The telephone channel is power limited as well, of course, since telephone companies are quite protective of their equipment. Generally, the limit on signaling over a line is 0.2 V rms; the interesting figure however is not the maximum signaling level but the overall signal-to-noise ratio of the line (i.e. the amount of unavoidable noise on the line *with respect to* the maximum signaling level). Nowadays, phone lines are extremely high-quality: a SNR of at least 28 dB can be assumed in all cases and one of 32-34 dB can be reasonably expected on a large percentage of individual connections.

12.2 Modem Design: The Transmitter

Data transmission over a physical medium is by definition analog; modern communication systems, however, place all of the processing in the digital domain so that the only interface with the medium is the final D/A converter at the end of the processing chain, following the signal processing paradigm of Section 9.7.

12.2.1 Digital Modulation and the Bandwidth Constraint

In order to develop a digital communication system over the telephone channel, we need to re-cast the problem in the discrete-time domain. To this end, it is helpful to consider a very abstract view of the data transmitter,

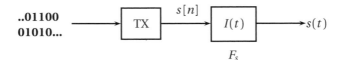

Figure 12.3 Abstract view of a digital transmitter.

as shown in Figure 12.3. Here, we neglect the details associated to the digital modulation process and concentrate on the digital-to-analog interface, represented in the picture by the interpolator $I(t)$; the input to the transmitter is some generic binary data, represented as a bit stream. The bandwidth constraints imposed by the channel can be represented graphically as in Figure 12.4. In order to produce a signal which "sits" in the prescribed frequency band, we need to use a D/A converter working at a frequency $F_s \geq 2f_{\max}$. Once the interpolation frequency is chosen (and we will see momentarily the criteria to do so), the requirements for the discrete-time signal $s[n]$ are set. The bandwidth requirements become simply

$$\omega_{\min,\max} = 2\pi \frac{f_{\min,\max}}{F_s}$$

and they can be represented as in Figure 12.5 (in the figure, for instance, we have chosen $F_s = 2.28 f_{\max}$).

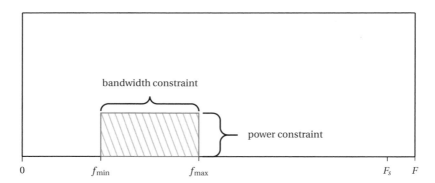

Figure 12.4 Analog specifications (positive frequencies) for the transmitter.

We can now try to understand how to build a suitable $s[n]$ by looking more in detail into the input side of the transmitter, as shown in Figure 12.6. The input bitstream is first processed by a *scrambler*, whose purpose is to randomize the data; clearly, it is a pseudo-randomization since this operation needs to be undone algorithmically at the receiver. Please note how the implementation of the transmitter in the digital domain allows for a seamless integration between the transmission scheme and more abstract data

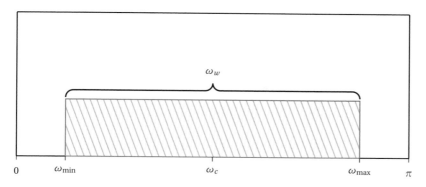

Figure 12.5 Discrete-time specifications (positive frequencies) for $F_s = 2.28\, f_{\max}$.

manipulation algorithms such as randomizers. The randomized bitstream could already be transmitted at this point; in this case, we would be implementing a binary modulation scheme in which the signal $s[n]$ varies between the two levels associated to a zero and a one, much in the fashion of telegraphic communications of yore. Digital communication devices, however, allow for a much more efficient utilization of the available bandwidth via the implementation of *multilevel* signaling. With this strategy, the bitstream is segmented in consecutive groups of M bits and these bits select one of 2^M possible signaling values; the set of all possible signaling values is called the *alphabet* of the transmission scheme and the algorithm which associates a group of M bits to an alphabet symbol is called the *mapper*. We will discuss practical alphabets momentarily; however, it is important to remark that the series of symbols *can be complex* so that all the signals in the processing chain up to the final D/A converter are complex signals.

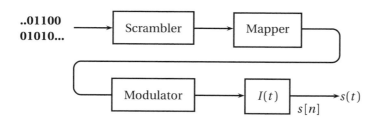

Figure 12.6 Data stream processing detail.

Spectral Properties of the Symbol Sequence. The mapper produces a sequence of symbols $a[n]$ which is the actual discrete-time signal which we need to transmit. In order to appreciate the spectral properties of this sequence consider that, if the initial binary bitstream is a maximum-information sequence (i.e. if the distribution of zeros and ones looks random and "fifty-fifty"), and with the scrambler appropriately randomizing the input bitstream, the sequence of symbols $a[n]$ can be modeled as a stochastic i.i.d. process distributed over the alphabet. Under these circumstances, the power spectral density of the random signal $a[n]$ is simply

$$P_A(e^{j\omega}) = \sigma_A^2$$

where σ_A depends on the design of the alphabet and on its distribution.

Choice of Interpolation Rate. We are now ready to determine a suitable rate F_s for the final interpolator. The signal $a[n]$ is a baseband, fullband signal in the sense that it is centered around zero and its power spectral density is nonzero over the entire $[-\pi, \pi]$ interval. If interpolated at F_s, such a signal gives rise to an analog signal with nonzero spectral power over the entire $[-F_s/2, F_s/2]$ interval (and, in particular, nonzero power at DC level). In order to fulfill the channel's constraints, we need to produce a signal with a bandwidth of $\omega_w = \omega_{\max} - \omega_{\min}$ centered around $\omega_c = \pm(\omega_{\max} + \omega_{\min})/2$. The "trick" is to *upsample* (and interpolate) the sequence $a[n]$, in order to narrow its spectral support.[1] Assuming ideal discrete-time interpolators, an upsampling factor of 2, for instance, produces a half-band signal; an upsampling factor of 3 produces a signal with a support spanning one third of the total band, and so on. In the general case, we need to choose an upsampling factor K so that:

$$\frac{2\pi}{K} \leq \omega_w$$

Maximum efficiency occurs when the available bandwidth is entirely occupied by the signal, i.e. when $K = 2\pi/\omega_w$. In terms of the analog bandwidth requirements, this translates to

$$K = \frac{F_s}{f_w} \qquad (12.2)$$

where $f_w = f_{\max} - f_{\min}$ is the effective positive bandwidth of the transmitted signal; since K must be an integer, the previous condition implies that we must choose an interpolation frequency *which is a multiple of the positive*

---

[1] A rigorous mathematical analysis of multirate processing of stochastic signals turns out to be rather delicate and beyond the scope of this book; the same holds for the effects of modulation, which will appear later on. Whenever in doubt, we may simply visualize the involved signals as a deterministic realization whose spectral shape mimics the power spectral density of their generating stochastic process.

passband width f_w. The two criteria which must be fulfilled for optimal signaling are therefore:

$$\begin{cases} F_s \geq 2f_{max} \\ F_s = Kf_w \quad K \in \mathbb{N} \end{cases} \tag{12.3}$$

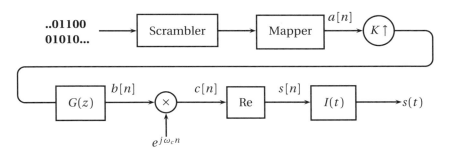

Figure 12.7 Complete digital transmitter.

The Baseband Signal. The upsampling by K operation, used to narrow the spectral occupancy of the symbol sequence to the prescribed bandwidth, must be followed by a lowpass filter, to remove the multiple copies of the upsampled spectrum; this is achieved by a lowpass filter which, in digital communication parlance, is known as the *shaper* since it determines the time domain shape of the transmitted symbols. We know from Section 11.2.1 that, ideally, we should use a sinc filter to perfectly remove all repeated copies. Since this is clearly not possible, let us now examine the properties that a practical discrete-time interpolator should possess in the context of data communications. The baseband signal $b[n]$ can be expressed as

$$b[n] = \sum_m a_{KU}[m] g[n-m]$$

where $a_{KU}[n]$ is the upsampled symbol sequence and $g[n]$ is the lowpass filter's impulse response. Since $a_{KU}[n] = 0$ for n not a multiple of K, we can state that:

$$b[n] = \sum_i a[i] g[n - iK] \tag{12.4}$$

It is reasonable to impose that, at multiples of K, the upsampled sequence $b[n]$ takes on the exact symbol value, i.e. $b[mK] = a[m]$; this translates to the following requirement for the lowpass filter:

$$g[mK] = \begin{cases} 1 & m = 0 \\ 0 & m \neq 0 \end{cases} \tag{12.5}$$

This is nothing but the classical interpolation property which we saw in Section 9.4.1. For realizable filters, this condition implies that the minimum frequency support of $G(e^{j\omega})$ cannot be smaller than $[-\pi/K, \pi/K]$.[2] In other words, there will always be a (controllable) amount of *frequency leakage* outside of a prescribed band with respect to an ideal filter.

To exactly fullfill (12.5), we need to use an FIR lowpass filter; FIR approximations to a sinc filter are, however, very poor, since the impulse response of the sinc decays very slowly. A much friendlier lowpass characteristic which possesses the interpolation property and allows for a precise quantification of frequency leakage, is the *raised cosine*. A raised cosine with nominal bandwidth ω_w (and therefore with nominal cutoff $\omega_b = \omega_w/2$) is defined over the positive frequency axis as

$$G(e^{j\omega}) = \begin{cases} 1 & \text{if } 0 < \omega < (1-\beta)\omega_b \\ 0 & \text{if } (1+\beta)\omega_b < \omega < \pi \\ \frac{1}{2} + \frac{1}{2}\cos\left(\pi\frac{\omega - (1-\beta)\omega_b}{2\beta\omega_b}\right) & \\ & \text{if } (1-\beta)\omega_b < \omega < (1+\beta)\omega_b \end{cases} \quad (12.6)$$

and is symmetric around the origin. The parameter β, with $0 < \beta < 1$, exactly defines the amount of frequency leakage as a percentage of the passband. The closer β is to one, the sharper the magnitude response; a set of frequency responses for $\omega_b = \pi/2$ and various values of β are shown in Figure 12.8. The raised cosine is still an ideal filter but it can be shown that its

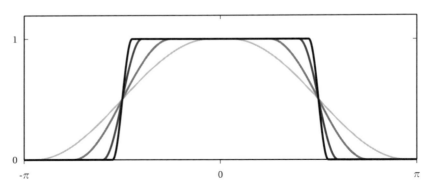

Figure 12.8 Frequency responses of a half-band raised-cosine filter for increasing values of β: from black to light gray, $\beta = 0.1$, $\beta = 0.2$, $\beta = 0.4$, $\beta = 0.9$.

---

[2] A simple proof of this fact can be outlined using multirate signal processing. Assume the spectrum $G(e^{j\omega})$ is nonzero only over $[-\omega_b, \omega_b]$, for $\omega_b < \pi/K$; $g[n]$ can therefore be subsampled by at least a factor of K without aliasing, and the support of the resulting spectrum is going to be $[-K\omega_b, K\omega_b]$, with $K\omega_b < \pi$. However, $g[Kn] = \delta[n]$, whose spectral support is $[-\pi, \pi]$.

impulse response decays as $1/n^3$ and, therefore, good FIR approximations can be obtained with a reasonable amount of taps using a specialized version of Parks-McClellan algorithm. The number of taps needed to achieve a good frequency response obviously increases as β approaches one; in most practical applications, however, it rarely exceeds 50.

The Bandpass Signal. The filtered signal $b[n] = g[n] * a_{KU}[n]$ is now a baseband signal with total bandwidth ω_w. In order to shift the signal into the allotted frequency band, we need to modulate[3] it with a sinusoidal *carrier* to obtain a complex bandpass signal:

$$c[n] = b[n] e^{j\omega_c n}$$

where the modulation frequency is the center-band frequency:

$$\omega_c = \frac{\omega_{\min} + \omega_{\max}}{2}$$

Note that the spectral support of the modulated signal is just the *positive* interval $[\omega_{\min}, \omega_{\max}]$; a complex signal with such a one-sided spectral occupancy is called an *analytic signal*. The signal which is fed to the D/A converter is simply the real part of the complex bandpass signal:

$$s[n] = \text{Re}\{c[n]\} \tag{12.7}$$

If the baseband signal $b[n]$ is real, then (12.7) is equivalent to a standard cosine modulation as in (12.1); in the case of a complex $b[n]$ (as in our case), the bandpass signal is the combination of a cosine *and* a sine modulation, which we will examine in more detail later. The spectral characteristics of the signals involved in the creation of $s[n]$ are shown in Figure 12.9.

Baud Rate vs Bit Rate. The *baud rate* of a communication system is the number of *symbols* which can be transmitted in one second. Considering that the interpolator works at F_s samples per second and that, because of upsampling, there are exactly K samples per symbol in the signal $s[n]$, the baud rate of the system is

$$B = \frac{F_s}{K} = f_w \tag{12.8}$$

where we have assumed that the shaper $G(z)$ is an ideal lowpass. As a general rule, *the baud rate is always smaller or equal to the positive passband of the channel*. Moreover, if we follow the normal processing order, we can

---

[3] See footnote (1).

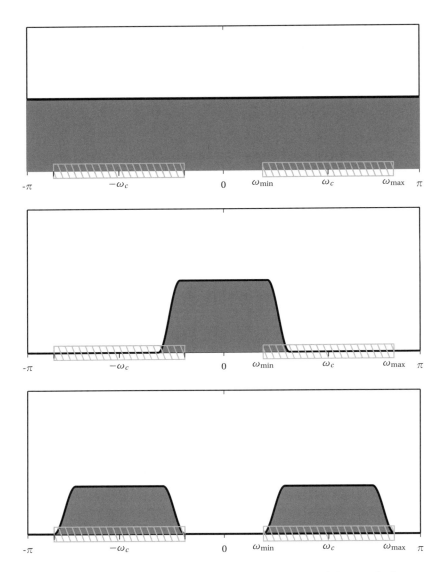

Figure 12.9 Construction of the modulated signal: PSD of the symbol sequence $a[n]$ (top panel); PSD of the upsampled and shaped signal $b[n]$ (middle panel); PSD of the real modulated signal $s[n]$ (bottom panel). The channel's bandwidth requirements are indicated by the dashed areas.

equivalently say that a symbol sequence generated at B symbols per second gives rise to a modulated signal whose positive passband is *no smaller than* B Hz. The effective bandwidth f_w depends on the modulation scheme and, especially, on the frequency leakage introduced by the shaper.

The total bit rate of a transmission system, on the other hand, is at most the baud rate times the log in base 2 of the number of symbols in the alphabet; for a mapper which operates on M bits per symbol, the overall bitrate is

$$R = MB \tag{12.9}$$

A Design Example. As a practical example, consider the case of a telephone line for which $f_{\min} = 450$ Hz and $f_{\max} = 3150$ Hz (we will consider the power constraints later). The baud rate can be at most 2700 symbols per second, since $f_w = f_{\max} - f_{\min} = 2700$ Hz. We choose a factor $\beta = 0.125$ for the raised cosine shaper and, to compensate for the bandwidth expansion, we deliberately reduce the actual baud rate to $B = 2700/(1+\beta) = 2400$ symbols per second, which leaves the effective positive bandwidth equal to f_w. The criteria which the interpolation frequency must fulfill are therefore the following:

$$\begin{cases} F_s \geq 2f_{\max} = 6300 \\ F_s = KB = 2400K \quad K \in \mathbb{N} \end{cases}$$

The first solution is for $K = 3$ and therefore $F_s = 7200$. With this interpolation frequency, the effective bandwidth of the discrete-time signal is $\omega_w = 2\pi(2700/7200) = 0.75\pi$ and the carrier frequency for the bandpass signal is $\omega_c = 2\pi(450+3150)/(2F_s) = \pi/2$. In order to determine the maximum attainable bitrate of this system, we need to address the second major constraint which affects the design of the transmitter, i.e. the power constraint.

12.2.2 Signaling Alphabets and the Power Constraint

The purpose of the mapper is to associate to each group of M input bits a value α from a given *alphabet* $\mathscr{A}$. We assume that the mapper includes a multiplicative factor G_0 which can be used to set the final gain of the generated signal, so that we don't need to concern ourselves with the absolute values of the symbols in the alphabet; the symbol sequence is therefore:

$$a[n] = G_0 \alpha[n], \qquad \alpha[n] \in \mathscr{A}$$

and, in general, the values α are set at integer coordinates out of convenience.

Transmitted Power. Under the above assumption of an i.i.d. uniformly distributed binary input sequence, each group of M bits is equally probable; since we consider only *memoryless* mappers, i.e. mappers in which no dependency between symbols is introduced, the mapper acts as the source of a random process $a[n]$ which is also i.i.d. The power of the output sequence can be expressed as

$$\sigma_a^2 = \mathrm{E}|a[n]|^2$$
$$= G_0^2 \sum_{\alpha \in \mathscr{A}} |\alpha|^2 p_a(\alpha) \qquad (12.10)$$
$$= G_0^2 \sigma_\alpha^2 \qquad (12.11)$$

where $p_a(\alpha)$ is the probability assigned by the mapper to symbol $\alpha \in \mathscr{A}$; the distribution over the alphabet $\mathscr{A}$ is one of the design parameters of the mapper, and is not necessarily uniform. The variance σ_α^2 is the intrinsic power of the alphabet and it depends on the alphabet size (it increases exponentially with M), on the alphabet structure, and on the probability distribution of the symbols in the alphabet. Note that, in order to avoid wasting transmission energy, communication systems are designed so that the sequence generated by the mapper is *balanced*, i.e. its DC value is zero:

$$\mathrm{E}[a[n]] = \sum_{\alpha \in \mathscr{A}} \alpha p_a(\alpha) = 0$$

Using (8.25), the power of the transmitted signal, after upsampling and modulation, is

$$\sigma_s^2 = \frac{1}{\pi} \int_{\omega_{\min}}^{\omega_{\max}} \frac{1}{2} |G(e^{j\omega})|^2 G_0^2 \sigma_\alpha^2 \qquad (12.12)$$

The shaper is designed so that its overall energy over the passband is $G^2 = 2\pi$ and we can express this as follows:

$$\sigma_s^2 = G_0^2 \sigma_\alpha^2 \qquad (12.13)$$

In order to respect the power constraint, we have to choose a value for G_0 and design an alphabet $\mathscr{A}$ so that:

$$\sigma_s^2 \leq P_{\max} \qquad (12.14)$$

where $P_{\max}$ is the maximum transmission power allowed on the channel. The goal of a data transmission system is to maximize the *reliable* throughput but, unfortunately, in this respect the parameters σ_α^2 and G_0 act upon conflicting priorities. If we use (12.9) and boost the transmitter's bitrate by increasing M, then σ_α^2 grows and we must necessarily reduce the gain G_0 to fulfill the power constraint; but, in so doing, we impair the reliability of the transmission. To understand why that is, we must leap ahead and consider both a practical alphabet and the mechanics of symbol decoding at the transmitter.

QAM. The simplest mapping strategies are one-to-one correspondences between binary values and signal values: note that in these cases the symbol sequence is uniformly distributed with $p_a(\alpha) = 2^{-M}$ for all $\alpha \in \mathscr{A}$. For example, we can assign to each group of M bits $(b_0, \ldots, b_{M-1})$ the signed binary number $b_0 b_1 b_2 \cdots b_{M-1}$ which is a value between -2^{M-1} and 2^{M-1} (b_0 is the sign bit). This signaling scheme is called *pulse amplitude modulation* (PAM) since the amplitude of each transmitted symbol is directly determined by the binary input value. The PAM alphabet is clearly balanced and the inherent power of the mapper's output is readily computed as[4]

$$\sigma_a^2 = \sum_{a=1}^{2^{M-1}} 2^{-M} \alpha^2 = \frac{2^M(2^M+3)+2}{24}$$

Now, a pulse-amplitude modulated signal prior to modulation is a baseband signal with positive bandwidth of, say, ω_0 (see Figure 12.9, middle panel); therefore, the *total* spectral support of the baseband PAM signal is $2\omega_0$. After modulation, the total spectral support of the signal actually doubles (Fig. 12.9, bottom panel); there is, therefore, some sort of redundancy in the modulated signal which causes an underutilization of the available bandwidth. The original spectral efficiency can be regained with a signaling scheme called *quadrature amplitude modulation* (QAM); in QAM the symbols in the alphabet are complex quantities, so that *two* real values are transmitted simultaneously at each symbol interval. Consider a complex symbol sequence

$$a[n] = G_0(\alpha_I[n] + j\alpha_Q[n]) = a_I[n] + j a_Q[n]$$

Since the shaper is a real-valued filter, we have that:

$$b[n] = (a_{I,KU} * g[n]) + j(a_{Q,KU} * g[n]) = b_I[n] + j b_Q[n]$$

so that, finally, (12.7) becomes:

$$s[n] = \mathrm{Re}\{b[n] e^{j\omega_c n}\}$$
$$= b_I[n]\cos(\omega_c n) - b_Q[n]\sin(\omega_c n)$$

In other words, a QAM signal is simply the linear combination of two pulse-amplitude modulated signals: a cosine carrier modulated by the real part of the symbol sequence and a sine carrier modulated by the imaginary part of the symbol sequence. The sine and cosine carriers are *orthogonal* signals, so that $b_I[n]$ and $b_Q[n]$ can be exactly separated at the receiver via a subspace projection operation, as we will see in detail later. The subscripts I

---

[4] A useful formula, here and in the following, is $\sum_{n=1}^{N} n^2 = N(N+1)(2N+1)/6$.

and Q derive from the historical names for the cosine carrier (the *in-phase* carrier) and the sine carrier which is the *quadrature* (i.e. the orthogonal carrier). Using complex symbols for the description of the internal signals in the transmitter is an abstraction which simplifies the overall notation and highlights the usefulness of complex discrete-time signal models.

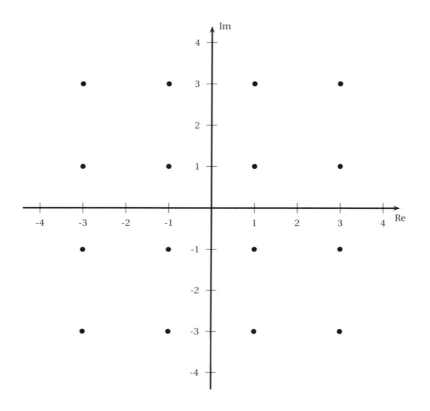

Figure 12.10 16-point QAM constellations ($M=4$).

Constellations. The 2^M symbols in the alphabet can be represented as points in the complex plane and the geometrical arrangement of all such points is called the signaling *constellation*. The simplest constellations are upright square lattices with points on the odd integer coordinates; for M even, the 2^M constellation points α_{hk} form a square shape with $2^{M/2}$ points per side:

$$\alpha_{hk} = (2h-1) + j(2k-1), \qquad -2^{M/2-1} < h, k \leq 2^{M/2-1}$$

Such square constellations are called *regular* and a detailed example is shown in Figure 12.10 for $M = 4$; other examples for $M = 2, 6, 8$ are shown in Figure 12.11. The nominal power associated to a regular, uniformly distributed constellation on the square lattice can be computed as the second moment of the points; exploiting the fourfold symmetry, we have

$$\sigma_\alpha^2 = 4 \sum_{h=1}^{2^{M/2-1}} \sum_{k=1}^{2^{M/2-1}} 2^{-M} \left[(2h-1)^2 + (2k-1)^2 \right]$$
$$= \frac{2}{3}(2^M - 1) \tag{12.15}$$

Square-lattice constellations exist also for alphabet sizes which are not perfect squares and examples are shown in Figure 12.12 for $M = 3$ (8-point constellation) and $M = 5$ (32-point). Alternatively, constellations can be defined on other types of lattices, either irregular or regular; Figure 12.13 shows an alternative example of an 8-point constellation defined on an irregular grid and a 19-point constellation defined over a regular hexagonal lattice. We will see later how to exploit the constellation's geometry to increase performance.

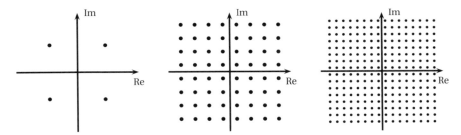

Figure 12.11 4-, 64- and 256-point QAM constellations (M bits/symbol for $M = 2$, $M = 6$, $M = 8$) respectively.

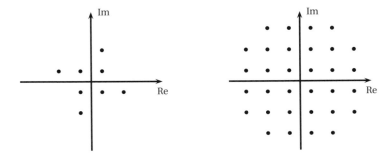

Figure 12.12 8- and 32-point square-lattice constellations.

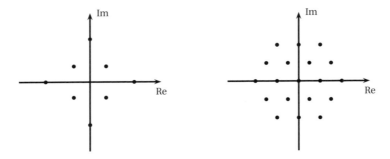

Figure 12.13 More exotic constellations: irregular low-power 8-point constellation (left panel) in which the outer point are at a distance of $1 + \sqrt{3}$ from the origin ; regular 19-point hexagonal-lattice constellation (right panel).

Transmission Reliability. Let us assume that the receiver has eliminated all the "fixable" distortions introduced by the channel so that an "almost exact" copy of the symbol sequence is available for decoding; call this sequence $\hat{a}[n]$. What no receiver can do, however, is eliminate all the additive noise introduced by the channel so that:

$$\hat{a}[n] = a[n] + \eta[n] \tag{12.16}$$

where $\eta[n]$ is a *complex* white Gaussian noise term. It will be clear later why the internal mechanics of the receiver make it easier to consider a complex representation for the noise; again, such complex representation is a convenient abstraction which greatly simplifies the mathematical analysis of the decoding process. The real-valued zero-mean Gaussian noise introduced by the channel, whose variance is σ_0^2, is transformed by the receiver into complex Gaussian noise whose real and imaginary parts are independent zero-mean Gaussian variables with variance $\sigma_0^2/2$. Each complex noise sample $\eta[n]$ is distributed according to

$$f_\eta(z) = \frac{1}{\pi \sigma_0^2} e^{-\frac{|z|^2}{\sigma_0^2}} \tag{12.17}$$

The magnitude of the noise samples introduces a shift in the complex plane for the demodulated symbols $\hat{a}[n]$ with respect to the originally transmitted symbols; if this displacement is too big, a decoding error takes place. In order to quantify the effects of the noise we have to look more in detail at the way the transmitted sequence is retrieved at the receiver. A bound on the probability of error can be obtained analytically if we consider a simple QAM decoding technique called *hard slicing*. In hard slicing, a value $\hat{a}[n]$ is associated to the most probable symbol $\alpha \in \mathscr{A}$ by choosing the alphabet symbol at the minimum Euclidean distance (taking the gain G_0 into account):

$$\mathscr{D}\{\hat{a}[n]\} = \arg\min_{\alpha \in \mathscr{A}} \{|\hat{a}[n] - G_0\alpha|^2\}$$

The hard slicer partitions the complex plane into decision regions centered on alphabet symbols; all the received values which fall into the decision region centered on α are mapped back onto α. Decision regions for a 16-point constellation, together with examples of correct and incorrect hard slicing are represented in Figure 12.14: when the error sample $\eta[n]$ moves the received symbol outside of the right decision region, we have a decoding error. For square-lattice constellations, this happens when either the real or the imaginary part of the noise sample is larger than the minimum distance between a symbol and the closest decision region boundary. Said distance is $d_{\min} = G_0$, as can be easily seen from Figure 12.10, and therefore the probability of error at the receiver is

$$p_e = 1 - \mathrm{P}\big[\big(\mathrm{Re}\{\eta[n]\} < G_0\big) \wedge \big(\mathrm{Im}\{\eta[n]\} < G_0\big)\big]$$
$$= 1 - \int_D f_\eta(z)\,dz$$

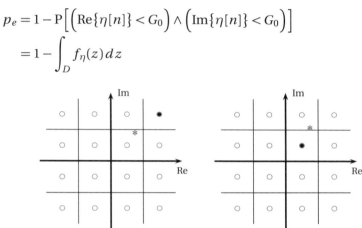

Figure 12.14 Decoding of noisy symbols: transmitted symbol is black dot, received value is the star. Correct decoding (left) and decoding error (right).

where $f_\eta(x)$ is the pdf of the additive complex noise and D is a square on the complex plane centered at the origin and $2d_{\min}$ wide. We can obtain a closed-form expression for the probability of error if we approximate the decision region D by the inscribed circle of radius $d_{\min}$ (Fig. 12.15), so:

$$p_e = 1 - \int_{|z|<G_0} f_\eta(z)\,dz$$
$$= 1 - \int_0^{2\pi} d\theta \int_0^{G_0} \frac{\rho}{\pi\sigma_0^2} e^{-\frac{\rho^2}{\sigma_0^2}}$$
$$= e^{-\frac{G_0^2}{\sigma_0^2}} \tag{12.18}$$

where we have used (12.17) and the change of variable $z = \rho\, e^{j\theta}$. The probability of error decreases exponentially with the gain and, therefore, with the power of the transmitter.

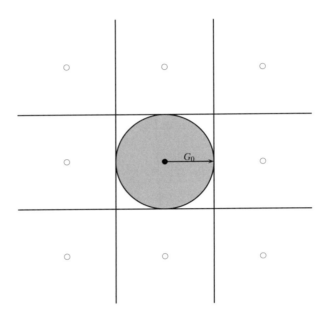

Figure 12.15 Decision region and its circular approximation.

The concept of "reliability" is quantified by the probability of error that we are willing to tolerate; note that this probability can never be zero, but it can be made arbitrarily low – values on the order of $p_e = 10^{-6}$ are usually taken as a reference. Assume that the transmitter transmits at the maximum permissible power so that the SNR on the channel is maximized. Under these conditions it is

$$\mathrm{SNR} = \frac{\sigma_s^2}{\sigma_0^2} = G_0^2 \frac{\sigma_a^2}{\sigma_0^2}$$

and from (12.18) we have

$$\mathrm{SNR} = -\ln(p_e)\sigma_a^2 \qquad (12.19)$$

For a regular square-lattice constellation we can use (12.15) to determine the maximum number of bits per symbol which can be transmitted at the given reliability figure:

$$M = \log_2\left(1 - \frac{3}{2}\frac{\mathrm{SNR}}{\ln(p_e)}\right) \qquad (12.20)$$

and this is how the power constraint ultimately affects the maximum achievable bitrate. Note that the above derivation has been carried out with very specific hypotheses on both the signaling alphabet and on the decoding algorithm (the hard slicing); the upper bound on the achievable rate on the channel is actually a classic result of information theory and is known under the name of Shannon's capacity formula. Shannon's formula reads

$$C = B \log_2 \left(1 + \frac{S}{N}\right)$$

where C is the absolute maximum capacity in bits per second, B is the available bandwidth in Hertz and S/N is the signal to noise ratio.

Design Example Revisited. Let us resume the example on page 341 by assuming that the power constraint on the telephone line limits the maximum achievable SNR to 22 dB. If the acceptable bit error probability is $p_e = 10^{-6}$, Equation (12.20) gives us a maximum integer value of $M = 4$ bits per symbol. We can therefore use a regular 16-point square constellation; recall we had designed a system with a baud rate of 2400 symbols per second and therefore the final reliable bitrate is $R = 9600$ bits per second. This is actually one of the operating modes of the V.32 ITU-T modem standard.[5]

12.3 Modem Design: the Receiver

The analog signal $s(t)$ created at the transmitter is sent over the telephone channel and arrives at the receiver as a distorted and noise-corrupted signal $\hat{s}(t)$. Again, since we are designing a purely digital communication system, the receiver's input interface is an A/D converter which, for simplicity, we assume, is operating at the same frequency F_s as the transmitter's D/A converter. The receiver tries to undo the impairments introduced by the channel and to demodulate the received signal; its output is a binary sequence which, in the absence of decoding errors, is identical to the sequence injected into the transmitter; an abstract view of the receiver is shown in Figure 12.16.

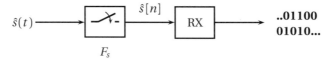

Figure 12.16 Abstract view of a digital receiver.

---

[5] ITU-T is the Standardization Bureau of the International Telecommunication Union.

12.3.1 Hilbert Demodulation

Let us assume for the time being that transmitter and receiver are connected back-to-back so that we can neglect the effects of the channel; in this case $\hat{s}(t) = s(t)$ and, after the A/D module, $\hat{s}[n] = s[n]$. Demodulation of the incoming signal to a binary data stream is achieved according to the block diagram in Figure 12.17 where all the steps in the modulation process are undone, one by one.

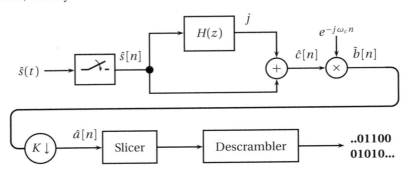

Figure 12.17 Complete digital receiver.

The first operation is retrieving the complex bandpass signal $\hat{c}[n]$ from the real signal $\hat{s}[n]$. An efficient way to perform this operation is by exploiting the fact that the original $c[n]$ is an analytic signal and, therefore, its imaginary part is completely determined by its real part. To see this, consider a complex analytic signal $x[n]$, i.e. a complex sequence for which $X(e^{j\omega}) = 0$ over the $[-\pi, 0]$ interval (with the usual 2π-periodicity, obviously). We can split $x[n]$ into real and imaginary parts:

$$x[n] = x_r[n] + jx_i[n]$$

so that we can write:

$$x_r[n] = \frac{x[n] + x^*[n]}{2}$$
$$x_i[n] = \frac{x[n] - x^*[n]}{2j}$$

In the frequency domain, these relations translate to (see (4.46)):

$$X_r(e^{j\omega}) = \frac{[X(e^{j\omega}) + X^*(e^{-j\omega})]}{2} \tag{12.21}$$

$$X_i(e^{j\omega}) = \frac{[X(e^{j\omega}) - X^*(e^{-j\omega})]}{2j} \tag{12.22}$$

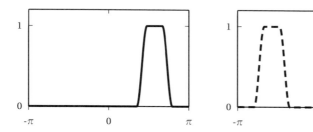

Figure 12.18 Magnitude spectrum of an analytic signal $x[n]$. $|X(e^{j\omega})|$ (left) and $|X^*(e^{-j\omega})|$ (right).

Since $x[n]$ is analytic, by definition $X(e^{j\omega}) = 0$ for $-\pi \leq \omega < 0$, $X^*(e^{-j\omega}) = 0$ for $0 < \omega \leq \pi$ and $X(e^{j\omega})$ does not overlap with $X^*(e^{-j\omega})$ (Fig. 12.18). We can therefore use (12.21) to write:

$$X(e^{j\omega}) = \begin{cases} 2X_r(e^{j\omega}) & \text{for } 0 \leq \omega \leq \pi \\ 0 & \text{for } -\pi < \omega < 0 \end{cases} \tag{12.23}$$

Now, $x_r[n]$ is a real sequence and therefore its Fourier transform is conjugate-symmetric, i.e. $X_r(e^{j\omega}) = X_r^*(e^{-j\omega})$; as a consequence

$$X^*(e^{-j\omega}) = \begin{cases} 0 & \text{for } 0 \leq \omega \leq \pi \\ 2X_r(e^{j\omega}) & \text{for } -\pi < \omega < 0 \end{cases} \tag{12.24}$$

By using (12.23) and (12.24) in (12.22) we finally obtain:

$$X_i(e^{j\omega}) = \begin{cases} -jX_r(e^{j\omega}) & \text{for } 0 \leq \omega \leq \pi \\ +jX_r(e^{j\omega}) & \text{for } -\pi < \omega < 0 \end{cases} \tag{12.25}$$

which is the product of $X_r(e^{j\omega})$ with the frequency response of a Hilbert filter (Sect. 5.6). In the time domain this means that the imaginary part of an analytic signal can be retrieved from the real part only via the convolution:

$$x_i[n] = h[n] * x_r[n]$$

At the demodulator, $\hat{s}[n] = s[n]$ is nothing but the real part of $c[n]$ and therefore the analytic bandpass signal is simply

$$\hat{c}[n] = \hat{s}[n] + j(h[n] * \hat{s}[n])$$

In practice, the Hilbert filter is approximated with a causal, $2L+1$-tap type III FIR, so that the structure used in demodulation is that of Figure 12.19.

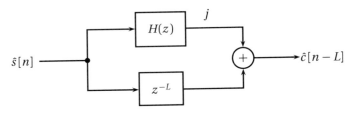

Figure 12.19 Retrieving the complex baseband signal with an FIR Hilbert filter approximation.

The delay in the bottom branch compensates for the delay introduced by the causal filter and puts the real and derived imaginary part back in sync to obtain:

$$\hat{c}[n] = \hat{s}[n-L] + j\left(h[n] * \hat{s}[n]\right)$$

Once the analytic bandpass signal is reconstructed, it can be brought back to baseband via a complex demodulation with a carrier with frequency $-\omega_c$:

$$\hat{b}[n] = \hat{c}[n] e^{-j\omega_c n}$$

Because of the interpolation property of the pulse shaper, the sequence of complex symbols can be retrieved by a simple downsampling-by-K operation:

$$\hat{a}[n] = \hat{b}[nK]$$

Finally, the slicer (which we saw in Section 12.2.2) associates a group of M bits to each received symbol and the descrambler reconstructs the original binary stream.

12.3.2 The Effects of the Channel

If we now abandon the convenient back-to-back scenario, we have to deal with the impairments introduced by the channel and by the signal processing hardware. The telephone channels affects the received signal in three fundamental ways:

- it adds noise to the signal so that, even in the best case, the signal-to-noise ratio of the received signal cannot exceed a maximum limit;
- it distorts the signal, acting as a linear filter;
- it delays the signal, according to the propagation time from transmitter to receiver.

Distortion and delay are obviously both linear transformations and, as such, their description could be lumped together; still, the techniques which deal with distortion and delay are different, so that the two are customarily kept separate. Furthermore, the physical implementation of the devices introduces an unavoidable lack of absolute synchronization between transmitter and receiver, since each of them runs on an independent internal clock. Adaptive synchronization becomes a necessity in all real-world devices, and will be described in the next Section.

Noise. The effects of noise have already been described in Section 12.2.2 and can be summed up visually by the plots in Figure 12.20 in each of which successive values of $\hat{a}[n]$ are superimposed on the same axes. The analog noise is transformed into discrete-time noise by the sampler and, as such, it leaks through the demodulation chain to the reconstructed symbols sequence $\hat{a}[n]$; as the noise level increases (or, equivalently, as the SNR decreases) the shape of the received constellation progressively loses its tightness around the nominal alphabet values. As symbols begin to cross the boundaries of the decision regions, more and more decoding errors, take place.

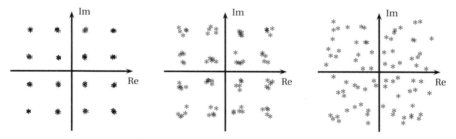

Figure 12.20 Noisy constellation for decreasing SNR.

Equalization. We saw previously that the passband of a communication channel is determined by the frequency region over which the channel introduces only linear types of distortion. The channel can therefore be modeled as a continuous-time linear filter $D_c(j\Omega)$ whose frequency response is unknown (and potentially time-varying). The received signal (neglecting noise) is therefore $\hat{S}(j\Omega) = D_c(j\Omega)S(j\Omega)$ and, after the sampler, we have

$$\hat{S}(e^{j\omega}) = D(e^{j\omega})S(e^{j\omega})$$

where $D(e^{j\omega})$ represents the combined effect of the original channel and of the anti-aliasing filter at the A/D converter. To counteract the channel distortion, the receiver includes an *adaptive equalizer* $E(z)$ right after the A/D converter; this is an FIR filter which is modified on the fly so that $E(z) \approx 1/D(z)$. While adaptive filter theory is beyond the scope of this book, the intuition behind adaptive equalization is shown in Figure 12.21. In fact, the demodulator contains an exact copy of the modulator as well; if we assume that the symbols produced by the slicer are error-free, a perfect copy of the transmitted signal $s[n]$ can be generated locally at the receiver. The difference between the equalized signal and the reconstructed original signal is used to adapt the taps of the equalizer so that:

$$d[n] = \hat{s}_e[n] - s[n] \longrightarrow 0$$

Clearly, in the absence of a good initial estimate for $D(e^{j\omega})$, the sliced values $\hat{a}[n]$ are nothing like the original sequence; this is obviated by having the transmitter send a pre-established *training sequence* which is known in advance at the receiver. The training sequence, together with other synchronization signals, is sent each time a connection is established between transmitter and receiver and is part of the modem's *handshaking protocol*. By using a training sequence, $E(z)$ can quickly converge to an approximation of $1/D(z)$ which is good enough for the receiver to start decoding symbols correctly and use them in driving further adaptation.

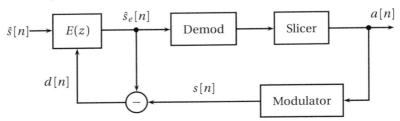

Figure 12.21 Adaptive equalization: based on the estimated symbols the receiver can synthesize the perfect desired equalizer output and use the difference to drive the adaptation.

Delay. The continuous-time signal arriving at the receiver can be modeled as

$$\hat{s}(t) = (s * v)(t - t_d) + \eta(t) \tag{12.26}$$

where $v(t)$ is the continuous-time impulse response of the channel, $\eta(t)$ is the continuous-time noise process and t_d is the *propagation delay*, i.e. the time it takes for the signal to travel from transmitter to receiver. After the sampler, the discrete-time signal to be demodulated is $\hat{s}[n] = \hat{s}(nT_s)$; if we neglect the noise and distortion, we can write

$$\hat{s}[n] = s(nT_s - t_d) = s((n - n_d)T_s - \tau T_s) \tag{12.27}$$

where we have split the delay as $t_d = (n_d + \tau)T_s$ with $n_d \in \mathbb{N}$ and $|\tau| \leq 1/2$. The term n_d is called the *bulk delay* and it can be estimated easily in a full-duplex system by the following handshaking procedure:

1. System A sends an impulse to system B at time $n = 0$; the impulse appears on the channel after a known processing delay t_{p1} seconds; let the (unknown) channel propagation delay be t_d seconds.

2. System B receives the impulse and sends an impulse back to A; the processing time t_{p2} (decoding of the impulse and generation of response) is known by design.

3. The response impulse is received by system A after t_d seconds (propagation delay is symmetric) and detected after a processing delay of t_{p3} seconds.

In the end, the total round-trip delay measured by system A is

$$t = 2t_d + t_{p1} + t_{p2} + t_{p3} = 2t_d + t_p$$

since t_p is known exactly in terms of the number of samples, t_d can be estimated *to within a sample*. The bulk delay is easily dealt with at the receiver, since it translated to a simple z^{-n_d} component in the channel's response. The fractional delay, on the other hand, is a more delicate entity which we will need to tackle with specialized machinery.

12.4 Adaptive Synchronization

In order for the receiver to properly decode the data, the discrete-time signals inside the receiver must be synchronous with the discrete-time signals generated by the transmitter. In the back-to-back operation, we could neglect synchronization problems since we assumed $\hat{s}[n] = s[n]$. In reality, we will need to compensate for the propagation delay and for possible clock differences between the D/A at the transmitter and the A/D at the receiver, both in terms of time offsets and in terms of frequency offsets.

12.4.1 Carrier Recovery

Carrier recovery is the modem functionality by which any phase offset between carriers is estimated and compensated for. Phase offsets between the transmitter's and receiver's carriers are due to the propagation delay and to the general lack of a reference clock between the two devices. Assume that the oscillator in the receiver has a phase offset of θ with respect to the transmitter; when we retrieve the baseband signal $\hat{b}[n]$ from $\hat{c}[n]$ we have

$$\hat{b}[n] = \hat{c}[n]\, e^{-j(\omega_c n - \theta)} = c[n]\, e^{-j(\omega_c n - \theta)} = b[n]\, e^{j\theta}$$

where we have neglected both distortion and noise and assumed $\hat{c}[n] = c[n]$. Such a phase offset translates to a rotation of the constellation points in the complex plane since, after downsampling, we have $\hat{a}[n] = a[n]\, e^{j\theta}$. Visually, the received constellation looks like in Figure 12.22, where $\theta = \pi/20 = 9°$. If we look at the decision regions plotted in Figure 12.22, it is clear that in the rotated constellation some points are shifted closer to the decision boundaries; for these, a smaller amount of noise is sufficient to cause slicing errors. An even worse situation happens when the receiver's carrier frequency is slightly different than the transmitter's carrier frequency; in this case the phase offset changes over time and the points in the constellation start to rotate with an angular speed equal to the difference between frequencies. In both cases, data transmission becomes highly unreliable: carrier recovery is then a fundamental part of modem design.

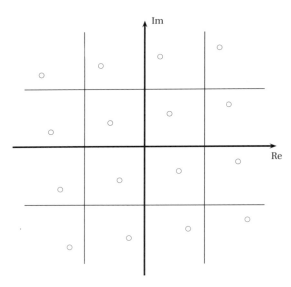

Figure 12.22 Rational effect of a phase offset on the received symbols.

The most common technique for QAM carrier recovery over well-behaved channels is a *decision directed loop*; just as in the case of the adaptive equalizer, this works when the overall SNR is sufficiently high and the distortion is mild so that the slicer's output is an almost error-free sequence of symbols. Consider a system with a phase offset of θ; in Figure 12.23 the

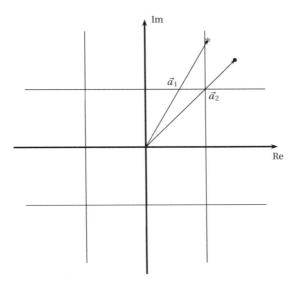

Figure 12.23 Estimation of the phase offset.

rotated symbol $\hat{a}$ (indicated by a star) is sufficiently close to the transmitted value α (indicated by a dot) to be decoded correctly. In the z plane, consider the two vectors $\vec{a}_1$ and $\vec{a}_2$, from the origin to $\hat{a}$ and α respectively; the magnitude of their vector product can be expressed as

$$|\vec{a}_1 \times \vec{a}_2| = \text{Re}\{\hat{a}\}\,\text{Im}\{\alpha\} - \text{Im}\{\hat{a}\}\,\text{Re}\{\alpha\} \tag{12.28}$$

Moreover, the angle between the vectors is θ and it can be computed as

$$|\vec{a}_1 \times \vec{a}_2| = |\vec{a}_1||\vec{a}_2|\sin(\theta) \tag{12.29}$$

We can therefore obtain an estimate for the phase offset:

$$\sin(\theta) = \frac{\text{Re}\{\hat{a}\}\,\text{Im}\{\alpha\} - \text{Im}\{\hat{a}\}\,\text{Re}\{\alpha\}}{|\vec{a}_1||\vec{a}_2|} \tag{12.30}$$

For small angles, we can invoke the approximation $\sin(\theta) \approx \theta$ and obtain a quick estimate of the phase offset. In digital systems, oscillators are realized using the algorithm we saw in Section 2.1.3; it is easy to modify such a routine to include a time-varying corrective term derived from the estimate of θ above so that the resulting phase offset is close to zero. This works also in the case of a slight frequency offset, with θ converging in this case to a nonzero constant. The carrier recovery block diagram is shown in Figure 12.24.

This decision-directed feedback method is almost always able to "lock" the constellation in place; due to the fourfold symmetry of regular square constellations, however, there is no guarantee that the final orientation of the locked pattern be the same as the original. This difficulty is overcome by a mapping technique called *differential encoding*; in differential encoding the first two bits of each symbol actually encode the *quadrant offset* of the symbol with respect to the previous one, while the remaining bits indicate the actual point within the quadrant. In so doing, the encoded symbol sequence becomes independent of the constellation's absolute orientation.

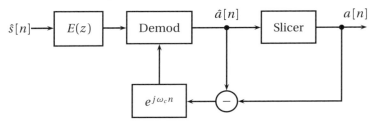

Figure 12.24 Carrier recovery by decision-directed loop.

12.4.2 Timing Recovery

Timing recovery is the ensemble of strategies which are put in place to recover the synchronism between transmitter and receiver at the level of discrete-time samples. This synchronism, which was one of the assumptions of back-to-back operation, is lost in real-world situations because of propagation delays and because of slight hardware differences between devices. The D/A and A/D, being physically separate, run on independent clocks which may exhibit small frequency differences and a slow drift. The purpose of timing recovery is to offset such hardware discrepancies in the discrete-time domain.

A Digital PLL. Traditionally, a Phase-Locked-Loop (PLL) is an analog circuit which, using a negative feedback loop, manages to keep an internal oscillator "locked in phase" with an external oscillatory input. Since the internal oscillator's parameters can be easily retrieved, PLLs are used to accurately measure the frequency and the phase of an external signal with respect to an internal reference.

In timing recovery, we use a PLL-like structure as in Figure 12.25 to compensate for sampling offsets. To see how this PLL works, assume that the discrete-time samples $\hat{s}[n]$ are obtained by the A/D converter as

$$\hat{s}[n] = \hat{s}(t_n) \tag{12.31}$$

where the sequence of sampling instants t_n is generated as

$$t_{n+1} = t_n + T[n] \tag{12.32}$$

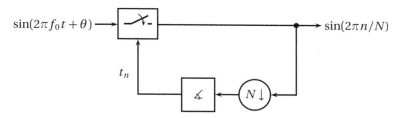

Figure 12.25 A digital PLL with a sinusoidal input.

Normally, the sampling period is a constant and $T[n] = T_s = 1/F_s$ but here we will assume that we have a special A/D converter for which the sampling period can be dynamically changed at each sampling cycle. Assume the input to the sampler is a zero-phase sinusoid of *known* frequency $f_0 = F_s/N$ for $N \in \mathbb{N}$ and $N \geq 2$:

$$x(t) = \sin(2\pi f_0 t)$$

If the sampling period is constant and equal to T_s and if the A/D is synchronous to the sinusoid, the sampled signal are simply:

$$x[n] = \sin\left(\frac{2\pi}{N} n\right)$$

We can test such synchronicity by downsampling $x[n]$ by N and we should have $x_{ND}[n] = 0$ for all n; this situation is shown at the top of Figure 12.26 and we can say that the A/D is *locked* to the reference signal $x(t)$.

If the local clock has a time lag τ with respect to the reference time of the incoming sinusoid (or, alternatively, if the incoming sinusoid is delayed by τ), then the discrete-time, downsampled signal is the constant:

$$x_{ND}[n] = \sin(2\pi f_0 \tau) \tag{12.33}$$

Note, the A/D is still locked to the reference signal $x(t)$, but it exhibits a phase offset, as shown in Figure 12.26, middle panel. If this offset is sufficiently small then the small angle approximation for the sine holds and $x_{ND}[n]$ provides a direct estimate of the corrective factor which needs to be injected into the A/D block. If the offset is estimated at time n_0, it will suffice to set

$$T[n] = \begin{cases} T_s - \tau & \text{for } n = n_0 \\ T_s & \text{for } n > n_0 \end{cases} \tag{12.34}$$

for the A/D to be locked to the input sinusoid.

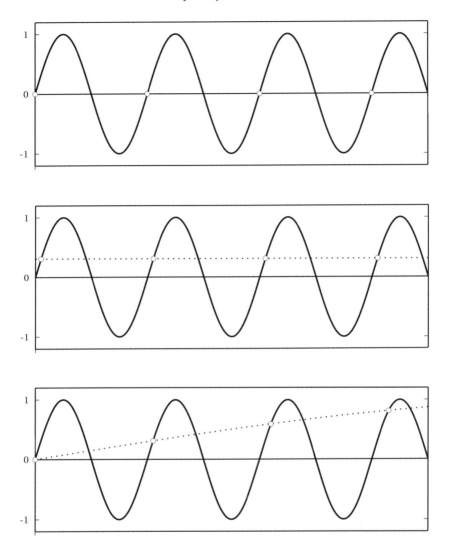

Figure 12.26 Timing recovery from a continuous-time sinusoid, with reference samples drawn as white circles: perfect locking (top); phase offset (middle) and frequency drift (bottom). All plots are in the time reference of the input sinusoid.

Suppose now that the the A/D converter runs slightly slower than its nominal speed or, in other words, that the effective sampling frequency is $F'_s = \beta F_s$, with $\beta < 1$. As a consequence the sampling period is $T'_s = T_s/\beta > T_s$ and the discrete-time, downsampled signal becomes

$$x_{ND}[n] = \sin\bigl((2\pi\beta)n\bigr) \tag{12.35}$$

i.e. it is a sinusoid of frequency $2\pi\beta$; this situation is shown in the bottom panel of Figure 12.26. We can use the downsampled signal to estimate β and we can re-establish a locked PLL by setting

$$T[n] = \frac{T_s}{\beta} \tag{12.36}$$

The same strategy can be employed if the A/D runs faster than normal, in which case the only difference is that $\beta > 1$.

A Variable Fractional Delay. In practice, A/D converters with "tunable" sampling instants are rare and expensive because of their design complexity; furthermore, a data path *from* the discrete-time estimators *to* the analog sampler would violate the digital processing paradigm in which all of the receiver works in discrete time and the one-way interface from the analog world is the A/D converter. In other words: the structure in Figure 12.25 is not a truly digital PLL loop; to implement a completely digital PLL structure, the adjustment of the sampling instants must be performed in discrete time via the use of a programmable fractional delay.

Let us start with the case of a simple time-lag compensation for a continuous-time signal $x(t)$. Of the total delay t_d, we assume that the bulk delay has been correctly estimated so that the only necessary compensation is that of a fractional delay τ, with $|\tau| \leq 1/2$. From the available sampled signal $x[n] = x(nT_s)$ we want to obtain the signal

$$x_\tau[n] = x(nT_s + \tau T_s) \tag{12.37}$$

using discrete-time processing only. Since we will be operating in discrete time, we can assume $T_s = 1$ with no loss of generality and so we can write simply:

$$x_\tau[n] = x(n+\tau)$$

We know from Section 9.7.2 that the "ideal" way to obtain $x_\tau[n]$ from $x[n]$ is to use a fractional delay filter:

$$x_\tau[n] = d_\tau[n] * x[n]$$

where $D_\tau(e^{j\omega}) = e^{j\omega\tau}$. We have seen that the problem with this approach is that $D_\tau(e^{j\omega})$ is an ideal filter, and that its impulse response is a sinc, whose slow decay leads to very poor FIR approximations. An alternative approach relies on the *local interpolation* techniques we saw in Section 9.4.2. Suppose $2N+1$ samples of $x[n]$ are available around the index $n = n_0$; we could easily build a local continuous-time interpolation around n_0 as

$$\hat{x}(n_0; t) = \sum_{k=-N}^{N} x[n_0 - k] L_k^{(N)}(t) \tag{12.38}$$

where $L_k^{(N)}(t)$ is the k-th Lagrange polynomial of order $2N$ defined in (9.14). The approximation

$$\hat{x}(n_0; t) \approx x(n_0 + t)$$

is good, at least, over a unit-size interval centered around n_0, i.e. for $|t| \leq 1/2$ and therefore we can obtain the fractionally delayed signal as

$$x_\tau[n_0] = \hat{x}_\tau(n_0; \tau) \qquad (12.39)$$

as shown in Figure 12.27 for $N = 1$ (i.e. for a three-point local interpolation).

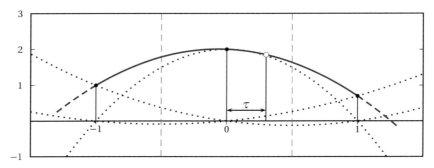

Figure 12.27 Local interpolation around $n_0 = 0$ and $T_s = 1$ for time lag compensation. The Lagrange polynomial components are plotted as dotted lines. The dashed lines delimit the good approximation interval. The white dot is the fractionally delayed sample for $n = n_0$.

Equation (12.39) can be rewritten in general as

$$x_\tau[n] = \sum_{k=-N}^{N} x[n-k] L_k^{(N)}(\tau) = \hat{d}_\tau[n] * x[n] \qquad (12.40)$$

which is the convolution of the input signal with a $(2N + 1)$-tap FIR whose coefficients are the values of the $2N + 1$ Lagrange polynomials of order $2N$ computed in $t = \tau$. For instance, for the above three-point interpolator, we have

$$\hat{d}_\tau[-1] = \tau \frac{\tau - 1}{2}$$
$$\hat{d}_\tau[0] = -(\tau + 1)(\tau - 1)$$
$$\hat{d}_\tau[1] = \tau \frac{\tau + 1}{2}$$

The resulting FIR interpolators are expressed in noncausal form purely out of convenience; in practical implementations an additional delay would make the whole processing chain causal.

The fact that the coefficients $\hat{d}_\tau[n]$ are expressed in closed form as a polynomial function of τ makes it possible to efficiently compensate for a time-varying delay by recomputing the FIR taps on the fly. This is actually the case when we need to compensate for a frequency drift between transmitter and receiver, i.e. we need to *resample* the input signal. Suppose that, by using the techniques in the previous Section, we have estimated that the actual sampling frequency is either higher or lower than the nominal sampling frequency by a factor β which is very close to 1. From the available samples $x[n] = x(nT_s)$ we want to obtain the signal

$$x_\beta[n] = x\left(\frac{nT_s}{\beta}\right)$$

using discrete-time processing only. With a simple algebraic manipulation we can write

$$x_\beta[n] = x\left(nT_s - n\frac{1-\beta}{\beta}T_s\right) = x(nT_s - n\tau T_s) \tag{12.41}$$

Here, we are in a situation similar to that of Equation (12.37) but in this case the delay term is linearly increasing with n. Again, we can assume $T_s = 1$ with no loss of generality and remark that, in general, β is very close to one so that it is

$$\tau = \frac{1-\beta}{\beta} \approx 0$$

Nonetheless, regardless of how small τ is, at one point the delay term $n\tau$ will fall outside of the good approximation interval provided by the local interpolation scheme. For this, a more elaborate strategy is put in place, which we can describe with the help of Figure 12.28 in which $\beta = 0.82$ and therefore $\tau \approx 0.22$:

1. We assume initial synchronism, so that $x_\beta[0] = x(0)$.

2. For $n = 1$ and $n = 2$, $0 < n\tau < 1/2$; therefore $x_\beta[1] = x_\tau[1]$ and $x_\beta[2] = x_{2\tau}[2]$ can be computed using (12.40).

3. For $n = 3$, $3\tau > 1/2$; therefore *we skip $x[3]$* and calculate $x_\beta[3]$ from a local interpolation around $x[4]$: $x_\beta[3] = x'_\tau[4]$ with $\tau' = 1 - 3\tau$ since $|\tau'| < 1/2$.

4. For $n = 4$, again, the delay 4τ makes $x_\beta[4]$ closer to $x[5]$, with an offset of $\tau' = 1 - 4\tau$ so that $|\tau'| < 1/2$; therefore $x_\beta[4] = x'_\tau[5]$.

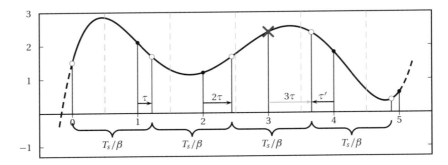

Figure 12.28 Sampling frequency reduction ($T_s = 1$, $\beta = 0.82$) in the discrete-time domain using a programmable fractional delay; white dots represent the resampled signal.

In general the resampled signal can be computed for all n using (12.40) as

$$x_\beta[n] = x_{\tau_n}[n + \gamma_n] \tag{12.42}$$

where

$$\tau_n = \text{frac}\left(n\tau + \frac{1}{2}\right) - \frac{1}{2} \tag{12.43}$$

$$\gamma_n = \left\lfloor n\tau + \frac{1}{2} \right\rfloor \tag{12.44}$$

It is evident that, τ_n is the quantity $n\tau$ "wrapped" over the $[-1/2, 1/2]$ interval[6] while γ_n is the number of samples skipped so far. Practical algorithms compute τ_n and $(n + \gamma_n)$ incrementally.

Figure 12.29 shows an example in which the sampling frequency is too slow and the discrete-time signal must be resampled at a higher rate. In the figure, $\beta = 1.28$ so that $\tau \approx -0.22$; the first resampling steps are:

1. We assume initial synchronism, so that $x_\beta[0] = x(0)$.

2. For $n = 1$ and $n = 2$, $-1/2 < n\tau$; therefore $x_\beta[1] = x_\tau[1]$ and $x_\beta[2] = x_{2\tau}[2]$ can be computed using (12.40).

3. For $n = 3$, $3\tau < -1/2$; therefore *we fall back on $x[2]$ and calculate $x_\beta[3]$* from a local interpolation around $x[2]$ once again: $x_\beta[3] = x'_\tau[2]$ with $\tau' = 1 + 3\tau$ and $|\tau'| < 1/2$.

---

[6] The frac function extracts the fractionary part of a number and is defined as $\text{frac}(x) = x - \lfloor x \rfloor$.

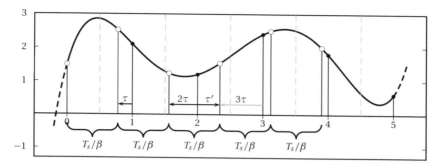

Figure 12.29 Sampling frequency increase ($T_s = 1$, $\beta = 1.28$) in the discrete-time domain using a programmable fractional delay.

4. For $n = 4$, the delay 4τ makes $x_\beta[4]$ closer to $x[3]$, with an offset of $\tau' = 1 + 4\tau$ so that $|\tau'| < 1/2$; therefore $x_\beta[4] = x'_\tau[5]$.

In general the resampled signal can be computed for all n using (12.40) as

$$x_\beta[n] = x_{\tau_n}[n - \gamma_n] \tag{12.45}$$

where τ_n and γ_n are as in (12.43) and (12.44).

Nonlinearity. The programmable delay is inserted in a PLL-like loop as in Figure 12.30 where $\mathscr{S}\{\cdot\}$ is a processing block which extracts a suitable sinusoidal component from the baseband signal.[7] Hypothetically, if the transmitter inserted an explicit sinusoidal component $p[n]$ in the baseband with a frequency equal to the baud rate and with zero phase offset with re-

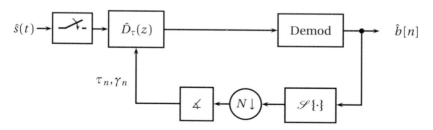

Figure 12.30 A *truly* digital PLL for timing recovery.

---

[7] Note that timing recovery is performed in the baseband signal since in baseband everything is slower and therefore easier to track; we also assume that equalization and carrier recovery proceed independently and converge before timing recovery is attempted.

spect to the symbol times, then this signal could be used for synchronism; indeed, from

$$p[n] = \sin\left(\frac{2\pi f_b}{F_s} n\right) = \sin\left(\frac{2\pi}{K} n\right)$$

we would have $p_{KD}[n] = 0$. If this component was present in the signal, then the block $\mathscr{S}\{\cdot\}$ would be a simple resonator $\mathscr{R}$ with peak frequencies at $\omega = \pm 2\pi/K$, as described in Section 7.3.1.

Now, consider more in detail the baseband signal $b[n]$ in (12.4); if we always transmitted the same symbol a, then $b[n] = a\sum_i g[n - iK]$ would be a periodic signal with period K and, therefore, it would contain a strong spectral line at $2\pi/K$ which we could use for synchronism. Unfortunately, since the symbol sequence $a[n]$ is a balanced stochastic sequence we have that:

$$E[b[n]] = E[a[n]]\sum_i g[n - iK] = 0 \qquad (12.46)$$

and so, even on average, no periodic pattern emerges.[8] The way around this impasse is to use a fantastic "trick" which dates back to the old days of analog radio receivers, i.e. we process the signal through a *nonlinearity* which acts like a diode. We can use, for instance, the square magnitude operator; if we process $b[n]$ with this nonlinearity, it will be

$$E\left[|b[n]|^2\right] = \sum_h \sum_i E[a[h]a^*[i]]\, g[n - hK]\, g[n - iK] \qquad (12.47)$$

Since we have assumed that $a[n]$ is an uncorrelated i.i.d. sequence,

$$E[a[h]a^*[i]] = \sigma_a^2\, \delta[h - i]$$

and, therefore,

$$E[\mathscr{S}\{b[n]\}] = \sigma_a^2 \sum_i (g[n - iK])^2 \qquad (12.48)$$

The last term in the above equation is periodic with period K and this means that, on average, the squared signal contains a periodic component at the frequency we need. By filtering the squared signal through the resonator above (i.e. by setting $\mathscr{S}\{x[n]\} = \mathscr{R}\{|x[n]|^2\}$), we obtain a sinusoidal component suitable for use by the PLL.

---

[8] Again, a rigorous treatment of the topic would require the introduction of cyclostationary analysis; here we simply point to the intuition and refer to the bibliography for a more thorough derivation.

Further Reading

Of course there are a good number of books on communications, which cover the material necessary for analyzing and designing a communication system like the modem studied in this Chapter. A classic book providing both insight and tools is J. M. Wozencraff and I. M. Jacobs's *Principles of Communication Engineering* (Waveland Press, 1990); despite its age, it is still relevant. More recent books include *Digital Communications* (McGraw Hill, 2000) by J. G. Proakis; *Digital Communications: Fundamentals and Applications* (Prentice Hall, 2001) by B. Sklar, and *Digital Communication* (Kluwer, 2004) by E. A. Lee and D. G. Messerschmitt.

Exercises

Exercise 12.1: Raised cosine. Why is the raised cosine an ideal filter? What type of linear phase FIR would you use for its approximation?

Exercise 12.2: Digital resampling. Use the programmable digital delay of Section 12.4.2 to design an *exact* sampling rate converter from CD to DVD audio (Sect. 11.3). How many different filters $h_\tau[n]$ are needed in total? Does this number depend on the length of the local interpolator?

Exercise 12.3: A quick design. Assume the specifications for a given telephone line are $f_{\min} = 300$ Hz, $f_{\max} = 3600$ Hz, and a SNR of at least 28 dB. Design a set of operational parameters for a modem transmitting on this line (baud rate, carrier frequency, constellation size). How many bits per second can you transmit?

Exercise 12.4: The shape of a constellation. One of the reasons for designing non-regular constellations, or constellation on lattices, different than the upright square grid, is that the energy of the transmitted signal is directly proportional to the parameter σ_a^2 as in (12.10). By arranging the same number of alphabet symbols in a different manner, we can sometimes reduce σ_a^2 and therefore use a larger amplification gain while keeping the total output power constant, which in turn lowers the probability of error. Consider the two 8-point constellations in the Figure 12.12 and Figure 12.13 and compute their intrinsic power σ_a^2 for uniform symbol distributions. What do you notice?

Index

A

A/D conversion, 283
aliasing, **255**, 250–259, 264
    in multirate, 297
allpass filter, 124
alphabet, 333
alternation theorem, 186
analog
    computer, 11
    transmission, 13
analytic signal, **133**, 337
anticausal filter, 114
aperiodic signal periodic, 31

B

bandlimited signal, **239**, 264
bandpass
    filter, 124, 131
    signal, 96, 137
bandwidth, 129, 138
    constraint, 328
    of telephone channel, 319
base-12, 4
baseband spectrum, 95
basis
    Fourier, 41
    orthonormal, 40
    sinc, 248
    span, 48
    subspace, 48
    (vector space), 40, 47–51
baud rate, 337
Bessel's inequality, 50

BIBO, 114
Blackman window, 178

C

carrier, 137, 337
    recovery, 353
cascade structure, 195
causal
    CCDE, 149
    filter, 113
CCDE, **134**, 134–136
    solving, 148
CD, 15, 102, 293, 311
    SNR, 281
CD to DVD conversion, 311
Chebyshev polynomials, 182
circulant matrix, 202
compander, 288
complex exponential, 24, 61
    aliasing, 33, 251
Constant-Coefficient Difference
    Equation, *see* CCDE
constellation, 342
continuous time vs. discrete time, 7–10
convolution, **110**
    as inner product, 112
    associativity, 111
    circular, 202
    in continuous time, 2
    of DTFTs, 112, 174
    theorem, **122**, 238
covariance, 221

368   Index

critical sampling, 256
cross-correlation, 222

D
D/A conversion, 286
DAT (Digital Audio Tape), 293
data compression rates, 15
decimation, 294
decision-directed loop, 354
delay, **26**, 117, 124, 125
    fractional, 125, **261**
demodulation, 138
DFS, 71–72
    properties, 85, 89
DFT, **64**, 63–71
    matrix form, 64
    properties, 86
    zero padding, 94
dichotomy paradox, *see* Zeno
differentiator
    approximate, 27
    exact, 260
digital
    computer, 11
    etymology of, 1
    frequency, **25**, 101–102
    revolution, 13
Dirac delta, **78**
    DTFT of, 80
    pulse train, 79
direct form, 197
Discrete Fourier Series, *see* DFS
Discrete Fourier Transform, *see* DFT
discrete time, **21**
    vs. continuous time, 7–10
Discrete-Time Fourier Transform,
        see DTFT
discrete-time
    signal, **19**
        finite-length, 29
        finite-support, 31
        infinite-length, 30
        periodic, 31
    vs. digital, 12
distortion nonlinear, 287
downsampling, **294**, 294–297

DTFT, **72**, 72–81
    from DFS, 81
    from DFT, 82
    of unit step, 103
    plotting, 91–92
    properties, 83, 88
DVD, 12, 293, 311, 319

E
eigenfunctions, 121
energy of a signal, 27
equiripple filter, 187
error correcting codes, 15

F
FFT, 93
    complexity, 203
    zero padding, 94
filter, **109**
    allpass, 205
    computational cost, 195
    delay, 117
    frequency response, 121
filter design, 165–170
    FIR, 171–190
        minimax, 179–190
        window method, 171–179
    IIR, 190
    specifications, 168
filter structures, 195–200
    cascade, 195
    direct forms, 197
    parallel, 196
finite-length signal, 29, 53
    filtering, 200
finite-support signal, 31
FIR, **113**
    linear phase, 154
    types, 180
    vs. IIR, 166
    zero locations, 181
first-order hold, 242
    (discrete-time), 308
Fourier basis, 41, 63
    series, 263
    transform (continuous time),
        238

fractional delay, 125, **261**
    variable, 359
frequency domain, 60
    response, 121
        magnitude, 123
        phase, 124

G
galena, 140
Gibbs phenomenon, 173
Goertzel filter, 204
group delay, **126**
    negative, 140

H
half-band filter, 101, **168**, 303, 321, 336
Hamming window, 178
highpass
    filter, 124, 131
    signal, 95
Hilbert
    demodulation, 349
    filter, **131**
    space, 41–46
        completeness, 45

I
ideal filter, 129
    bandpass, 131
    highpass, 131
    Hilbert, 131
    lowpass, 129
IIR, **113**
    vs. FIR, 166
impulse, 23
    response, 113–115
infinite-length signal, 30, 54
inner product, **39**
    approximation by, 46
    Euclidean distance, 46
    for functions, 237
integrator
    discrete-time, 26
    leaky, 117
    planimeter, 8
    RC network, 8

interpolation, 236, 240–246
    in multirate, 307
    local, 241
        first-order, 242
        Lagrange, 244
        zero-order, 241
    sinc, 246

K
Kaiser formula, 188
Karplus-Strong, 207

L
Lagrange interpolation, **243**, 244, 359
    polynomials, 244
leaky integrator, 117–120, 127, 157, 192
Leibnitz, 5
linear phase, **125**
    generalized, 179
linearity, 110
Lloyd-Max, 282
lowpass
    filter, 124, 129
    signal, 95

M
magnitude response, 123
mapper, 333
Matlab, 17
modem, 320
modulation
    AM, 329
    AM radio, 137
    theorem, 122
moving average, 115–116, 126, 156, 192
μ-law, 288

N
negative frequency, 61
Nile floods, 2
noise
    amplification, 14
    complex Gaussian, 344
    floor, 328
    thresholding, 14

nonlinear processing, 140, 364
notch, 192

O
optimal filter design, *see* filter design, FIR, minimax
orthonormal basis, 40
oversampling, 255
    in A/D conversion, 311
    in D/A conversion, 314

P
parallel structure, 196
Parks-McClellan, *see* filter design, FIR, minimax
Parseval's identity, 50
    for the DFS, 85
    for the DFT, 90
    for the DTFT, 84
passband, 124, **168**
periodic
    extension, 31
    signal, 53
        filtering, 201
periodization, **33**, 76, 263, 265
phase
    linear, **125**
    response, 124
    zero, 97
phonograph, 7
planimeter, *see* integrator
PLL (phase locked loop), 356
Poisson sum formula, 263
pole-zero plot, 153
power
    constraint, 328, 347
    of a signal, 28
    spectral density, **226**
product of signals, 26
programmable delay, 359
PSD, *see* power spectral density
Pythagoras, 4

Q
QAM, 341–344
    constellation, 342

quadrature amplitude modulation, *see* QAM
quantization, 10–12, **276**, 275–276
    nonuniform, 289
    reconstruction, 277
    uniform, 278–281

R
raised cosine, 336
random
    variable, 217–219
    vector, 219–220
        Gaussian, 220
    process, 221–227
        covariance, 221
        cross-correlation, 222
        ergodic, 222
        filtered, 229
        Gaussian, 223
        power spectral density, 224–226
        stationary, 222
        white noise, 227
rational sampling rate change, 310
rect, **130**, 239
rectangular window, 173
region of convergence, *see* ROC
reproducing formula, 27
resonator, 192
ROC, 150
roots
    of complex polynomial, 158, 265
    of transfer function, 153
    of unity, **63**

S
sampling, 236
    frequency, 236, 250
    theorem, 10, **249**
scaling, 26
sequence, *see* discrete-time signal
shift, 26
sinc, **130**, 240, 266
    interpolation, 246
        (discrete-time), 308
slicer, 344

SNR, 281, 313, 328
    of a CD, 281
    of a telephone channel, 331
spectrogram, 98
spectrum
    magnitude, 95
    overlap, 138
    periodicity, **75**
    phase, 96
stability, 114
    criteria, 114, 152
stationary process, 222
stopband, 124, **168**
sum of signals, 26
system LTI, 109–110

T

telephone channel, 330
    bandwidth, 319, 331
    SNR, 319, 331
thermograph, 7
time domain, 60
time-invariance, 110
timing recovery, 356–364
Toeplitz matrix, 232
transatlantic cable, 15
transfer function, **148**, 152
    ROC and stability, 152
    roots, 153
    sketching, 155

transmission reliability, 346
triangular window, 176

U

unit step, 23
    DTFT, 103
upsampling, 304

V

vector, **39**
    distance, 39
    norm, 39
    orthogonality, 39
    space, 38–40
        basis, 40
vinyl, 12, 247, 324

W

wagon wheel effect, 33
white noise, 227
Wiener filter, 231

Z

Zeno, 4
zero initial conditions, 119, 136
zero-order hold, 241
    (discrete-time), 308
z-transform, **147**, 147–152
    of periodic signals, 158
    ROC, 150